Table of Contents

Preface

This *Study Guide* is designed to supplement the first eleven chapters of *Calculus Early Transcendentals*, 7th edition, by James Stewart. It may also be used with *Single Variable Calculus Early Transcendentals*, 7th edition. If you later go on to multivariable calculus, you will want to obtain the multivariable volume of this *Study Guide*.

Your text is well written in a very complete and patient style. This *Study Guide* is not intended to replace it. You should read the relevant sections of the text and work problems, lots of problems. Calculus is learned by doing it.

What this *Study Guide* does do is capture the main points and formulas of each section of your text and provide complete examples and short, concise questions that will help you understand the essential concepts. Every question has an explained answer. Some of our solutions begin with parenthetical comments offset by < . . . > and in italics to explain the approach to take to solve the problem. The two-column format allows you to cover the answer portion of a question while working on it and then uncover the given answer to check your solution. Working in this fashion leads to greater success than simply perusing the solutions. Students have found this *Study Guide* helpful for reviewing for examinations.

Technology, such as graphing calculators and computer algebra systems, can help the understanding of calculus concepts by drawing accurate graphs, solving or approximating solutions to equations, doing numerically intensive calculations, and performing symbolic manipulations. A number of sections have "Technology Plus" exercises at the conclusion of the section to help you master the calculus.

As a quick check of your understanding of a section we have included a page of On Your Own questions located toward the back of the *Study Guide*. These are nearly 400 multiple choice type questions—the kind you might see on an exam in a calculus class. You are "on your own" in the sense that an answer, but no solution, is provided for each question.

I hope that you find this *Study Guide* helpful in understanding the concepts and solving the exercises in *Calculus Early Transcendentals*, 7th edition.

Richard St. Andre

Study Guide

for

Stewart's

Single Variable Calculus

Early Transcendentals

Seventh Edition

Richard St. Andre
Central Michigan University

BROOKS/COLE
CENGAGE Learning™

Australia • Brazil • Japan • Korea • Mexico • Singapore • Spain • United Kingdom • United States

BROOKS/COLE
CENGAGE Learning™

For product information and technology assistance, contact us at **Cengage Learning Customer & Sales Support, 1-800-354-9706**

For permission to use material from this text or product, submit all requests online at **www.cengage.com/permissions**
Further permissions questions can be emailed to **permissionrequest@cengage.com**

ISBN-13: 978-0-8400-5420-3
ISBN-10: 0-8400-5420-3

Brooks/Cole
20 Davis Drive
Belmont, CA 94002-3098
USA

Cengage Learning is a leading provider of customized learning solutions with office locations around the globe, including Singapore, the United Kingdom, Australia, Mexico, Brazil, and Japan. Locate your local office at: **www.cengage.com/global**

Cengage Learning products are represented in Canada by Nelson Education, Ltd.

To learn more about Brooks/Cole, visit **www.cengage.com/brookscole**

Purchase any of our products at your local college store or at our preferred online store **www.cengagebrain.com**

Printed in the United States of America
1 2 3 4 5 6 7 15 14 13 12 11

Chapter 1 — Functions and Models

Section 1.1 Four Ways to Represent a Function

This section defines what is a function and several concepts that go along with that definition. The concept of a function is central to all of calculus.

Concepts to Master

A. Definition of a function; Function value; Domain; Range; Independent variable; Dependent variable

B. Four ways to describe a function; Tables of values; Graphs

C. Implied domain

D. Piecewise defined function

E. Symmetry (even and odd functions)

F. Increasing; Decreasing

Summary and Focus Questions

A. A function f is a rule that associates pairs of numbers: to each number x in a set (the **domain**) there is associated another real number, denoted $f(x)$, the **value** of f at x.

The set of all images (that is, the set of all $f(x)$ values) is the **range** of f. The variable x is the **independent variable** and $y = f(x)$ is the **dependent variable.** When we say "H is a function of t," that means H is the name of the function and we can write $H(t)$ for the values of the function.

Example: The function S, "to each nonnegative number, associate its square root," has domain $[0, \infty)$ and rule of association given by:

$$S(x) = \sqrt{x}$$

The value of S at $x = 25$ is 5 because $S(25) = \sqrt{25} = 5$. S has no value at -7 since $\sqrt{-7}$ is not a real number. Another way to say this is "-7 is not in the domain of S."

1

Example: The function T, given by the table at the right, has domain $\{-1, 1, 3, 5\}$ and range of $\{2, 4, 6\}$. One possible rule to describe T is $T(x) = |x| + 1$.

x	y
-1	2
1	2
3	4
5	6

To determine whether a given graph is the graph of a function use this *Vertical Line Test*: If every vertical line passes through the graph in at most one point, the graph represents a function. In other words, no vertical line intersects the graph more than once.

1) Sometimes, Always, Never:
Both $(2, 5)$ and $(2, 7)$ could be pairs of a function f.

Never. $f(2)$ cannot be both 5 and 7.

2) True or False:
$x = y^2$ defines y as a function of x.

<Look to see whether there is an x-value with multiple corresponding y-values.>

False, since $(4, 2)$ and $(4, -2)$ satisfy the equation. Note that x is a function of y.

3) A track is in the shape of two semicircular ends and straight sides as in the figure. Find the distance around the track as a function of the radius.

<Determine the length of each part of the track in terms of r and add them up.>

The length of each straight side is $3r$ and each semicircular end is πr. The distance d is $d(r) = 2(3r + \pi r) = (6 + 2\pi)r$.

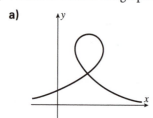

4) Which of these is the graph of a function?

<Apply the Vertical Line Test.>

a)

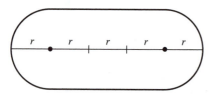

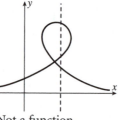

Not a function.

b)

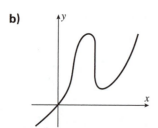

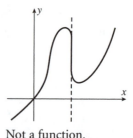

Not a function.

c)

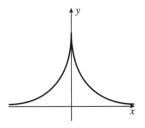

Is a function.

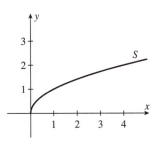

Page
12

B. The four ways to describe the rule of association for a function are

    i) in words, a verbal description.

    ii) a table of values which lists various domain values and corresponding range values.

    iii) the **graph** of $y = f(x)$; that is, all points (x, y) in the Cartesian plane that make $y = f(x)$ true. The y coordinate of a point on the graph of function f is the value of f associated with the x coordinate.

    iv) an explicit algebraic formula or equation.

It is important to remember that if (a, b) is a point on the graph of $y = f(x)$, then $b = f(a)$.

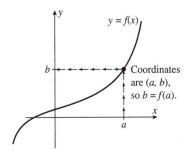

$y = f(x)$

Coordinates are (a, b), so $b = f(a)$.

Example: Here are four ways to represent the rule of association for the square root function.

    i) "to each nonnegative real number, associate its square root."

    ii)

x	0	1	2	3	...
y	0	1	$\sqrt{2}$	$\sqrt{3}$	...

    (Note: For this function with infinite domain, we can only list a few x values with corresponding y values and hope that the meaning is clear.)

    iii)

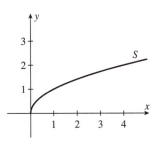

With an infinite domain, we can only represent a portion of the function.

    iv) $S(x) = \sqrt{x}$, where $x \geq 0$.

5) Write 3 other ways to describe each function:

   a) "to each real number, associate one more than its square."

   i) By table of values

x	y
−3	10
−2	5
−1	2
0	1
1	2
2	5
3	10

(A partial list of values)

   ii) *<Plot points and draw a curve through them; it helps to recognize the graph is a parabola.>*

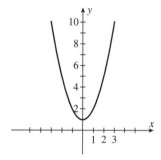

   iii) $f(x) = x^2 + 1$, where x is a real number.

b)

x	1	2	3	4	5
y	5	4	3	2	1

   i) "to each of $x = 1, 2, 3, 4, 5$ associate 6 minus the value of x."

   ii)

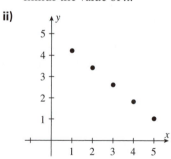

Note: Unless we know the domain is all real numbers, we should not "connect the dots" to form a line.

   iii) For $x = 1, 2, 3, 4, 5$, let $f(x) = 6 - x$.

6) Given the graph of f below, what is the value of f for each of these?

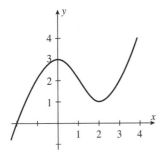

a) $f(2)$

b) $f(0)$

c) $f(3)$

d) $f(f(0))$ (Hint: First determine $f(0)$.)

e) $f(7)$

7) Sketch the graph of $f(x) = \sqrt{x^2 - 16}$.

<Locate each x-value on the x-axis and use the curve to determine the corresponding y-value.>

1 (since the pair $(2, 1)$ is on the graph).

3.

2.

Since $f(0)$ is 3, $f(f(0)) = f(3) = 2$.

7 is not pictured on the x-axis; we can only surmise from the trend of the graph that $f(7)$ will be a large positive number.

<Plot points and recognize the equation as a hyperbola.>

Here are several computed values: $(4, 0)$, $(5, 3)$, $(6, \sqrt{20})$, $(-4, 0)$, $(-5, 3)$.
Squaring both sides of $y = \sqrt{x^2 - 16}$ gives $y^2 = x^2 - 16$, or $x^2 - y^2 = 16$. The graph is the top half of a hyperbola.

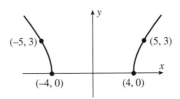

Note: About all we can do now to sketch graphs is plot points and recognize the general shape of certain graphs from the form of the equation. Calculus will help us draw better graphs.

C. The domain of a function f, if unspecified, is understood to be the largest possible set of the real numbers x for which $f(x)$ exists.

Example: $f(x) = \dfrac{1}{\sqrt{x-5}}$ has implied domain $(5, \infty)$ because $\sqrt{x-5}$ is defined only for $x \geq 5$. Since $\sqrt{x-5}$ cannot be zero, $x > 5$. The range of $f(x)$ is $(0, \infty)$.

---

8) Find the domain of $f(x) = \sqrt{x^2 - 16}$.

<Look for where the square root can be calculated.>

For $f(x)$ to exist, $x^2 - 16 \geq 0$. Thus $(x-4)(x+4) \geq 0$. The solution set is $x \leq -4$ or $x \geq 4$. The domain is $(-\infty, -4] \cup [4, \infty)$.

9) Find the domain of $f(x) = \dfrac{1}{|x| + 1}$.

<Look for where the denominator is not zero.>

Since $|x| \geq 0$ for all x, $|x| + 1$ is never 0. The domain is all real numbers.

10) What is the domain and range of the function whose graph is given below?

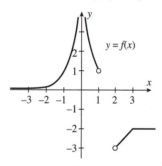

<Describe all the x-values used by points on the graph and all the y-values used.>

The domain appears to be $(-\infty, 0) \cup (0, 1) \cup (2, \infty)$. This is all the x values for which there is a point (x, y) on the graph. The range is $(-3, -2] \cup (0, \infty)$—the set of all y-coordinates used in the graph.

11) Write a rule for a function whose unspecified domain is all real numbers except 3 and 6.

<Make sure that $f(3)$ and $f(6)$ cannot be calculated.>

One such function is
$$f(x) = \frac{1}{(x-3)(x-6)}.$$ There are many other functions, such as
$$g(x) = \frac{x^3}{(x-3)^2(6-x)}.$$

D. A piecewise defined function has its domain divided into disjoint (nonoverlapping) parts and uses a different formula on each part. For example:

Page 16

$$f(x) = \begin{cases} 3x & \text{if } x < 1 \\ \dfrac{1}{x} & \text{if } x \geq 1 \end{cases}$$

has a domain of all real numbers split up into $(-\infty, 1)$ and $[1, \infty)$. Use the appropriate rule to draw the graph over each part.

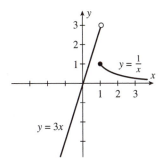

12) The phone company charges a flat rate for local calls of $22.60 per month with an additional charge of $0.09 per call after 55 calls. Express the monthly bill as a function of the number of local calls.

<Separately find the rates with less than and with more than 55 calls.>

Let $x =$ the number of local calls. Then

$$M(x) = \begin{cases} 22.60 & \text{if } x \in [0, 55] \\ 22.60 + 0.09(x - 55) & \text{if } x \in (55, \infty) \end{cases}$$

13) Sketch a graph of

$$f(x) = \begin{cases} 1 - x & \text{if } x \leq 1 \\ 3 & \text{if } 1 < x \leq 2. \\ 7 - x^2 & \text{if } x > 2 \end{cases}$$

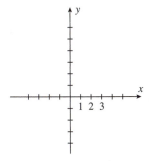

<Sketch pieces of each of the three functions $y = 1 - x$, $y = 3$, and $y = 7 - x^2$.>

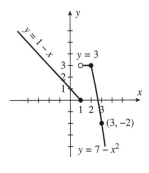

E. Sometimes one part of a graph of a function will be a reflected image of another part:

Page 17

$y = f(x)$ is **even** means $f(x) = f(-x)$ for all x in the domain.

An even function is symmetric about the y-axis—if the point (a, b) is on the graph of f, then so is the point $(-a, b)$.

$y = f(x)$ is **odd** means $f(x) = -f(-x)$ for all x in the domain.

An odd function is symmetric about the origin—if the point (a, b) is on the graph of f, then so is the point $(-a, -b)$.

14) If $(7, 3)$ is on the graph of an odd function, what other point must also be on the graph?

$(-7, -3)$. This is because $3 = f(7) = -f(-7)$, so $f(-7) = -3$.

15) Is $f(x) = 10 - 2x$ even or odd?

<See whether the definitions of odd and even hold.>

i) $f(x) = 10 - 2x$
$f(-x) = 10 - 2(-x) = 10 + 2x$
$f(x)$ is *not even* since $f(x) \neq f(-x)$.

ii) $-f(-x) = -(10 + 2x) = -10 - 2x$
$f(x)$ is *not odd* since $f(x) \neq -f(-x)$.

16) Complete the graph of $y = f(x)$ given below,

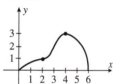

a) if f is even

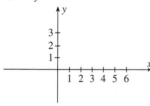

<Reflect the graph about the y-axis.>

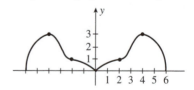

b) if f is odd

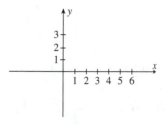

<Reflect the graph about the origin $(0, 0)$.>

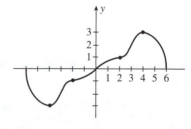

Page 19

F. A function f is **increasing** on an interval I if for all $x_1, x_2 \in I$, $x_1 < x_2$ implies $f(x_1) < f(x_2)$. In other words, as x gets larger, so do the corresponding $f(x)$ values.

f is **decreasing** means that $x_1 < x_2$ implies $f(x_1) > f(x_2)$ for all $x_1, x_2 \in I$. In other words, as x gets larger, $f(x)$ gets smaller.

17) Given the graph of f below, on what intervals is f increasing? decreasing?

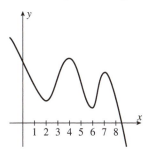

<Look for intervals on the x-axis where the corresponding curve is going upward and intervals where it is going downward.>

f is decreasing on each of these intervals: $(-\infty, 2], [4, 6], [7, \infty)$.

f is increasing on $[2, 4]$ and on $[6, 7]$.

18) Is $f(x) = x^2 + 4x$ increasing on $[-1, 2]$?

<Graph the function and observe whether it is increasing or decreasing.>

Yes. The graph of f is given below. As x increases from -1 to 2, $f(x)$ increases from -3 to 12.

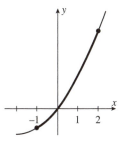

19) Find where $f(x) = 1 - |x|$ is increasing and where it is decreasing.

<Sketch the graph.>

The graph of $f(x) = 1 - |x|$ is

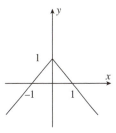

$f(x)$ is increasing on $(-\infty, 0]$ and decreasing on $[0, \infty)$.

20) Can a function be both increasing and decreasing on an interval?

No, not as we have defined the terms: If $x_1 < x_2$, we cannot have both $f(x_1) < f(x_2)$ and $f(x_1) > f(x_2)$.

21) A pizza frozen at $-10°C$ is cooked in a microwave oven until it reaches $45°C$, then allowed to cool to a room temperature of $24°C$. Draw a graph of the temperature (T) as a function of time (t).

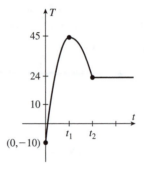

At time t_1, the temperature reaches $45°C$. At t_2, the temperature reaches $24°C$.

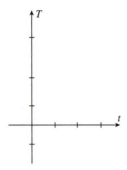

Section 1.2   **Mathematical Models**

This section describes several types of functions that are useful in describing (modeling) real-world phenomenon.

Concepts to Master

A. Characteristics of these modeling functions: linear, quadratic, cubic, polynomial, power, root, rational, algebraic, trigonometric, exponential, logarithmic, and transcendental

B. Fit a math model to a real-world phenomenon; Interpolation; Extrapolation

Summary and Focus Questions

Page 23

A. A mathematical model is an abstraction of a real-world phenomenon into symbols and equations. Our models relate two variables with a function. For example, $C = \frac{5}{9}(F - 32)$ models the relationship between temperature measured in degrees Fahrenheit (F) and degrees Celsius (C). Thus, when the temperature is 50°F, the temperature is $C(50) = \frac{5}{9}(50 - 32) = 10°C$.

Example: If a hardware store makes a profit of $1.38 for each latch hasp it sells, then $P(x) = 1.38x$ models the profit on selling x such hasps.

Example: The radius (r) of a spherical balloon is related to the volume (V) of air in the balloon by $V = \frac{4}{3}\pi r^3$. The volume of a balloon with radius 3 inches is $V(3) = \frac{4}{3}\pi(3)^3 = 36\pi$ in.^{3}

Many types of functions can serve as models; some include:

Type	Description	Examples
Linear	$f(x) = mx + b$	$f(x) = 5x$ $f(x) = -\pi x + 3$
Quadratic	$f(x) = ax^2 + bx + c, a \neq 0$	$f(x) = 2x^2 - x + 7$
Cubic	$f(x) = ax^3 + bx^2 + cx + d, a \neq 0$	$f(x) = -3x^3 + 2x^2 - 5x + 1$
Polynomial	$f(x) = a_n x^n + \dots + a_1 x + a_0$ (n is the degree)	$f(x) = 6x^2 + 3x + 1$ $f(x) = -5x^{12} + 7x$
Power	$f(x) = x^n$ $n = 1, 2, 3, \dots$	$f(x) = x^7$ $f(x) = x^1$
Root	$f(x) = x^a$ where $a = \frac{1}{2}, \frac{1}{3}, \frac{1}{4}, \dots$	$f(x) = \sqrt{x} \ (= x^{1/2})$ $f(x) = x^{1/10}$

Rational	$f(x) = \frac{P(x)}{Q(x)}$, where P, Q are polynomials	$f(x) = \frac{7x + 6}{x^2 + 10}$ $f(x) = \frac{3}{2 - x}$
Algebraic	$f(x)$ is obtained from polynomials, using algebra $(+, -, \cdot, /,$ roots$)$	$f(x) = \sqrt{16 - x^2}$ $f(x) = \sqrt[3]{\frac{x^3}{x + 1}}$
Trigonometric	sine, cosine, tangent, cotangent, secant, cosecant	$f(x) = \tan x$ $f(x) = \cos 3x$
Exponential	$f(x) = a^x$ (for $a > 0$)	$f(x) = 2^x$ $f(x) = (1/3)^{x + 1}$
Logarithmic	$f(x) = \log_a(x)$ (for $a > 0$)	$f(x) = \ln x$ $f(x) = \log_4(x + 1)$

A special case of a rational function is the **reciprocal** function $f(x) = \dfrac{1}{x}$.

1) True or False:
$f(x) = x^3 + 6x + x^{-1}$ is a polynomial.

False, (because of the x^{-1} term).

2) True or False:
$f(x) = x + \dfrac{1}{x}$ is a rational function.

<Rewrite in the form $\dfrac{P(x)}{Q(x)}$.>

True. $\left(x + \dfrac{1}{x} = \dfrac{x^2 + 1}{x}\right)$.

3) True or False:
  a) Every rational function is an algebraic function.

True.

  b) Every linear function is a polynomial.

True.

4) $f(x) = 6x^{10} + 12x^4 + x^{11}$ has degree _____.

<Look for the largest exponent.>
11.

5) True or False:
$f(x) = 2^{\pi}$ is an exponential function.

False. 2^{π} is a constant (approximately 8.825).

6) Which functions are not algebraic?
  a) $f(x) = \sin 2x$
  b) $g(x) = \sqrt{1 - x}$
  c) $h(x) = 2^x + x^2$

a) and c).

7) True or False:
$p(x) = 3$ is linear.

<Determine if the function fits the form $f(x) = mx + b$.>
True. $(m = 0, b = 3)$

B. Success fitting models to given problems comes with practice and gaining insight into how variables and data values are related. Sometimes the graph of a phenomenon initially given in tabular data will look rather like a certain type of model. For example, if the data appears to be periodic, a trigonometric function will likely be involved. Something that seems to grow very rapidly may involve an exponential or a polynomial of a high degree.

Example: A tenant pays a non-refundable $250 deposit upon leasing an apartment and $650 per month. Rent is paid in advance. A linear model for the total amount paid to the landlord after x months is:

$$C(x) = 250 + 650x.$$

Example: The amounts of sulfur dioxide particulates in the air around Rose City are given in the table at the right. A scatter plot of the data suggests that a model for the amount of particulates after x days is a quadratic (dashed curve is added).

day	parts per million
1	48
2	32
3	28
5	27
6	33
7	47

Using a calculator or software, a quadratic that best fits the data (quadratic regression) is $y = 2.55x^2 - 20.48x + 65.03$.

For the missing day, $x = 4$, we estimate that the amount of sulfur dioxide is $y(4) = 2.55(4)^2 - 20.48(4) + 65.03 \approx 23.91$ parts per million. Estimations for values of x much beyond day 7 would involve a great deal of uncertainty.

8) Which type of function model is associated with each scatter diagram?

a)

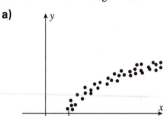

<Know the general shapes of the graphs of the various functions in this section.>

Logarithmic, such as $y = \log_2 x$, for $x \geq 1$.

b)

Rational, such as $y = 10 + \frac{1}{x} = \frac{10x + 1}{x}$, for $x > 0$.

c)

Linear, such as $y = \frac{1}{2}x + 5$.

d)

There does not appear to be a functional relationship between x and y.

9) To determine whether pennies change weight as they age (from factors such as wear from handling) the average weight of a sample of 100 pennies was calculated for each of these years 2002, 2003, 2005, 2007, 2008, and 2009.

Year	Average weight in sample
2002	1.912
2003	1.913
2005	1.914
2007	1.914
2008	1.915
2009	1.916

a) Draw a scatter diagram and determine whether the model seems linear.

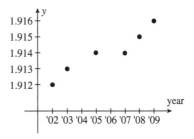

The model appears to be rather linear.

b) Assuming a linear model, find the model that passes through the 2002 and 2008 data points.

<Find the slope and then the equation of the line.>

The slope of the line through (2002, 1.912) and (2008, 1.915) is

$$\frac{1.915 - 1.912}{2008 - 2002} = \frac{.003}{6} = .0005.$$

The equation is

$$y - 1.915 = .0005(x - 2008)$$

or

$$y = .0005x + .911.$$

c) Estimate the average weight of a 2004 penny and a 2012 penny.

For $x = 2004$,
$y = .0005(2004) + .911 = 1.913$.
For $x = 2012$,
$y = .0005(2012) + .911 = 1.917$.

d) Does using the model to estimate the weight of a 2025 penny with the function in part **b)** seem reasonable?

No. Pennies in the future will probably have the same weight as newer ones do today.

10) Karen lives 10 miles from work in a busy city. In light traffic it takes her about 20 minutes to get home from work. Sketch a graph of a model of the duration of her trip home from work (in minutes) as a function of the starting time of the trip—4:00, 4:30, 5:00, 5:30, 6:00, 6:30.

<*It takes longer to get home during rush hour so the graph must show those trip starting times correspond to larger trip durations.*>

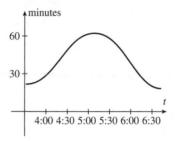

We have assumed that it takes her about 20 minutes in light traffic at 4:00 and 6:30 and nearly an hour in heavy rush hour traffic to drive from work to home (5:00). The model is not linear.

Section 1.3 New Functions from Old Functions

This section describes several ways to combine simple functions to produce one with a more complex expression, and conversely, writing a "complicated" function in terms of ones with simpler expressions. Knowing how to do this will be essential for calculus in the next chapter.

Concepts to Master

A. Translations and stretchings of functions; Reflection

B. Combining functions (sum, difference, product, quotient, and composition)

Summary and Focus Questions

Page 36

A. For $c > 0$, adding or subtracting the constant c to either the independent or dependent variable of $y = f(x)$ shifts the graph either left, right, up, or down.

Function	Graph of the new function
$y = f(x - c)$	Shift graph of $y = f(x)$ *right* by c units
$y = f(x + c)$	Shift graph of $y = f(x)$ *left* by c units
$y = f(x) + c$	Shift graph of $y = f(x)$ *upward* by c units
$y = f(x) - c$	Shift graph of $y = f(x)$ *downward* by c units

Example: The function $f(x) = x^2$ is translated by 2 units:

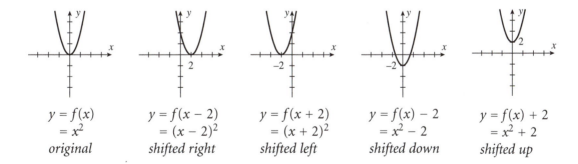

$y = f(x)$
$= x^2$
original

$y = f(x - 2)$
$= (x - 2)^2$
shifted right

$y = f(x + 2)$
$= (x + 2)^2$
shifted left

$y = f(x) - 2$
$= x^2 - 2$
shifted down

$y = f(x) + 2$
$= x^2 + 2$
shifted up

For $c > 1$, multiplying either the independent or dependent variable of $y = f(x)$ by the constant c will stretch or compress a graph either horizontally or vertically.

Function	Graph of the new function
$y = cf(x)$	Stretch *vertically* graph of $y = f(x)$ by a factor of c.
$y = \dfrac{1}{c}f(x)$	Compress *vertically* graph of $y = f(x)$ by a factor of c units.
$y = f(cx)$	Compress *horizontally* graph of $y = f(x)$ by a factor of c units.
$y = f\left(\dfrac{x}{c}\right)$	Stretch *horizontally* graph of $y = f(x)$ by a factor of c units.

Example: The function $f(x) = x^2$ is stretched/compressed by 2 units:

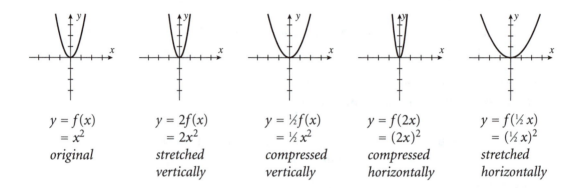

$y = f(x)$
$= x^2$
original

$y = 2f(x)$
$= 2x^2$
*stretched
vertically*

$y = \frac{1}{2}f(x)$
$= \frac{1}{2}x^2$
*compressed
vertically*

$y = f(2x)$
$= (2x)^2$
*compressed
horizontally*

$y = f(\frac{1}{2}x)$
$= (\frac{1}{2}x)^2$
*stretched
horizontally*

Changing the sign of either the independent or dependent variable of $y = f(x)$ will reflect the graph about an axis.

Function	Graph of the new function
$y = -f(x)$	Reflect graph of $y = f(x)$ about the *x*-axis.
$y = f(-x)$	Reflect graph of $y = f(x)$ about the *y*-axis.

Example: The function $f(x) = x^2 - 4x$ is reflected about each axis:

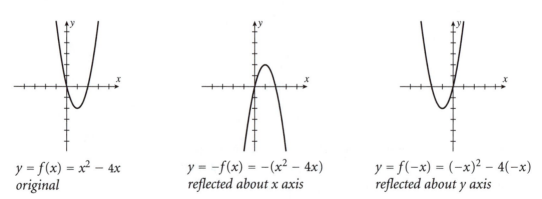

$y = f(x) = x^2 - 4x$
original

$y = -f(x) = -(x^2 - 4x)$
reflected about x axis

$y = f(-x) = (-x)^2 - 4(-x)$
reflected about y axis

1) Given the graph below, sketch each:

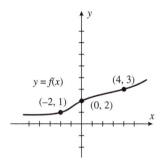

a) $y = 2f(x)$

<Stretch in y-direction by a factor of 2.>

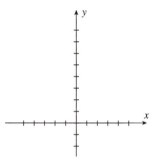

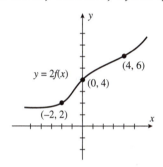

b) $y = f(x) - 2$

<Lower graph by 2 units.>

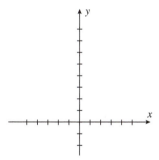

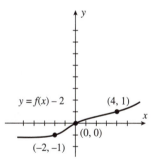

c) $y = f(-x)$

<Reflect graph about y-axis.>

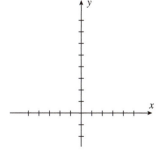

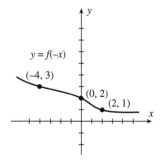

d) $y = \frac{1}{2}f(x) + 3$

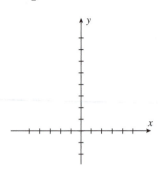

<Compress graph by half and then raise by 3 units.>

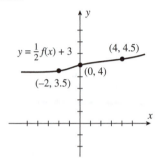

$y = \frac{1}{2}f(x) + 3$
(4, 4.5)
(0, 4)
(−2, 3.5)

Page 39

B. Given two functions $f(x)$ and $g(x)$:

The **sum** $f + g$ associates to each x the number $f(x) + g(x)$.

The **difference** $f - g$ associates to each x the number $f(x) - g(x)$.

The **product** fg associates to each x the number $f(x) \cdot g(x)$.

The **quotient** $\frac{f}{g}$ associates to each x the number $\frac{f(x)}{g(x)}$, provided that $g(x) \neq 0$.

For example, the sum, difference, product, and quotient of $f(x) = x^2$ and $g(x) = 2x - 8$ are:

$$(f + g)(x) = x^2 + 2x - 8$$
$$(f - g)(x) = x^2 - 2x + 8$$
$$fg(x) = x^2(2x - 8)$$
$$\frac{f}{g}(x) = \frac{x^2}{2x - 8}, \text{ for } x \neq 4.$$

It is important to recognize the components of a function so that it can be broken down into simpler functions.

Example: The function $h(x) = \dfrac{x^2(x + 1)}{2 + x}$ may be written as $h = \dfrac{fg}{k}$, where

$$f(x) = x^2,$$
$$g(x) = x + 1,$$
$$\text{and } k(x) = 2 + x.$$

For $f(x)$ and $g(x)$, the **composition** of f and g, written $f \circ g$, is the function that associates to each x the same number that f associates to $g(x)$, that is:

$$(f \circ g)(x) = f(g(x))$$

To compute $(f \circ g)(x)$, first compute $g(x)$, then compute f of that result.

Example: If $f(x) = x^2 + 1$ and $g(x) = 3x$, then:

for $x = 5$, $(f \circ g)(5) = f(g(5)) = f(3(5)) = f(15) = 15^2 + 1 = 226,$

for $x = -2$, $(f \circ g)(-2) = f(g(-2)) = f(-6) = (-6)^2 + 1 = 37$,
and, in general, $(f \circ g)(x) = f(g(x)) = f(3x) = (3x)^2 + 1$.

To determine the components f and g of a composite function,
$h(x) = (f \circ g)(x)$, requires an examination of which function is applied first.

Example: For $h(x) = \sqrt{x + 5}$, we may use $g(x) = x + 5$ (it is applied first,
before the square root) and $f(x) = \sqrt{x}$. Then

$$h(x) = \sqrt{x + 5} = \sqrt{g(x)} = f(g(x)).$$

2) Find $f + g$ and $\dfrac{f}{g}$ for $f(x) = 8 + x$,

$g(x) = \sqrt{x}$. What is the domain of each?

$(f + g)(x) = 8 + x + \sqrt{x}$ with
domain $= [0, \infty)$.

$\dfrac{f}{g}(x) = \dfrac{8 + x}{\sqrt{x}}$ with

domain $= (0, \infty)$.

3) Write $f(x) = \dfrac{2\sqrt{x + 1}}{x^2(x + 3)}$ as a combination of

simpler functions.

*<Look for simple components like $x + 1$ and
$x + 3$.>*

There are many answers to this question.

One is $f = \dfrac{hk}{mn}$ where $h(x) = 2$,

$k(x) = \sqrt{x + 1}$, $m(x) = x^2$,

$n(x) = x + 3$.

4) Find $f \circ g$ where:

a) $f(x) = x - 2x^2$ and $g(x) = x + 1$.

b) $f(x) = \sin x$ and $g(x) = \sin x$.

<Use the definition of composition.>

$(f \circ g)(x) = f(g(x)) = f(x + 1)$
$\qquad = (x + 1) - 2(x + 1)^2$.

$(f \circ g)(x) = f(g(x)) = f(\sin x) = \sin(\sin x)$.
This is not the same as $\sin^2(x)$.

5) Rewrite $h(x) = (x^2 + 3x)^{2/3}$ as a
composition, $f \circ g$.

*<Look for basic components; the expression
$x^2 + 3x$ is one.>*

Let $f(x) = x^{2/3}$ and $g(x) = x^2 + 3x$.
$(f \circ g)(x) = f(g(x)) = f(x^2 + 3x)$
$\qquad = (x^2 + 3x)^{2/3}$.

Thus $h = f \circ g$.
There are other answers. This one is the best.

6) Find $f \circ f$ where
$f(x) = x^2 + 2x$.

<Use the definition of composition.>

$(f \circ f)(x) = f(f(x)) = f(x^2 + 2x)$
$\qquad = (x^2 + 2x)^2 + 2(x^2 + 2x)$.

Section 1.4 **Graphing Calculators and Computers**

Concepts to Master

A. Determining an appropriate viewing rectangle

B. Technology Plus

Summary and Focus Questions

Page 44

A. A **viewing rectangle** for a function $y = f(x)$ is a rectangle:

$$\{(x, y) | a \le x \le b, c \le y \le d\}$$

corresponding to a calculator or computer screen within which a portion of the graph of $y = f(x)$ is displayed. Selecting the correct window coordinates $(a, b, c, d,)$ can be tricky—choose a, b, c, and d so that the display brings out the most salient features of the graph.

Example: Choose an appropriate viewing window for $f(x) = 12 - x^2$.

Since the graph will have no y values greater than 12 ((0, 12) is a high point—a salient feature), $d = 13$ is a good choice. The graph is symmetric about the y-axis and crosses the x-axis (other salient features) between 3 and 4 and between -4 and -3. Good choices for a and b are $a = -5$ and $b = 5$. $f(5) = -13$, so $c = -15$ will work. The graph and viewing rectangle are shown below:

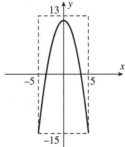

window: $[-5, 5]$ by $[-15, 13]$

1) Find an appropriate viewing rectangle for each:

   a) $y = f(x)$ whose graph is below:

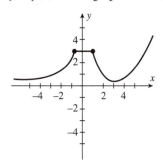

Your choices may vary. A good viewing window is $-4 \le x \le 5$, $-1 \le y \le 5$, or $[-4, 5]$ by $[-1, 5]$.

   b)

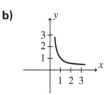

Your choices may vary. A good viewing window is $0 \le x \le 4$, $0 \le y \le 4$, or $[0, 4]$ by $[0, 4]$.

2) Is $[-50, 50]$ by $[-100, 100]$ an appropriate viewing rectangle for $y = \sin x$?

No. Since $-1 \le y \le 1$, the y scale is too large. Also, since sine is periodic, only a few periods need to be displayed to give a good idea of what it looks like. A window such as $[-12, 12]$ by $[-2, 2]$ would be more appropriate.

3) Find where the two graphs $y = x^2$ and $y = 2 - x$ intersect.

<Graph both functions on a calculator or CAS and use the "intersect" or "intersection" command twice.>

The graphs are:

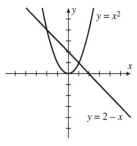

Using a calculator, the two points of intersection are $(1, 1)$ and $(-2, 4)$. Of course, these could have been found by solving the equation $x^2 = 2 - x$, or $x^2 + x - 2 = 0$.

B. Technology Plus. Use a computer algebra system or a graphing calculator to solve.

T-1) Use a graphing device to graph
$f(x) = ax - x^2$, for $a = 0, 1, 2, 3, 4, 5$.
Use $-2 \le x \le 7$.

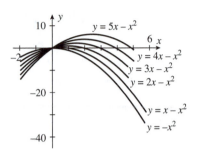

T-2) Use a graphing device to graph $f(x)$, $2f(x)$,
$4f(x)$, $-f(x)$, $-2f(x)$, and $-4f(x)$ where
$f(x) = 2x^2 - 4x$.
Use $-1 \le x \le 3$.

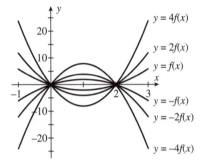

T-3) Draw a graph of

$$f(x) = \begin{cases} x + 6 & \text{if } x < -1 \\ x^2 - 2x + 2 & \text{if } -1 \le x \le 2 \\ 6 - 2x & \text{if } x > 2 \end{cases}$$

Using a calculator, the graph is:

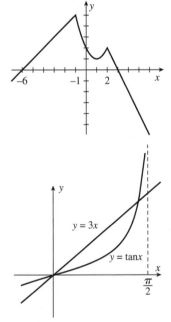

T-4) Solve graphically:
$3x = \tan x$ for $x \in \left[0, \frac{\pi}{2}\right]$.

Graph both $y = 3x$ and $y = \tan x$ for
$0 \le x \le \frac{\pi}{2}$. Note than $\tan x$ is not defined
at $x = \frac{\pi}{2}$. One solution is $x = 0$. Zooming in
on the other, we see that $x \approx 1.32$.

Section 1.5   Exponential Functions

An exponential function contains a variable in an exponent of the formula for the function. This section defines exponential functions and gives some of their properties. Exponential functions may be used to model many types of phenomenon, including growth and decay of a population over time.

Concepts to Master

A. Definition, properties and graph of an exponential function

B. Model growth and decay with an exponential function

C. Definition of e

Summary and Focus Questions

Page 51

A. Let a be a positive constant. The **exponential function with base a** is $f(x) = a^x$ and is defined as follows:

Type of exponent x	Definition of a^x	Example
$x = n$, a positive integer	$a^n = a \cdot a \cdot a \ldots$ (n times)	$4^3 = 4 \cdot 4 \cdot 4$
$x = 0$	$a^0 = 1$	$5^0 = 1$
$x = -n$ (n, a positive integer)	$a^{-n} = \frac{1}{a^n}$	$3^{-2} = \frac{1}{3^2}$
$x = \frac{p}{q}$, a rational number	$a^{p/q} = \sqrt[q]{a^p}$	$5^{2/3} = \sqrt[3]{5^2}$
x, irrational	a^x is approximately a^r, where r is a rational number near a.	$3^{\sqrt{2}} \approx 3^{\frac{1414}{1000}}$ $= 3^{1.414}$

The shape of the graph of $y = a^x$ depends on whether $a > 1$ or $0 < a < 1$:

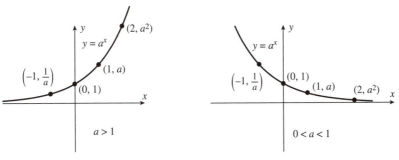

Properties of exponentials include:

$$a^{x+y} = a^x a^y \qquad a^{x-y} = \frac{a^x}{a^y} \qquad a^{xy} = (a^x)^y \qquad (ab)^x = a^x b^x$$

1) By definition, $2^{4/3} = $ _____ and $2^{-5} = $ _____.

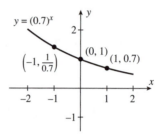

<Use the definition of a^x.>
$\sqrt[3]{2^4}$ and $\dfrac{1}{2^5}$.

2) The rational number 1.72 is near $\sqrt{3}$. An approximation of $3^{\sqrt{3}}$ is _____.

$3^{\sqrt{3}} \approx 3^{1.72} \approx 6.617.$

3) Graph $y = (0.7)^x$.

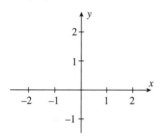

<Note that .7 < 1. Be familiar with the shapes of exponential functions.>

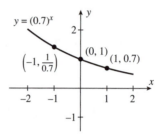

4) True or False:

   a) $(x + y)^a = x^a + y^a.$

<Know the properties of exponentials.>
False.

   b) $x^{a-b+c} = \dfrac{x^a x^c}{x^b}.$

True.

   c) $(a^b)^c = a^{b^c}.$

False.

5) For large positive values of x, which is larger, $100x^4$ or $\dfrac{4^x}{100}$?

<Know that an exponential function a^x (for $a > 1$) will always eventually have values larger than any given polynomial function.>
For $x \geq 15, \dfrac{4^x}{100} > 100x^4.$

Page 54

B. Exponential functions can be used to model growth and decay of the number of individuals in a population as a function of time.

Example: A tribble colony is known to double its size every 60 days. For an initial population of 100 tribbles, what is the population after 1 year (360 days)?

If $P(t)$ is the population after t days,
$$P(0) = 100$$
$$P(60) = 200 = 100 \cdot 2$$
$$P(120) = 400 = 100 \cdot 2^2$$
and in general $P(t) = 100 \cdot 2^{t/60}$.
Thus, $P(360) = 100 \cdot 2^{360/60} = 100 \cdot 2^6 = 6{,}400$.

6) The half life of a certain radioactive isotope is 100 years. To what size has a 12 mg sample disintegrated after 500 years?

<Find the formula for P(t); then use t = 500.>

Let $P(t)$ be the mass after t years. We are given $P(0) = 12$. $P(100) = 6 = 12 \cdot 2^{-1}$. $P(200) = 3 = 12 \cdot 2^{-2}$ and, in general, $P(t) = 12 \cdot 2^{-t/100}$.
$P(500) = 12 \cdot 2^{-500/100} = \dfrac{12}{2^5} = \dfrac{3}{8}$ mg.

7) The price of Macrosoft stock at the close of each day for a week is given in this table.

Mon	Tue	Wed	Thu	Fri
2.30	3.96	6.57	11.33	19.26

If x represents the number of days since Monday, which exponential model best describes the stock prices?

A) $y = 2.3^x$

B) $y = 2.3(1.7)^x$

C) $y = (2.3)^{-x}$

D) $y = (2.3)(1.7)^{-x}$

<Be familiar with the shapes of exponential functions.>

B) Here is a table of values for $y = 2.3(1.7)^x$:

x	0	1	2	3	4
y	2.30	3.91	6.65	11.30	19.21

Note: Functions C and D cannot be correct because they are decreasing. Function A grows too quickly.

C. The number *e* is that unique real number for which the line tangent to $y = e^x$ at $(0, 1)$ has slope one. In chapter 3 we will see that the value of *e*, correct to five decimal places, is

$$e \approx 2.71828.$$

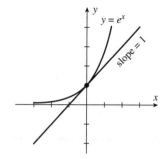

8) For the graphs below, which is the graph of $y = e^x$ and which is $y = 3^x$?

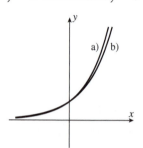

<Be familiar with the shapes of exponential functions. The greater the base, the steeper the graph.>

Since $e < 3$, a) is the graph of $y = 3^x$ and b) is graph of $y = e^x$.

9) Write the following as *e* to a power:

$$\frac{(e^x \cdot e^3)^2}{e}.$$

<Use laws of exponents.>

$$\frac{(e^x \cdot e^3)^2}{e} = \frac{(e^{x+3})^2}{e} = \frac{e^{2x+6}}{e} = e^{2x+6-1}$$
$$= e^{2x+5}.$$

10) If the slope of the tangent line to $y = a^x$ at $(0, 1)$ is 1, then $a = $ ____.

<Use the definition of e.>

e.

11) Slope of tangent line to the graph of $y = a^x$ at $(0, 1)$ is -1. What is *a*?

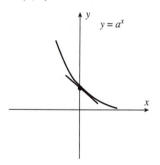

$a = e^{-1}$. This graph is a reflection of the graph of $f(x) = e^x$ about the *y*-axis. Hence $y = f(-x) = e^{-x} = (e^{-1})^x$.

Section 1.6   Inverse Functions and Logarithms

If y is a function of x described in any one of the several ways (such as by rule, by table of values, or by graph), then there are times when the same description may be used with the roles of x and y reversed, so that x is a function of y.

For example, if C is the circumference of a circle and r is the radius then $C = 2\pi r$ may be used to define C as a function of r. The same equation may be rewritten to define r as a function of C: $r = \frac{C}{2\pi}$. The two functions are inverses of each other.

This section defines inverses and sets conditions when the inverse is a function. For $a > 0$ and $a \neq 1$, $y = \log_a x$ is defined as the inverse of the exponential function $y = a^x$.

Concepts to Master

A. One-to-one functions; Horizontal Line Test

B. Inverse of a function

C. Definition, properties and graph of $y = \log_a x$; natural logarithm

D. Inverse trigonometric functions

E. Technology Plus

Summary and Focus Questions

Page 59

A. A function f is **one-to-one** means that for all x_1, x_2 in the domain of f,

$$\text{if } x_1 \neq x_2, \text{ then } f(x_1) \neq f(x_2).$$

Examples: a) $f(x) = 2x + 5$ is one-to-one because if $x_1 \neq x_2$ then $2x_1 + 5 \neq 2x_2 + 5$.
b) The function $g(x) = x^2$ is not one-to-one $(4 \neq -4$, yet $g(4) = 4^2 = 16^2 = (-4)^2 = g(-4))$.

Increasing and decreasing functions are one-to-one.

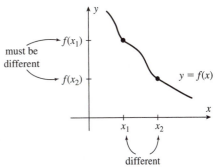

Horizontal Line Test

A function f is one-to-one if no horizontal line intersects the graph of f more than once.

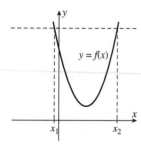

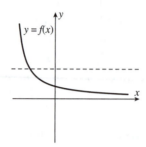

Horizontal Line Test fails:
$x_1 \neq x_2$, but $f(x_1) = f(x_2)$.
$y = f(x)$ is not one-to-one.

Horizontal Line Test succeeds:
$y = f(x)$ is one-to-one because no horizontal line intersects the graph more than once.

---

1) Is $f(x) = \sin x$ one-to-one?

<Determine whether two different x-values correspond to the same y-value.>

No. For example, $0 \neq \pi$, yet $\sin 0 = 0 = \sin \pi$.

2) Is the function $y = g(x)$ given by this table one-to-one?

x	g(x)
1	10
2	9
3	8
4	7
5	8

No, $3 \neq 5$ but both $g(3)$ and $g(5)$ are 8.

3) True, False: If $y = f(x)$ is one-to-one, then $y = f(x)$ is either increasing or decreasing.

False. This function f is one-to-one but is neither increasing nor decreasing:

$$f(x) = \begin{cases} \dfrac{1}{x} & \text{if } x < 0 \\ x+1 & \text{if } x \geq 0 \end{cases}$$

4) Is the given figure the graph of a one-to-one function?

a)

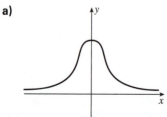

<Pass horizontal lines through the graph. See if they touch the graph in more than one place.>

The figure is the graph of a function but the function is not one-to-one.

b)

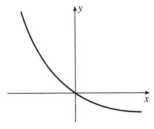

The figure is the graph of a one-to-one function.

c)

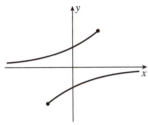

While the figure is not the graph of a function and hence not the graph of a one-to-one function.

Page 60

B. If f is one-to-one with domain A and range B then the **inverse function** f^{-1} has domain B and range A, and

$$f^{-1}(b) = a \text{ if and only if } f(a) = b.$$

Therefore:

The pair (a, b) satisfies $y = f(x)$ if and only if the pair (b, a) satisfies $y = f^{-1}(x)$.

The pair (a, b) is in the table that defines f if and only if the pair (b, a) is in the table that defines f^{-1}.

The point (a, b) is on the graph of f if and only if the point (b, a) is on the graph of f^{-1}.

These inverse properties hold

$f^{-1}(f(x)) = x$ for all x in the domain of f.
$f(f^{-1}(x)) = x$ for all x in the domain of f^{-1}.

To find the rule for $y = f^{-1}(x)$,

i) write $y = f(x)$

ii) solve the equation $y = f(x)$ for x

iii) interchange x and y.

Example: Find the inverse of the function $f(x) = x^3 - 4$.

i) $y = x^3 - 4$.

ii) Solve $y = x^3 - 4$ for x:

$$y = x^3 - 4$$
$$y + 4 = x^3$$
$$x = \sqrt[3]{y + 4}.$$

iii) Interchange x and y:

$$y = \sqrt[3]{x + 4}, \text{ so } f^{-1}(x) = \sqrt[3]{x + 4}.$$

5) Is the graph below the graph of a function with an inverse function?

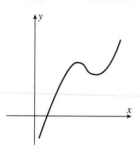

<Use the Horizontal Line Test.>

No. The function is not one-to-one.

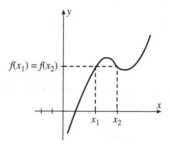

$x_1 \neq x_2$, but $f(x_1) = f(x_2)$.

6) Given the following graph of f, sketch the graph of f^{-1} on the same axis.

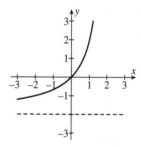

<Reflect the graph about the line $y = x$.>

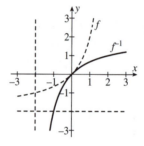

7) If f^{-1} is a function and $f(5) = 8$, then $f^{-1}(8) =$ _____.

<Use the definition of f^{-1}.>

5.

8) Find the inverse function of
$f(x) = 6x + 30$.

i) $y = 6x + 30$.
ii) Solve for x:
$$y = 6x + 30$$
$$y - 30 = 6x$$
$$x = \tfrac{1}{6}y - 5$$

iii) Interchange y and x:
$$y = \tfrac{1}{6}x - 5, \text{ so } f^{-1}(x) = \tfrac{1}{6}x - 5.$$

9) True, False:
If f is one-to-one with domain A and range B, then $f(f^{-1}(x)) = x$ for all $x \in A$.

False. $f(f^{-1}(x)) = x$ for all $x \in \underline{B}$.

10) Is $f^{-1}(x) = \dfrac{1}{f(x)}$?

No.

Page 62

C. The **logarithmic function** $y = \log_a x$ is defined as the inverse of the exponential $y = a^x$. For $a > 0$ and $a \ne 1$, $\log_a x = y$ means $a^y = x$. This definition requires $x > 0$ since $a^y > 0$ for all y. You should think of $\log_a x$ as the exponent that you put on a to get x. Thus, the function $y = \log_a x$ has these properties:

$$\log_a(a^x) = x, \text{ for all } x.$$
$$a^{\log_a x} = x, \text{ for } x > 0.$$
$$\log_a(xy) = \log_a x + \log_a y.$$
$$\log_a\left(\frac{x}{y}\right) = \log_a x - \log_a y.$$
$$\log_a x^c = c \log_a x, \text{ for all } c.$$

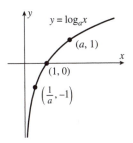

For $a > 1$, the graph of $y = \log_a x$ is given at the right.

In the special case where $a = e$, we use the notation **ln x** to mean $\log_e x$.

Thus, $\ln e = \log_e e = 1$ and $\log_a x = \dfrac{\ln x}{\ln a}$.

---

11) $\log_4 8 = $ _____.

<Use the definition of $\log_4 x$.>

Let $y = \log_4 8$.
Then $4^y = 8$
$2^{2y} = 2^3$
$2y = 3$
$y = \dfrac{3}{2}$.

12) True or False:

  a) $\log_a x^3 = 3 \log_a x$.

<Know all the properties of logarithms.>

True.

  b) $\log_2(10) = \log_2(-5) + \log_2(-2)$.

False. $\log_2(-5)$ and $\log_2(-2)$ are not defined.

  c) $\ln(a - b) = \ln a - \ln b$.

False.

  d) $\ln e = 1$.

True. In $\log_e$ notation, this is $\log_e e = 1$.

  e) $\ln\left(\frac{a}{b}\right) = -\ln\left(\frac{b}{a}\right)$.

True. $\ln\left(\frac{a}{b}\right) = \ln\left(\frac{b}{a}\right)^{-1} = -\ln\left(\frac{b}{a}\right)$.

  f) $\log_a(-x) = -\log_a x$

False.

13) Write $2 \ln x + 3 \ln y$ as a single logarithm.

$2 \ln x + 3 \ln y = \ln x^2 + \ln y^3 = \ln(x^2 y^3)$.

14) For $x > 0$, simplify using properties of logarithms: $\log_2\left(\frac{4x^3}{\sqrt{2}}\right)$.

$\log_2\left(\frac{4x^3}{\sqrt{2}}\right) = \log_2 4x^3 - \log_2 \sqrt{2}$
$= \log_2 4 + \log_2 x^3 - \log_2 2^{1/2}$
$= \log_2 4 + 3 \log_2 x - \frac{1}{2} \log_2 2$
$= 2 + 3 \log_2 x - \frac{1}{2}(1)$
$= \frac{3}{2} + 3 \log_2 x$.

15) Solve for x:

<Use the definition of ln to rewrite the equation and then solve for x.>

a) $e^{x+1} = 10$.

By definition $\log_e x = \ln x$, this means $\ln 10 = x + 1$, so $x = \ln 10 - 1$.

b) $\ln(x + 5) = 2$.

By definition of $\log_e x = \ln x$, this means $e^2 = x + 5$, so $x = e^2 - 5$.

c) $2^{3x} = 7$.

Take $\log_2$ of both sides and simplify:
$\log_2 2^{3x} = \log_2 7$
$3x = \log_2 7$
$x = \frac{1}{3} \log_2 7$.

16) $\log_2 10 = \dfrac{\ln \underline{\quad}}{\ln \underline{\quad}}$.

$\log_2 10 = \dfrac{\ln 10}{\ln 2}$

17) The magnitude of an earthquake on the Richter scale is $\log_{10}\left(\frac{I}{S}\right)$, where I is the intensity of the quake and S is the intensity of a "standard" quake. The Mexico City and San Francisco quakes were 6.4 and 7.1 on the Richter scale, respectively. How many times more powerful was the San Francisco quake than the Mexico City quake?

<Use properties of logarithms and algebra to simplify and solve.>

If the Mexico City quake intensity is I_M and the San Francisco quake intensity is I_S then we need to find a value k for which $I_S = k \cdot I_M$.

$7.1 = \log_{10}\left(\frac{I_S}{S}\right) = \log_{10}\left(\frac{kI_M}{S}\right)$
$= \log_{10} k + \log_{10}\left(\frac{I_M}{S}\right)$
$= \log_{10} k + 6.4$.

Thus $7.1 = \log_{10} k + 6.4$.

$\log_{10} k = 0.7$.

$k = 10^{0.7} \approx 5$ times as powerful.

D. None of the six trigonometric functions are one-to-one. However, when the domain of a trigonometric function is restricted to a certain interval, that restricted trigonometric function is one-to-one and thus has an inverse.

Example: The sine function is not one-to-one, but sine with domain $\left[-\frac{\pi}{2}, \frac{\pi}{2}\right]$ has an inverse function $\sin^{-1}$. Thus, for example,

$$y = \sin^{-1}\left(\frac{1}{2}\right) \text{ means } \sin(y) = \frac{1}{2} \text{ for } -\frac{\pi}{2} \le y \le \frac{\pi}{2}; \text{ therefore } y = \frac{\pi}{6}.$$

The domains and ranges of the six inverse trigonometric functions are:

Function	Domain	Range
$\sin^{-1}$	$[-1, 1]$	$\left[-\frac{\pi}{2}, \frac{\pi}{2}\right]$
$\cos^{-1}$	$[-1, 1]$	$[0, \pi]$
$\tan^{-1}$	all reals	$\left(-\frac{\pi}{2}, \frac{\pi}{2}\right)$
$\cot^{-1}$	all reals	$(0, \pi)$
$\sec^{-1}$	$(-\infty, -1] \cup [1, \infty)$	$\left[0, \frac{\pi}{2}\right) \cup \left[\pi, \frac{3\pi}{2}\right)$
$\csc^{-1}$	$(-\infty, -1] \cup [1, \infty)$	$\left(0, \frac{\pi}{2}\right] \cup \left(\pi, \frac{3\pi}{2}\right]$

The first three functions are used frequently; their graphs are:

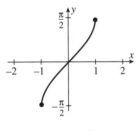

$y = \sin^{-1} x$

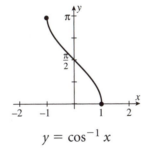

$y = \cos^{-1} x$

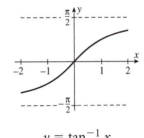

$y = \tan^{-1} x$

18) True or False:

a) $\sin^{-1}\left(\frac{3}{2}\right)$ is not defined.

b) For what x is $\tan^{-1} x = \frac{\pi}{2}$.

<Use the definition of the inverse trigonometric function in each.>
True. $\frac{3}{2}$ is not in the domain of $\sin^{-1}$.

No such x exists since $\frac{\pi}{2}$ is not in the range of $\tan^{-1}$.

19) Sometimes, Always, or Never:

    a) $\cos(\cos^{-1} x) = x$

                                                                    Sometimes. True for $-1 \le x \le 1$.

    b) $\tan(\tan^{-1} x) = x$

                                                                     Always.

    c) $\tan^{-1}(\tan x) = x$

                                                                  Sometimes. True for $-\frac{\pi}{2} < x < \frac{\pi}{2}$.

20) Determine each:

    a) $\cos^{-1}\left(\frac{1}{2}\right)$

                                                                    Let $y = \cos^{-1}\left(\frac{1}{2}\right)$. Then $0 \le y \le \pi$ and $\frac{1}{2} = \cos(y)$. From your knowledge of trigonometry, $y = \frac{\pi}{3}$.

    b) $\tan^{-1}(-\sqrt{3})$

                                                                    Let $y = \tan^{-1}(-\sqrt{3})$. Then $-\frac{\pi}{2} < y < \frac{\pi}{2}$ and $-\sqrt{3} = \tan y$. Thus $y = -\frac{\pi}{3}$.

    c) $\sin^{-1}\left(\sin \frac{3\pi}{4}\right)$

                                                                    Since $\sin \frac{3\pi}{4} = \frac{1}{\sqrt{2}}$, this problem asks you to find $y = \sin^{-1}\left(\frac{1}{\sqrt{2}}\right)$. Then $-\frac{\pi}{2} \le y \le \frac{\pi}{2}$ and $\sin y = \frac{1}{\sqrt{2}}$. Thus $y = \frac{\pi}{4}$.

    d) $\sin^{-1}\left(\sin \frac{\pi}{3}\right)$

                                                                     For $-\frac{\pi}{2} \le x \le \frac{\pi}{2}$, $\sin^{-1}(\sin x) = x$. Thus $\sin^{-1}\left(\sin \frac{\pi}{3}\right) = \frac{\pi}{3}$.

    e) $\sec^{-1}\left(\sin \frac{\pi}{7}\right)$

                                                                     This does not exist because $-1 \le \sin \frac{\pi}{7} \le 1$; $\sin \frac{\pi}{7}$ is not in the domain of $\sec^{-1}$.

21) For $0 < x < 1$, simplify $\tan(\cos^{-1} x)$.

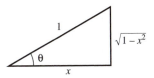

<Drawing and labeling a right triangle will help.>

Let $\theta = \cos^{-1} x$. Then $0 < \theta < \frac{\pi}{2}$ since $0 < x < 1$.

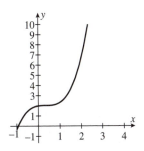

If $\cos \theta = x$ where $0 < \theta < \frac{\pi}{2}$ then

$\theta = \cos^{-1} x$ and $\tan \theta = \dfrac{\sqrt{1 - x^2}}{x}$.

E. Technology Plus. Use a computer algebra system or a graphing calculator to solve.

T-1) Graph $f(x) = x^3 - x^2 + 0.335x + 2$ in the window $[-1, 3]$ by $[-1, 10]$. Is the function one-to-one?

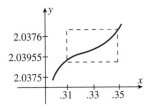

In this window it is difficult to tell. Zooming in near $x = 0.33$ we see that f is indeed always increasing and hence one-to-one.

T-2) Let $f(x) = x^3 + x$ and

$$g(x) = \sqrt[3]{\frac{x + \sqrt{x^2 + \frac{4}{27}}}{2}} + \sqrt[3]{\frac{x - \sqrt{x^2 + \frac{4}{27}}}{2}}.$$

Draw the graphs of $y = f(x)$, $y = g(x)$, and $y = x$ on the same axes. What can you conclude about f and g?

From the graphs below, $g = f^{-1}$.

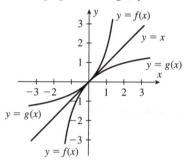

Chapter 2 — Limits and Derivatives

Section 2.1   The Tangent and Velocity Problems

This section illustrates two types of problems to which calculus may be applied:
- finding the slope of a line tangent to a curve, and
- finding the velocity of a moving object.

These are two examples of instantaneous rate of change that we will calculate as a "limit" of average rates of change.

Concepts to Master

A. Slope of secant line to the graph of a function; Slope of tangent line

B. Interpretation of slope as instantaneous velocity

Summary and Focus Questions

Page 82

A. A **secant line** at $P(a, b)$ for the graph of $y = f(x)$ is a line joining P and another point Q also on the graph. If Q has coordinates (c, d), then the slope of the secant line is

$$\frac{d - b}{c - a}.$$

Example: For $f(x) = x^3$, the secant line through

$P(2, 8)$ and $Q(3, 27)$ has slope $\dfrac{27 - 8}{3 - 2} = 19.$

$P(2, 8)$ and $Q(2.5, 15.625)$ has slope $\dfrac{15.625 - 8}{2.5 - 2} = 15.25.$

$P(2, 8)$ and $Q(2.1, 9.261)$ has slope $\dfrac{9.261 - 8}{2.1 - 2} = 12.61.$

$P(2, 8)$ and $Q(2.01, 8.1206)$ has slope $\dfrac{8.1206 - 8}{2.01 - 2} = 12.06.$

A **tangent line** to a graph of a function touches the graph at a point much like a tangent line to a circle touches the circle at a point.

To find the slope of the tangent line at P, we repeatedly select points Q closer and closer to P; the slopes of the secant lines become better and better estimates of the slope of the tangent line at P. Finally, our guess for the slope of the tangent line is the number that the slopes of the secant lines seem to be approaching as points Q get closer and closer to P.

Example: For $f(x) = x^3$ and $P(2, 8)$, the slopes of the secant lines in the above example are 19, 15.25, 12.61, 12.06 as Q gets closer to P. We estimate the slope of the tangent line to f at P to be 12.

Once we have determined the slope m of the tangent line to the curve, then using the point P as a point on the line, the equation of the tangent line is

$$y - b = m(x - a).$$

1) Let P be the point $(3, 10)$ on the graph of $y = f(x)$. Each point Q in the table below is also on the graph of f. Find the slopes of each secant line PQ.

<Calculate $\frac{d-b}{c-a}$ for each point Q.>

Q	Slope of PQ
(6, 18)	
(5, 15)	
(4, 12.3)	
(3.5, 11.1)	
(3.1, 10.21)	

Q	Slope of PQ
(6, 18)	$\frac{18-10}{6-3} = 2.67$
(5, 15)	$\frac{15-10}{5-3} = 2.5$
(4, 12.3)	$\frac{12.3-10}{4-3} = 2.3$
(3.5, 11.1)	$\frac{11.1-10}{3.5-3} = 2.2$
(3.1, 10.21)	$\frac{10.21-10}{3.1-3} = 2.1$

2) What is your guess for the slope of the tangent line to $y = f(x)$ at $x = 3$ in question 1?

The slopes of the secant lines appear to be approaching 2.

3) Let $P(1, 5)$ be a point on the graph of $f(x) = 6x - x^2$. Let $Q(x, f(x))$ be on the graph. Find the slope of the secant line PQ for each given x value for Q.

x	f(x)	Slope of PQ
3		
2		
1.5		
1.01		

x	f(x)	Slope of PQ
3	9	$\frac{9-5}{3-1} = 2$
2	8	$\frac{8-5}{2-1} = 3$
1.5	6.75	$\frac{6.75-5}{1.5-1} = 3.5$
1.01	5.0399	$\frac{5.0399-5}{1.01-1} = 3.99$

4) Use your answer to question 3 to guess the slope of the tangent to $f(x)$ at P.

The values appear to be approaching 4.

5) What is the equation of the tangent line to $f(x)$ at P in question 3?

<Use the given point $P(1, 5)$ and the slope (4) from question 3 in the point-slope form for the equation of a line.>

$y - 5 = 4(x - 1)$.

Page 84

B. Suppose an object is moving along an axis (a number line). If $f(x)$ represents the distance the object is located from its initial starting point at time x, then

the slope of the secant line between points P and Q is the **average velocity** of the object as it travels from P to Q.

the slope of the tangent line at the point P is the **instantaneous velocity** of the object at P.

Thus, given a distance function:

To find an average velocity over an interval of time, calculate the slope of a secant line.

To find an instantaneous velocity at a particular time, calculate the slope of a tangent line.

6) Let $f(x) = x^2 + 3x$ be the distance in feet a race car has traveled from its starting point after x seconds.

a) How far has the race car traveled after 2 seconds? After 4 seconds?

<Calculate function values.>
At $x = 2$, $f(2) = 2^2 + 3(2) = 10$ ft.
At $x = 4$, $f(4) = 4^2 + 3(4) = 28$ ft.

b) What is the average velocity over the time interval $x = 2$ to $x = 4$?

<Calculate the slope of the secant line.>
From part **a)**, $f(2) = 10$ and $f(4) = 28$.
$$\frac{f(4) - f(2)}{4 - 2} = \frac{28 - 10}{4 - 2} = \frac{18}{2} = 9 \text{ ft/sec.}$$

c) What is the instantaneous velocity when $x = 2$?

<Calculate the slopes of several secant lines.>
Choose some x values approaching 2:

x	$f(x)$	Slope of Secant
3	18	$\frac{18 - 10}{3 - 2} = 8$
2.5	13.75	$\frac{13.75 - 10}{2.5 - 2} = 7.5$
2.1	10.71	$\frac{10.71 - 10}{2.1 - 2} = 7.1$

We guess the slope of the tangent line, and thus the instantaneous velocity, is 7 ft/sec.

7) Let $f(x) = mx + b$ be a linear position function. Sometimes, Always, or Never:

The average velocity for $f(x)$ from P to Q is the same as the instantaneous velocity at P.

Always. The tangent line and all secant lines are the same: $y = mx + b$.

8) The distance that a runner is from the starting point after t seconds is given in this table:

t (seconds)	0	1	2	3	4
d (meters)	0	4	9	16	24

a) Find the average velocity over the time intervals $[1, 4]$, $[1, 3]$, $[1, 2]$, and $[0, 1]$.

Interval	Average Velocity
$[1, 4]$	$\frac{24-4}{4-1} = \frac{20}{3} = 6.67$ m/s
$[1, 3]$	$\frac{16-4}{3-1} = \frac{12}{2} = 6$ m/s
$[1, 2]$	$\frac{9-4}{2-1} = \frac{5}{1} = 5$ m/s
$[0, 1]$	$\frac{4-0}{1-0} = \frac{4}{1} = 4$ m/s

b) Estimate the instantaneous velocity at $t = 1$.

The intervals $[0, 1]$ and $[1, 2]$ give average velocities of 4 m/s and 5 m/s, respectively. We estimate the instantaneous velocity at $t = 1$ to be about 4.5 m/s, the average of the two average velocities.

Section 2.2   The Limit of a Function

This section introduces the concept of the limit of a function $y = f(x)$—the number, if any, that the values of $f(x)$ cluster about for x values near a particular number a. (Section 2.4 will provide a precise definition.) We will estimate the value of a limit by calculating several functional values or by observing trends in a graph. We will also look at limits from just one side of the limiting value— considering functional values just for x less than a, or just for x greater than a. Finally, we will see that some limits do not exist because the functional values get larger and larger without bound.

Concepts to Master

A. Limit of a function; Numerical estimation of a limit; Estimates of limit from a graph

B. Right- and left-hand limits; Relationships between limits and one-sided limits

C. Infinite limits; Vertical asymptotes

D. Technology Plus

Summary and Focus Questions

Page
87

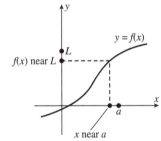

A. Suppose a function f is defined near a point a (that is, $f(x)$ exists in an open interval about a, except perhaps at $x = a$). $\lim\limits_{x \to a} f(x) = L$ means that as x gets closer and closer to the number a (but not equal to a), the corresponding $f(x)$ values get closer and closer to the number L.

Sometimes we will write "as $x \to a$, $f(x) \to L$" to mean $\lim\limits_{x \to a} f(x) = L$.

Important: For now, ignore the value $f(a)$, regardless of whether it exists, in determining $\lim\limits_{x \to a} f(x)$.

Example: $\lim\limits_{x \to 4} (x + 3) = 7$ because whenever x is near 4, then $x + 3$ will be near 7.

$\lim\limits_{x \to a} f(x) = L$ is like a guarantee: if you choose x close enough to a on the x-axis, then $f(x)$ is guaranteed to be close to L on the y-axis.

To say that "$\lim\limits_{x \to a} f(x)$ exists" means that there is some number L such that $\lim\limits_{x \to a} f(x) = L$.

To evaluate $\lim\limits_{x \to a} f(x)$ from the graph of $y = f(x)$, simply look for a number on the y-axis approached by $f(x)$ values as you choose x values closer and closer to a on the x-axis.

Another way to calculate $\lim\limits_{x \to a} f(x)$ is to select several values of x near a and observe whether there is a pattern for the corresponding $f(x)$ values. If so, then guess the value of the limit. Your answer may vary depending on the particular values of x you choose and whether you experience rounding errors from your calculator or computer. Computer algebra systems and some calculators include a "limit" command that performs the calculations and make a guess for you.

1) Find each:

a) $\lim\limits_{x \to 2} (x + 8) =$ _____ .

10. If x is near 2, $x + 8$ is near 10.

b) $\lim\limits_{x \to 4} (2x + 6) =$ _____ .

14.

c) As $x \to 2$, $x^3 \to$ _____ .

8.

d) $\lim\limits_{x \to 3} 7 =$ _____ .

7. For the constant function $f(x) = 7$, the limit is 7 because all functional values are 7. The intuitive description "as x gets closer to 3, then 7 gets closer to 7" does not make much sense when taken literally. If x is near 3, then indeed the function values (7) are near (and actually equal to) 7.

2) Answer each using the graph below:

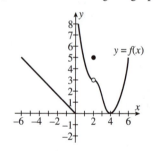

$y = f(x)$

<For x values near the limiting value on the x-axis, use the graph to see what the corresponding $f(x)$ values are near on the y-axis.>

a) $\lim\limits_{x \to -5} f(x) =$ _____ .

4. (If x is near -5, then $f(x)$ is near 4.)

b) $\lim\limits_{x \to 2} f(x) =$ _____ .

3. (It does not matter that $f(2) = 5$.)

c) $\lim\limits_{x \to 0} f(x) =$ _____ .

Does not exist.
If x is near 0 and negative, $f(x)$ is near 0. But if x is near 0 and positive then $f(x)$ is a large positive number.

d) $\lim\limits_{x \to 4} f(x) =$ _____ .

0.

3) Find $\lim\limits_{x\to 2} |x - 5|$ by making a table of values for x near 2.

x	$|x - 5|$
3	2
2.1	2.9
1.99	3.01
2.0003	2.9997

We guess that $\lim\limits_{x\to 2} |x - 5| = 3$.

4) Sketch three different functions, each of which has $\lim\limits_{x\to 4} f(x) = 2$.

<Draw a graph so that if x is near 4 then f(x) is near 2. The rest of the graph is unimportant.>

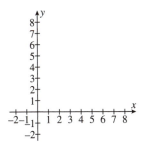

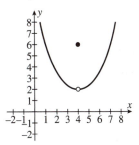

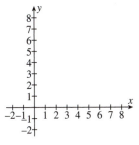

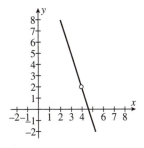

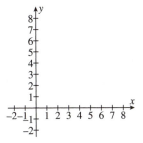

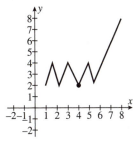

Your answers will probably not look much like the ones above except near the point $(4, 2)$—for x near 4, the $f(x)$ value must be near 2. ($f(4)$ may be any value, or not exist.)

5) Estimate $\lim\limits_{x \to 1} \dfrac{1}{1-x}$.

<Compute $\dfrac{1}{1-x}$ for x near 1.>

We make a table of values:

x	$\dfrac{1}{1-x}$
1.5	-2
1.01	-100
0.99	100
1.0004	-2500

The values of $f(x)$ do not seem to congregate near a particular value. We guess that $\lim\limits_{x \to 1} \dfrac{1}{1-x}$ does not exist.

Page 91

B. $\lim\limits_{x \to a^-} f(x) = L$, the **left-hand limit of *f* at *a*,** means that $f(x)$ gets closer to L as x approaches a and $x < a$. It is called "left-hand" because it only concerns values of x less than a (to the left of a on the x-axis).

$\lim\limits_{x \to a^+} f(x) = L$, the **right-hand limit,** is a similar concept but only considers values of x greater than a.

The relationships between limits and one-sided limits are:

If $\lim\limits_{x \to a^-} f(x)$ and $\lim\limits_{x \to a^+} f(x)$ both exist and are the same number L, then $\lim\limits_{x \to a} f(x)$ exists and is L.

If either left or right hand limit does not exist, or if they both exist but are different numbers, then $\lim\limits_{x \to a} f(x)$ does not exist.

If $\lim\limits_{x \to a} f(x) = L$, then both one-sided limits must exist and are equal to L.

6) Answer each using the graph of $y = f(x)$ below:

<For $\lim\limits_{x \to a^+} f(x)$, restrict your inspection of the graph to the right of a. For $\lim\limits_{x \to a^-} f(x)$, restrict your inspection of the graph to the left of a.>

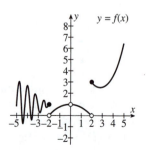

a) $\lim\limits_{x \to -2^-} f(x) =$ _____.

1.

b) $\lim\limits_{x \to -2^+} f(x) =$ _____.

0.

c) $\lim\limits_{x \to -2} f(x) = $ _____.

<Use parts **a**) *and* **b**).>

Does not exist.

d) $\lim\limits_{x \to 2^+} f(x) = $ _____.

3.

e) $\lim\limits_{x \to 2^-} f(x) = $ _____.

0.

f) $\lim\limits_{x \to 2} f(x) = $ _____.

<Use parts **d**) *and* **e**).>

Does not exist.

g) $\lim\limits_{x \to 0^-} f(x) = $ _____.

1.

h) $\lim\limits_{x \to 0^+} f(x) = $ _____.

1.

i) $\lim\limits_{x \to 0} f(x) = $ _____.

<Use parts **g**) *and* **h**).>

1.

7) $\lim\limits_{x \to 3^-} \dfrac{3 - x}{|x - 3|} = $ _____.

1. For x near 3 with $x < 3$, $x - 3 < 0$.

$|x - 3| = -(x - 3) = 3 - x.$

Therefore, $\lim\limits_{x \to 3^-} \dfrac{3 - x}{|x - 3|} = \lim\limits_{x \to 3^-} \dfrac{3 - x}{3 - x} = 1.$

8) Find $\lim\limits_{x \to 2} f(x)$, where

$$f(x) = \begin{cases} 3x + 1 & \text{if } x < 2 \\ 8 & \text{if } x = 2 \\ x^2 + 3 & \text{if } x > 2 \end{cases}$$

<Compute the right- and left-hand limits and see whether they are equal.>

$\lim\limits_{x \to 2^+} f(x) = \lim\limits_{x \to 2^+} (x^2 + 3) = 7.$

$\lim\limits_{x \to 2^-} f(x) = \lim\limits_{x \to 2^-} (3x + 1) = 7.$

Since both one-sided limits are 7,

$\lim\limits_{x \to 2} f(x) = 7.$

9) Sometimes, Always, or Never:

If $\lim\limits_{x \to 2} f(x)$ does not exist, then $\lim\limits_{x \to 2^+} f(x)$ does not exist.

Sometimes. True in the case $f(x) = \dfrac{1}{x - 2}$.

False in the case

$f(x) = \dfrac{x - 2}{|x - 2|}$ $\left(\lim\limits_{x \to 2^+} \dfrac{x - 2}{|x - 2|} = 1 \right).$

10) Sometimes, Always, or Never:

If $\lim\limits_{x\to 2^+} f(x)$ does not exist, then $\lim\limits_{x\to 2} f(x)$ does not exist.

Always.

11) The spherical helium balloons sold at the Four Ring Circus are known to burst when inflated to diameters of 80 cm or greater. Describe the volume of a balloon, V, as a function of the diameter (x) and find $\lim\limits_{x\to 80^-} V(x)$.

The volume of a sphere is $\frac{4}{3}\pi(\text{radius})^3$, so

$$V(x) = \frac{4}{3}\pi\left(\frac{x}{2}\right)^3 = \frac{\pi}{6}x^3 \text{ for } 0 \leqslant x < 80.$$

$\lim\limits_{x\to 80^-} V(x) = \lim\limits_{x\to 80^-} \frac{\pi}{6}x^3$. For x near 80, $V(x)$

is near $\dfrac{\pi(80)^3}{6} \approx 268{,}083 \text{ cm}^3.$

C. Sometimes $\lim\limits_{x\to a} f(x)$ does not exist because as x is assigned values that approach a, the corresponding $f(x)$ values grow larger without bound. In this case we say

$$\lim\limits_{x\to a} f(x) = \infty.$$

Remember that $\lim\limits_{x\to a} f(x) = \infty$ still means $\lim\limits_{x\to a} f(x)$ does not exist, but it fails to exist in this special way.

Example: For $f(x) = \dfrac{1}{(x-2)^2}$, $\lim\limits_{x\to a} f(x) = \infty$ because the graph "shoots upward" for values of x approaching 2.

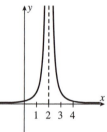

$\lim\limits_{x\to a} f(x) = -\infty$ means $f(x)$ becomes smaller without bound as x approaches a.

Just like with ordinary limits:

For $\lim\limits_{x\to a^+} f(x) = \infty$ and $\lim\limits_{x\to a^+} f(x) = -\infty$ consider only $x > a$.

For $\lim\limits_{x\to a^-} f(x) = \infty$ and $\lim\limits_{x\to a^-} f(x) = -\infty$ consider only $x < a$.

A **vertical asymptote for $y = f(x)$** is a vertical line $x = a$ for which at least one of these limits hold:

$$\lim\limits_{x\to a^-} f(x) = \infty \qquad \lim\limits_{x\to a^-} f(x) = -\infty \qquad \lim\limits_{x\to a^+} f(x) = \infty \qquad \lim\limits_{x\to a^+} f(x) = -\infty$$

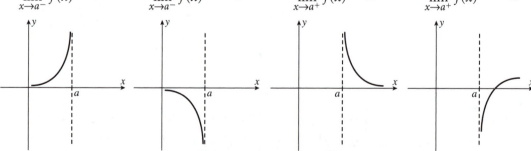

A function may have any number of vertical asymptotes or none at all. Some functions, such as $y = \tan x$, have an infinite number.

12) Evaluate the limits given this graph of
$y = f(x)$:

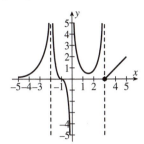

a) $\lim\limits_{x \to -2^+} f(x) = $ _____.

∞.

b) $\lim\limits_{x \to -2^-} f(x) = $ _____.

∞.

c) $\lim\limits_{x \to -2} f(x) = $ _____.

∞. (Use parts **a)** and **b)**.)

d) $\lim\limits_{x \to 3^-} f(x) = $ _____.

∞.

e) $\lim\limits_{x \to 3^+} f(x) = $ _____.

0.

f) $\lim\limits_{x \to 3} f(x) = $ _____.

Does not exist. (Use parts **d)** and **e)**.)

g) $\lim\limits_{x \to 0^+} f(x) = $ _____.

∞.

h) $\lim\limits_{x \to 0^-} f(x) = $ _____.

−∞.

i) $\lim\limits_{x \to 0} f(x) = $ _____.

Does not exist. (Use parts **g)** and **h)**.)

13) $\lim\limits_{x \to 3} \dfrac{1}{|x - 3|} = $ _____.

∞. For x near 3, but not equal to 3, $|x - 3|$ is a small positive number. Therefore, $\dfrac{1}{|x - 3|}$ is a large positive number.

14) $\lim\limits_{x \to 0^+} \ln x = $ _____.

<Know the shape of the graph of ln x near $x = 0$.>

−∞. The graph of $y = \ln x$ is:

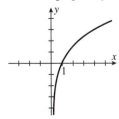

15) True or False: If $\lim\limits_{x \to a} f(x) = \infty$ and $\lim\limits_{x \to a} g(x) = \infty$, then $\lim\limits_{x \to a} (f(x) - g(x)) = 0$.

<You should be skeptical because we cannot treat ∞ as a number.>

False. For example, let $f(x) = \dfrac{3}{x^2}$ and $g(x) = \dfrac{2}{x^2}$. Then $\lim\limits_{x \to 0} \dfrac{3}{x^2} = \infty$, $\lim\limits_{x \to 0} \dfrac{2}{x^2} = \infty$, but $\lim\limits_{x \to 0} \left(\dfrac{3}{x^2} - \dfrac{2}{x^2} \right) = \lim\limits_{x \to 0} \dfrac{1}{x^2} = \infty$ (not 0).

16) Find $\lim\limits_{x \to 3} \dfrac{x}{3 - x}$.

<Determine one-sided limits first.>

For x near 3, $\dfrac{x}{3 - x}$ is the result of a number near 3 divided by a number near 0. Find one-sided limits:

$\lim\limits_{x \to 3^+} \dfrac{x}{3 - x} = -\infty$ because the denominator is negative.

$\lim\limits_{x \to 3^-} \dfrac{x}{3 - x} = \infty$. Thus, $\lim\limits_{x \to 3} \dfrac{x}{3 - x}$ does not exist.

17) What are the vertical asymptotes for the function in question 12?

$x = -2, x = 0$ and $x = 3$.

18) Find the vertical asymptotes of $f(x) = \dfrac{x}{|x| - 1}$ and sketch the graph.

<Good places to check are the x values where f(x) is not defined: x = 1 and x = −1.>

$\lim\limits_{x \to 1^+} \dfrac{x}{|x| - 1} = \infty$, and $\lim\limits_{x \to 1^-} \dfrac{x}{|x| - 1} = -\infty$, so $x = 1$ is a vertical asymptote.

$\lim\limits_{x \to -1^+} \dfrac{x}{|x| - 1} = \infty$ and $\lim\limits_{x \to -1^-} \dfrac{x}{|x| - 1} = -\infty$, so $x = -1$ is another vertical asymptote.

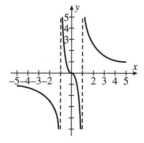

Later we will see that $y = f(x)$ has "horizontal" asymptotes as well.

D. Technology Plus. Use a computer algebra system or a graphing calculator to solve.

T-1) a)   Use a graphing device to graph

$$f(x) = \frac{x^4 - 1}{x - 1}.$$

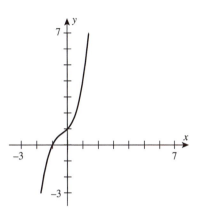

b)   As you zoom in on the graph near $x = 1$, what does the limit appear to be? Describe the nature of the graph.

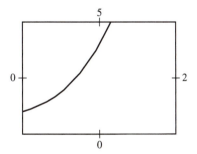

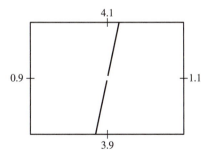

It appears that $\lim\limits_{x \to 1} f(x) = 4$.

The graph appears more linear.

For some systems a small gap will appear because $f(x)$ is not defined when $x = 1$.

c) Is $f(x)$ continuous at $x = 1$? What does the graphing device appear to tell you versus what you know about

$$f(x) = \frac{x^4 - 1}{x - 1} \text{ at } x = 1?$$

$f(x)$ is not continuous at $x = 1$. Some devices will produce a continuous looking line; others will show a gap, as mentioned above.

T-2) Graph the function

$$f(x) = \frac{(x + 1)(x - 2)}{(x^2 - 4)(x - 3)}.$$

What are the vertical asymptotes?

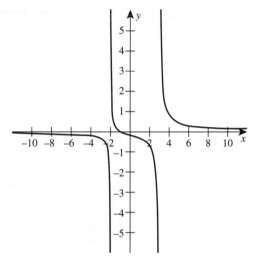

Vertical asymptotes at $x = -2$ and $x = 3$.

Section 2.3   Calculating Limits Using Limit Laws

This section will give you some useful "limit laws" for evaluating some limits without resorting to calculating several estimates for the limit value. Once we know the limits of some basic functions we will use the limit laws to evaluate limits of more complex functions. For example, if you are given the problem of trying to find $\lim_{x \to a} (f(x) + g(x))$, you do so by first finding $\lim_{x \to a} f(x)$ and $\lim_{x \to a} g(x)$, then adding the results.

Concepts to Master

A. Evaluation of limits of rational functions at points in their domain

B. Limit laws (theorems) for evaluating limits

C. Algebraic manipulations prior to using the limit laws

D. Squeeze Theorem

Summary and Focus Questions

Page 101

A. Some limits of some functions at some points are easy to find:

> If $f(x)$ is a rational function and a is in the domain of f, then $\lim_{x \to a} f(x) = f(a)$.

Since polynomials are rational functions, the above applies to them as well.

Examples: $\lim_{x \to 3} \dfrac{x}{x + 2} = \dfrac{3}{3 + 2} = \dfrac{3}{5}$ since 3 is in the domain of the rational function $f(x) = \dfrac{x}{x + 2}$.

$\lim_{x \to 2} (4x^3 - 6x^2 + 10x + 2) = 4(2)^3 - 6(2)^2 + 10(2) + 2 = 30.$

1) Find $\lim_{x \to 6} (x^2 - 10x + 3)$.

<First note that this is the limit of a polynomial function, so you may substitute directly.>

Since $x^2 - 10x + 3$ is a polynomial, the limit is $(6)^2 - 10(6) + 3 = -21.$

2) Find $\lim\limits_{x \to 1} \dfrac{x^2 - 6x + 5}{x - 1}$.

Since 1 is not in the domain of $f(x) = \dfrac{x^2 - 6x + 5}{x - 1}$, we cannot substitute $x = 1$ to find this limit. The substitution yields $\frac{0}{0}$ which will be a clue that the limit might still exist but must be found by a different method (see question 12).
For $x = 1.001$,

$$f(x) = \frac{(1.001)^2 - 6(1.001) + 5}{1.001 - 1} = -3.999.$$

The limit appears to be -4.

3) Find $\lim\limits_{x \to 2} \dfrac{x^2 - 6x + 5}{x - 1}$.

Unlike question 2, this one is straightforward.

$$\lim_{x \to 2} \frac{x^2 - 6x + 5}{x - 1} = \frac{2^2 - 6(2) + 5}{2 - 1} = -3,$$

because 2 is in the domain of the rational function $f(x) = \dfrac{x^2 - 6x + 5}{x - 1}$.

Page 99

B. The eleven basic limit laws of this section are:

1. The limit of a sum is the sum of the limits.

2. The limit of a difference is the difference of the limits.

3. The limit of a constant times a function is the constant times the limit of the function.

4. The limit of a product is the product of the limits.

5. The limit of a quotient is the quotient of the limits (provided that the limit of the denominator is not zero).

6. $\lim\limits_{x \to a} [f(x)]^n = [\lim\limits_{x \to a} f(x)]^n$ where n is a positive integer.

7. $\lim\limits_{x \to a} c = c.$

8. $\lim\limits_{x \to a} x = a.$

9. $\lim\limits_{x \to a} x^n = a^n$ where n is a positive integer.

10. $\lim\limits_{x \to a} \sqrt[n]{x} = \sqrt[n]{a}$ where n is a positive integer. (If n is even, we assume $a \geq 0$.)

11. $\lim\limits_{x \to a} \sqrt[n]{f(x)} = \sqrt[n]{\lim\limits_{x \to a} f(x)}$ where n is a positive integer. (If n is even, we assume that $\lim\limits_{x \to a} f(x) \geq 0$).

Note: Limit laws also hold for one-sided limits.

When applicable, the limit laws allow you to evaluate the limit of a complex function by rewriting the complex limit as an algebraic combination of simpler limits whose values you can find.

Example:

$$\lim_{x\to3}\left(\sqrt{x^2+x}+\frac{4x}{3x-8}+2\right) = \lim_{x\to3}\sqrt{x^2+x}+\lim_{x\to3}\frac{4x}{3x-8}+\lim_{x\to3}2 \text{ (by limit law 1)}$$

$$= \lim_{x\to3}\sqrt{x^2+x}+\lim_{x\to3}\frac{4x}{3x-8}+2 \text{ (by limit law 7)}$$

$$= \lim_{x\to3}\sqrt{x^2+x}+\frac{4(3)}{3(3)-8}+2 \left(\text{because } \frac{4x}{3x-8} \text{ is a rational function}\right)$$

$$= \lim_{x\to3}\sqrt{x^2+x}+14$$

$$= \sqrt{\lim_{x\to3}(x^2+x)}+14 \text{ (by limit law 11)}$$

$$= \sqrt{12}+14 \text{ (because } x^2+x \text{ is a polynomial function).}$$

The limit laws fail when a zero denominator results (see part **C** of this section for what to try in such cases).

4) True or False:
$$\lim_{x\to a}(f(x)g(x)) = \left(\lim_{x\to a}f(x)\right)\left(\lim_{x\to a}g(x)\right).$$

True, provided that both $\lim_{x\to a}f(x)$ and $\lim_{x\to a}g(x)$ exist. (False of one or both of the limits does not exist.)

5) True or False:
$$\lim_{x\to a}f(g(x)) = \left(\lim_{x\to a}f(x)\right)\left(\lim_{x\to a}g(x)\right).$$

False. $f(g(x))$ is not $f(x)g(x)$.

6) Evaluate $\lim_{x\to5}\sqrt[3]{\frac{4x+44}{6x-29}}$.

$$\lim_{x\to5}\sqrt[3]{\frac{4x+44}{6x-29}} = \sqrt[3]{\lim_{x\to5}\frac{4x+44}{6x-29}} = \sqrt[3]{\frac{64}{1}} = 4.$$

7) Evaluate $\lim_{x\to1}\frac{x^2(6x+3)(2x-7)}{(x^3+4)(x+17)}$.

Applying the limit laws, this limit is

$$\frac{\left(\lim_{x\to1}x^2\right)\left(\lim_{x\to1}(6x+3)\right)\left(\lim_{x\to1}(2x-7)\right)}{\left(\lim_{x\to1}(x^3+4)\right)\left(\lim_{x\to1}(x+17)\right)}$$

$$= \frac{(1)(9)(-5)}{(5)(18)} = -\frac{1}{2}.$$

8) Find $\lim_{x\to3}f(x)$ where

$$f(x) = \begin{cases} 6x-4 & \text{if } x<3 \\ x^2 & \text{if } x\geq3 \end{cases}.$$

Because $f(x)$ is defined piecewise, we need to consider one-sided limits.

$$\lim_{x\to3^+}f(x) = \lim_{x\to3}x^2 = 9.$$
$$\lim_{x\to3^-}f(x) = \lim_{x\to3}6x-4 = 14.$$

Since the one-sided limits do not agree, $\lim_{x\to3}f(x)$ does not exist.

9) Given the graphs below find $\lim\limits_{x \to 2} (f + g)(x)$.

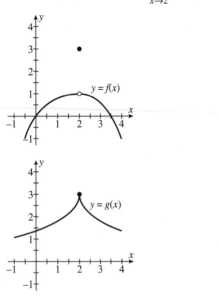

y = f(x)

y = g(x)

<Use limit law 1.>

From the graphs, we have $\lim\limits_{x \to 2} f(x) = 1$ and $\lim\limits_{x \to 2} g(x) = 3$. Thus

$$\lim\limits_{x \to 2} (f + g)(x) = 1 + 3 = 4.$$

10) $\lim\limits_{x \to 4} \sqrt{17 + 8x} = \underline{\quad\quad}.$

$$\lim\limits_{x \to 4} \sqrt{17 + 8x} = \sqrt{\lim\limits_{x \to 4} (17 + 8x)}$$

$$= \sqrt{17 + \lim\limits_{x \to 4} 8x} = \sqrt{17 + 32} = 7.$$

Page
101

C. To find a limit, first try the limit laws. Sometimes they fail, as in

$$\lim\limits_{x \to 4} \frac{x^2 - 16}{x - 4} \left(\text{gives } \tfrac{0}{0}, \text{ no information}\right).$$

You should next try algebraic techniques to rewrite the function in a form for which the limit theorems do work.

"Division by zero" is one common reason why the limit laws fail. Try to rewrite the function in a form that will not give a zero denominator. There are several techniques for this and it may be that more than one technique is required in a given problem. The most common techniques are:

i) *Cancellation:*

When substitution of $x = a$ in $\lim\limits_{x \to a} \dfrac{f(x)}{g(x)}$ results in $\tfrac{0}{0}$, it may be that $x - a$ is a factor of both $f(x)$ and $g(x)$ and can be canceled.

Example: Find $\lim\limits_{x \to 2} \dfrac{2x^2 - 4x}{x^2 - 4}$.

Using $x = 2$ gives $\tfrac{0}{0}$ but $\dfrac{2x^2 - 4x}{x^2 - 4} = \dfrac{2x(x - 2)}{(x + 2)(x - 2)} = \dfrac{2x}{x + 2}$, provided $x \neq 2$.

Thus, $\lim\limits_{x \to 2} \dfrac{2x^2 - 4x}{x^2 - 4} = \lim\limits_{x \to 2} \dfrac{2x}{x + 2} = \dfrac{2 \cdot 2}{2 + 2} = 1.$

ii) *Fraction Manipulation:*

Some limits involve functions expressed as complex fractions. Rewriting the fraction in simpler terms may allow the limit laws to be used.

Example: Find $\displaystyle\lim_{x \to 1} \frac{\frac{1}{x} - x}{\frac{1}{x} - 1}$.

Using $x = 1$ gives $\frac{0}{0}$. We rewrite the function by first finding a simpler numerator and denominator and then using cancellation:

$$\frac{\frac{1}{x} - x}{\frac{1}{x} - 1} = \frac{\frac{1}{x} - \frac{x^2}{x}}{\frac{1}{x} - \frac{x}{x}} = \frac{\frac{1 - x^2}{x}}{\frac{1 - x}{x}} = \frac{1 - x^2}{1 - x} = \frac{(1 - x)(1 + x)}{1 - x} = 1 + x, \text{ for } x \neq 1.$$

Thus, $\displaystyle\lim_{x \to 1} \frac{\frac{1}{x} - x}{\frac{1}{x} - 1} = \lim_{x \to 1} (1 + x) = 2.$

iii) *Rationalizing an Expression:*

If two radicals appear in the function, a process similar to using the conjugate of a complex number may sometimes be used to rewrite part of the function without radicals. For functions with expressions of the form

$$\sqrt{a} - \sqrt{b}, \text{ use } \frac{\sqrt{a} + \sqrt{b}}{\sqrt{a} + \sqrt{b}}.$$

The result may be simplified using cancellation.

Example: Find $\displaystyle\lim_{x \to 0} \frac{\sqrt{2 + x} - \sqrt{2}}{x}$.

The expression $\frac{0}{0}$ results when you try $x = 0$, but if $\sqrt{2 + x} - \sqrt{2}$ is multiplied by $\sqrt{2 + x} + \sqrt{2}$, no radicals will remain in the numerator:

$$\frac{\sqrt{2 + x} - \sqrt{2}}{x} \cdot \frac{\sqrt{2 + x} + \sqrt{2}}{\sqrt{2 + x} + \sqrt{2}} = \frac{2 + x - 2}{x(\sqrt{2 + x} + \sqrt{2})} = \frac{1}{\sqrt{2 + x} + \sqrt{2}},$$

for $x \neq 0$. We end up with radicals in the denominator, but now we can use the limit laws.

$$\lim_{x \to 0} \frac{\sqrt{2 + x} - \sqrt{2}}{x} = \lim_{x \to 0} \frac{1}{\sqrt{2 + x} + \sqrt{2}} = \frac{1}{\sqrt{2} + \sqrt{2}} = \frac{1}{2\sqrt{2}}.$$

11) Evaluate $\displaystyle\lim_{x \to 6} \frac{2x - 12}{x^2 - x - 30}$.

<Note that both the numerator and denominator may be factored.>

$$\frac{2x - 12}{x^2 - x - 30} = \frac{2(x - 6)}{(x - 6)(x + 5)}$$

$$= \frac{2}{x + 5}, \text{ for } x \neq 6.$$

$$\lim_{x \to 6} \frac{2}{x + 5} = \frac{2}{11}.$$

12) Evaluate $\lim\limits_{x \to 1} \dfrac{x^2 - 6x + 5}{x - 1}$.

<Note that the numerator may be factored.>

$$\frac{x^2 - 6x + 5}{x - 1} = \frac{(x - 1)(x - 5)}{x - 1}$$
$$= x - 5, \text{ for } x \neq 1.$$

$$\lim\limits_{x \to 1} (x - 5) = -4.$$

(Go back and look at how question 2 was answered.)

13) Evaluate $\lim\limits_{x \to 3} \left[\dfrac{2x^2}{x - 3} + \dfrac{6x}{3 - x} \right]$.

<First combine the fractions to produce a single fraction.>

$$\frac{2x^2}{x - 3} + \frac{6x}{3 - x} = \frac{2x^2}{x - 3} - \frac{6x}{x - 3}$$
$$= \frac{2x^2 - 6x}{x - 3} = \frac{2x(x - 3)}{x - 3} = 2x, \text{ for } x \neq 3.$$

$$\lim\limits_{x \to 3} 2x = 6.$$

14) Evaluate $\lim\limits_{x \to 0} \dfrac{\dfrac{1}{4 + x} - \dfrac{1}{4}}{x}$.

<First simplify the fraction.>

$$\frac{\dfrac{1}{4 + x} - \dfrac{1}{4}}{x} = \frac{\dfrac{4}{(4 + x)4} - \dfrac{(4 + x)}{(4 + x)4}}{x}$$

$$= \frac{\dfrac{4 - (4 + x)}{(4 + x)4}}{x} = \frac{\dfrac{-x}{(4 + x)4}}{x} = \frac{-1}{(4 + x)4}, \text{ for } x \neq 0.$$

$$\lim\limits_{x \to 0} \frac{-1}{(4 + x)4} = -\frac{1}{16}.$$

15) Evaluate $\lim\limits_{x \to 3} \dfrac{\sqrt{x} - \sqrt{3}}{x - 3}$.

$$\frac{\sqrt{x} - \sqrt{3}}{x - 3} \cdot \frac{\sqrt{x} + \sqrt{3}}{\sqrt{x} + \sqrt{3}} = \frac{x - 3}{(x - 3)(\sqrt{x} + \sqrt{3})}$$

$$= \frac{1}{\sqrt{x} + \sqrt{3}}, \text{ for } x \neq 3.$$

$$\lim\limits_{x \to 3} \frac{1}{\sqrt{x} + \sqrt{3}} = \frac{1}{2\sqrt{3}}.$$

16) Find $\lim\limits_{x \to 2} f(x)$, where

$$f(x) = \begin{cases} \dfrac{x^2 - 4}{x - 2} & \text{for } x < 2 \\ \dfrac{x^2 - x - 2}{x^2 - 4} & \text{for } x \geq 2 \end{cases}.$$

We need to compare one-sided limits:

$$\lim_{x \to 2^-} f(x) = \lim_{x \to 2^-} \frac{x^2 - 4}{x - 2}$$

$$= \lim_{x \to 2^-} \frac{(x - 2)(x + 2)}{x - 2} = \lim_{x \to 2^-} (x + 2) = 4.$$

$$\lim_{x \to 2^+} f(x) = \lim_{x \to 2^+} \frac{x^2 - x - 2}{x^2 - 4}$$

$$= \lim_{x \to 2^+} \frac{(x - 2)(x + 1)}{(x - 2)(x + 2)} = \lim_{x \to 2^+} \frac{x + 1}{x + 2} = \frac{3}{4}.$$

Since the one-sided limits $\left(4 \text{ and } \dfrac{3}{4}\right)$ do not agree, $\lim\limits_{x \to 2} f(x)$ does not exist.

D. Two theorems helps us find limit values:

Suppose $g(x)$ is in between $f(x)$ and $h(x)$; that is, $f(x) \leq g(x) \leq h(x)$ for all x near a, except perhaps when $x = a$.

Then

$$\lim_{x \to a} f(x) \leq \lim_{x \to a} g(x) \leq \lim_{x \to a} h(x)$$

if each limit exists.

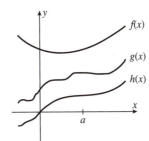

The Squeeze Theorem:

Suppose $f(x) \leq g(x) \leq h(x)$ for all x

near a, except $x = a$ and $\lim\limits_{x \to a} f(x)$ and $\lim\limits_{x \to a} h(x)$

exist and are equal. Then $\lim\limits_{x \to a} g(x)$ also exists

and $\lim\limits_{x \to a} f(x) = \lim\limits_{x \to a} g(x) = \lim\limits_{x \to a} h(x)$.

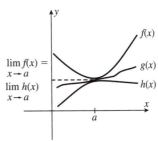

Example: For $f(x) = 4x$ and $h(x) = (x + 1)^2$,
$f(x) \leq h(x)$ for all real numbers. At $x = 1$, $\lim\limits_{x \to 1} f(x) = 4$ and $\lim\limits_{x \to 1} h(x) = 4$.
Thus, if g is any function between f and h, then $\lim\limits_{x \to 1} g(x) = 4$.

Page
105

17) True or False: If $f(x) < g(x)$ for all x and $\lim\limits_{x \to a} f(x)$ and $\lim\limits_{x \to a} g(x)$ exist, then $\lim\limits_{x \to a} f(x) < \lim\limits_{x \to a} g(x)$.

False. Let $a = 0$, $f(x) = -x^2$ and

$$g(x) = \begin{cases} x^2 & \text{if } x \neq 0 \\ 1 & \text{if } x = 0 \end{cases}.$$

Then $f(x) < g(x)$ for all x. But $\lim\limits_{x \to 0} f(x) = \lim\limits_{x \to 0} -x^2 = 0$ and $\lim\limits_{x \to 0} g(x) = \lim\limits_{x \to 0} x^2 = 0.$

18) For $0 \leq x \leq 1$, $x + 1 \leq 3^x \leq 2x + 1$. Use this to find $\lim\limits_{x \to 0^+} 3^x$.

Since $\lim\limits_{x \to 0^+} (x + 1) = 1$ and $\lim\limits_{x \to 0^+} (2x + 1) = 1$, by the Squeeze Theorem, $\lim\limits_{x \to 0^+} 3^x = 1$.

19) Sometimes, Always, or Never:

If $\lim\limits_{x \to a} f(x)$ and $\lim\limits_{x \to a} h(x)$ both exist and $f(x) < g(x) < h(x)$ for all x near a, then $\lim\limits_{x \to a} g(x)$ exists.

Sometimes. The statement is true for $f(x) = -x^2 - 1$, $g(x) = \sin x$, $h(x) = x^2 + 1$, and $a = 0$. In this case $\lim\limits_{x \to 0} \sin x = 0$. The statement is false for $f(x) = -x^2 - 1$, $g(x) = \sin \frac{1}{x}$, $h(x) = x^2 + 1$, and $a = 0$. In this case $\lim\limits_{x \to 0} \sin \frac{1}{x}$ does not exist.

20) Using the fact that if $0 < x < 1$, then $0 < x^x < 1$, what can be said about $\lim\limits_{x \to 0^+} x^x$?

If the limit exists, then $0 \leq \lim\limits_{x \to 0^+} x^x \leq 1$. This is because the function is in between the constant functions $f(x) = 0$ and $h(x) = 1$ for $0 < x < 1$. (Later we shall see that this limit exists and is 1.)

Section 2.4   The Precise Definition of a Limit

This section gives a precise definition of $\lim\limits_{x \to a} f(x) = L$ using a sentence with ϵ and δ symbols. The definition is easier to remember if you think of ϵ and δ as measures of distances between points. The limit definition is like a promise— $f(x)$ is guaranteed to be within ϵ of L when you specify an x within δ of a.

Concepts to Master

A. Epsilon-delta (ϵ-δ) definition of $\lim\limits_{x \to a} f(x) = L$

B. Epsilon-delta proof of $\lim\limits_{x \to a} f(x) = L$

C. Definitions of infinite limits

D. Technology Plus

Summary and Focus Questions

Page 108

A. The precise definition of **limit** is:

$\lim\limits_{x \to a} f(x) = L$ means for any given real number $\epsilon > 0$ there exists another real number $\delta > 0$ such that if $0 < |x - a| < \delta$, then $|f(x) - L| < \epsilon$.

This definition will be easier to remember if you can see how it is consistent with the intuitive definition of limit given in section 2.2. You should think of

  i) ϵ and δ as small positive numbers representing distances,

  ii) $0 < |x - a| < \delta$ says the distance between x and a is less than δ and $x \neq a$,

  iii) $|f(x) - L| < \epsilon$ says the distance between $f(x)$ and L is less than ϵ.

Thus, the definition says that $f(x)$ will always be within ϵ units of L (that is $|f(x) - L| < \epsilon$) whenever x is within δ units of a (that is, $|x - a| < \delta$). In other words, if x is close to a, then $f(x)$ is close to L.

There is no single technique for determining δ in terms of ϵ. Often this may be accomplished by starting with $|f(x) - L| < \epsilon$ and replacing it with equivalent statements until finally the statement $|x - a| < \delta$ results.

Reducing $|f(x) - L| < \epsilon$ down to $|x - a| < \delta$ may require several rewritings of $|f(x) - L| < \epsilon$ using factoring and cancellation. You may also need properties of absolute value, the most common being

$$|a \cdot b| = |a| \, |b| \text{ and}$$

$$|a| < b \text{ is equivalent to } -b < a < b.$$

Example: Given $\lim\limits_{x \to 3} (6x - 5) = 13$ and $\epsilon > 0$, find a suitable δ.

We start with $|f(x) - L| < \epsilon$ and simplify:

$|f(x) - L| = |(6x - 5) - 13| = |6x - 18| = |6(x - 3)| = |6| \, |x - 3| = 6|x - 3|$.

Thus, $|6x - 5 - 13| < \epsilon$ is equivalent to $6|x - 3| < \epsilon$.

Divide by 6: $|x - 3| < \frac{\epsilon}{6}$. Our choice for δ is $\frac{\epsilon}{6}$.

Then, if $0 < |x - 3| < \delta$, we can conclude $|(6x - 5) - 3| < \epsilon$.

When $f(x)$ is nonlinear, a simplified $|f(x) - L| < \epsilon$ may be in the form $|M(x)| \, |x - a| < \epsilon$ where $M(x)$ is some expression involving x. The procedure here is to find a constant K such that $|M(x)| < K$ for all x such that $|x - a| < p$ for some positive number p (often $p = 1$). The choice of δ is then the smaller of p and $\frac{\epsilon}{K}$, written $\delta = \min\left\{p, \frac{\epsilon}{K}\right\}$.

Example: Given $\lim\limits_{x \to 4} (x^2 - 1) = 15$ and $\epsilon > 0$, find a corresponding $\delta > 0$.

$|f(x) - L| = |(x^2 - 1) - 15| = |x^2 - 16| = |(x + 4)(x - 4)| = |x + 4| \, |x - 4|$.

For $p = 1$, if $|x - 4| < p$, then

$|x - 4| < 1$
$-1 < x - 4 < 1$
$3 < x < 5$
$7 < x + 4 < 9$.

Thus, $|x + 4| < 9$, when $|x - 4| < 1$. Therefore, using $K = 9$, we have $|f(x) - L| = |x + 4| \, |x - 4| < 9|x - 4|$ and $9|x - 4| < \epsilon$ if $|x - 4| < \frac{\epsilon}{9}$.

Thus choose $\delta = \min\left\{1, \frac{\epsilon}{9}\right\}$.

Then, if $0 < |x - 4| < \delta$, we can conclude $|(x^2 - 1) - 15| < \epsilon$.

1) Fill in the blanks in the definition of $\lim\limits_{x \to a} f(x) = L$:

For all _____ > 0 there exists _____ > 0 such that if _____ $< \delta$ then _____ $< \epsilon$.

$\epsilon, \delta, 0 < |x - a|, |f(x) - L|$.

2) What part of the definition of $\lim\limits_{x \to a} f(x) = L$ guarantees that $x = a$ is not considered?

$0 < |x - a|$.

3) Alter the $\epsilon - \delta$ definition of $\lim\limits_{x \to a} f(x) = L$ to define $\lim\limits_{x \to a^+} f(x) = L$.

<Take into account that x must be greater than a.>

For any given real number $\epsilon > 0$ there exists another real number $\delta > 0$ such that if $|x - a| < \delta$ and $x > a$, then $|f(x) - L| < \epsilon$.

4) The choice of δ usually (does, does not) depend upon the size of ϵ.

Does. For most limits, the smaller the ϵ given, the smaller must be the corresponding choice of δ.

5) For the function in the figure below, what is an appropriate value of δ for the given ϵ in $\lim_{x \to a} f(x) = L$?

A) $\delta = p$

B) $\delta = q$

C) $\delta = p + q$

D) None of these

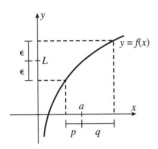

A) δ should be p or smaller, for then if x is within δ units of a, $f(x)$ will be within ϵ units of L.

6) For arbitrary $\epsilon > 0$, find a $\delta > 0$, in terms of ϵ such that the definition of limit is satisfied for $\lim_{x \to 4} (2x + 1) = 9$.

<Rewrite $|f(x) - L| < \epsilon$ so that eventually the left side of the inequality is $|x - a| < \delta$.>

$|f(x) - L| < \epsilon$ (Start)

$|(2x + 1) - 9| < \epsilon$ (Replacement)

$|2x - 8| < \epsilon$ (Simplification)

$2|x - 4| < \epsilon$ (Factoring)

$|x - 4| < \frac{\epsilon}{2}$ (Divide by 2).

The desired δ is $\delta = \frac{\epsilon}{2}$.

7) Find $\delta > 0$ for an arbitrary $\epsilon > 0$ in $\lim_{x \to 3} (x^2 - 4) = 5$.

$$|f(x) - L| = |(x^2 - 4) - 5|$$
$$= |x^2 - 9| = |(x + 3)(x - 3)|$$
$$= |x + 3| \, |x - 3|.$$

If $p = 1$ and $|x - 3| < 1$, then

$$-1 < x - 3 < 1$$
$$2 < x < 4$$
$$5 < x + 3 < 7.$$

So $|x + 3| < 7$ when $|x - 3| < 1$.

Thus, $|x + 3| \, |x - 3| < 7|x - 3| < \epsilon$ if $|x - 3| < \frac{\epsilon}{7}$.

Let $\delta = \min\left\{1, \frac{\epsilon}{7}\right\}$.

Page
112

B. A proof that $\lim\limits_{x \to a} f(x) = L$ consists of a verification of the limit definition. Start by assuming you have $\epsilon > 0$ and then determine a number δ so that the statement $|f(x) - L| < \epsilon$ can be deduced from the statement $0 < |x - a| < \delta$.

Your proof then consists of

i) Assuming $\epsilon > 0$ and stating your choice of $\delta > 0$.

ii) Showing that $0 < |x - a| < \delta$.

iii) Concluding that $\lim\limits_{x \to a} f(x) = L$.

Example: Prove $\lim\limits_{x \to 3} (6x - 5) = 13$.

For $\epsilon > 0$, the scratchwork to determine a suitable $\delta > 0$ is in part **A** of this section.

Proof:

i) Let $\epsilon > 0$. Choose $\delta = \frac{\epsilon}{6}$.

ii) Then for $0 < |x - 3| < \delta$
$$|x - 3| < \frac{\epsilon}{6}$$
$$6|x - 3| < \epsilon$$
$$|6x - 18| < \epsilon$$
$$|(6x - 5) - 13| < \epsilon.$$
In summary, if $0 < |x - 3| < \delta$, then $|(6x - 5) - 13| < \epsilon$.

iii) Thus, $\lim\limits_{x \to 3} (6x - 5) = 13$.

8) Prove that $\lim\limits_{x \to 4} (2x + 1) = 9$.
(See question 6.)

For $\epsilon > 0$, we know from question 6 that
$\delta = \frac{\epsilon}{2}$ is suitable.

Proof:

i) Let $\epsilon > 0$. Choose $\delta = \frac{\epsilon}{2}$.

ii) Then for $0 < |x - 4| < \delta$
$$|x - 4| < \frac{\epsilon}{2}$$
$$2|x - 4| < \epsilon$$
$$|2x - 8| < \epsilon$$
$$|(2x + 1) - 9| < \epsilon.$$
In summary, if $0 < |x - 4| < \delta$,
then $|(2x + 1) - 9| < \epsilon$.

iii) Thus, $\lim\limits_{x \to 4} (2x + 1) = 9$.

9) Prove that $\lim\limits_{x \to 3} (x^2 - 4) = 5$.
(See question 7.)

From question 7, we see $\delta = \min\left\{1, \frac{\epsilon}{7}\right\}$.

Proof:

i) Let $\epsilon > 0$. Choose $\delta = \min\left\{1, \frac{\epsilon}{7}\right\}$.

ii) Then for $0 < |x - 3| < \delta$, we first have

$$|x - 3| < 1$$
$$-1 < x - 3 < 1$$
$$5 < x + 3 < 7$$

Thus $|x + 3| < 7$.

We also have $|x - 3| < \frac{\epsilon}{7}$,

$$7|x - 3| < \epsilon$$

Therefore $|x + 3| \, |x - 3| < \epsilon$

$$|x^2 - 9| < \epsilon$$
$$|(x^2 - 4) - 5| < \epsilon.$$

In summary, if $0 < |x - 3| < \delta$, then $|(x^2 - 4) - 5| < \epsilon$.

iii) Thus, $\lim\limits_{x \to 3} (x^2 - 4) = 5$.

C. Recall that $\lim\limits_{x \to a} f(x) = \infty$ means that as x is assigned values that approach a, the corresponding $f(x)$ values grow larger without bound. The precise definition of this infinite limit is

Page 115

$$\lim\limits_{x \to a} f(x) = \infty \text{ means for all positive } M \text{ there is a } \delta > 0 \text{ such that}$$
$$\text{if } 0 < |x - a| < \delta \text{ then } f(x) > M.$$

Likewise,

$$\lim\limits_{x \to a} f(x) = -\infty \text{ means for all negative numbers } N \text{ there is a}$$
$$\delta > 0 \text{ such that if } 0 < |x - a| < \delta \text{ then } f(x) < N.$$

Note: The limit laws might not hold for infinite limits because we cannot treat ∞ and $-\infty$ as real numbers.

10) Prove that $\lim\limits_{x \to 2} \dfrac{1}{|4 - 2x|} = \infty$.

<Just like with ordinary limits, work "backwards" from $f(x) > M$ to find δ.>

For $M > 0$, $f(x) > M$ is $\dfrac{1}{|4 - 2x|} > M$.

$$\frac{1}{M} > |4 - 2x|$$
$$|4 - 2x| < \frac{1}{M}$$
$$|(-2)(x - 2)| < \frac{1}{M}$$

$$2|x - 2| < \frac{1}{M}$$
$$|x - 2| < \frac{1}{2M}.$$
Choose $\delta = \frac{1}{2M}.$

Proof:

i) Let $M > 0$; choose $\delta = \frac{1}{2M}.$

ii) Then if $0 < |x - 2| < \delta,$

$$|x - 2| < \frac{1}{2M}$$
$$2|x - 2| < \frac{1}{M}$$
$$|2x - 4| < \frac{1}{M}$$
$$|4 - 2x| < \frac{1}{M}$$
$$M < \frac{1}{|4 - 2x|}$$
$$\frac{1}{|4 - 2x|} > M.$$

Thus, if $0 < |x - 2| < \delta$, then $\dfrac{1}{|4 - 2x|} > M.$

iii) Therefore $\lim\limits_{x \to 2} \dfrac{1}{|4 - 2x|} = \infty.$

D. Technology Plus. Use a computer algebra system or a graphing calculator to solve.

T-1) A manufacturer of computer memory chips knows that selling x (hundred) chips will result in a revenue of $700x - .2x^2$ dollars.

a) Find $\lim\limits_{x \to 100} (700x - .2x^2).$

$$\lim_{x \to 100} (700x - .2x^2) = 700(100) - .2(100)^2$$
$$= \$68{,}000.$$

b) If x varies by 1 unit, what is the range of the revenue?

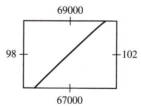

At $x = 99$, the revenue is \$67,339.80.
At $x = 101$, the revenue is \$68,659.80.
The range is from \$67,339.80 to \$68,659.80.

c) Question b) may be written in terms of the definition of a limit. What are each of these?

$f(x), a, L, \delta, \epsilon$

$f(x) = 700x - .2x^2,$

$a = 100, L = 68{,}000, \delta = 1,$

$\epsilon = \max\{|67{,}339.8 - 68{,}000|,$
$\qquad\qquad |68{,}659.8 - 68{,}000|\}$
$\quad = \max\{660.2, 659.8\} = 660.2.$

Section 2.5   Continuity

Suppose you could draw a graph of a function without lifting your pencil as you move from one end of an interval domain to the other. We would say that the function is "continuous"—meaning that the graph is unbroken over its interval domain. This section gives a precise description of the concept of an unbroken graph.

Concepts to Master

A. Continuity of a function at a point and on an interval

B. Intermediate Value Theorem

C. Technology Plus

Summary and Focus Questions

Page 118

A. A function *f* is **continuous at *a*** means both numbers $f(a)$ and $\lim_{x \to a} f(x)$ exist and are equal; that is, $\lim_{x \to a} f(x) = f(a)$.

If *f* is continuous at *a*, then the calculation of $\lim_{x \to a} f(x)$ is easy—just calculate $f(a)$.

For the type of functions we will consider, if the function *f* is continuous at the number *a*, then the graph of *f* is unbroken as it passes through $(a, f(a))$.

All functions of the following types are continuous at every real number in their domains.

Type	Example	Continuous for
Polynomial	$f(x) = 6x^2 - 5x$	all reals
Rational	$f(x) = \dfrac{x}{(x + 1)(x - 2)}$	all reals except $x = -1$, $x = 2$
Root	$f(x) = \sqrt{x + 1}$	$x \geq -1$
Trigonometric	$f(x) = \cot x$	all reals except multiples of π
Inverse Trigonometric	$f(x) = \cos^{-1} x$	$-1 \leq x \leq 1$
Exponential	$f(x) = 3^x$	all reals
Logarithmic	$f(x) = \ln x$	$x > 0$

Example: Where is the function $f(x) = \dfrac{x-2}{x^2-5x+6}$ continuous?

Since f is a rational function, f is continuous at each point in its domain, which is all x such that $x^2 - 5x + 6$ is not zero. From $x^2 - 5x + 6 = (x-2)(x-3)$, the domain of f is all real numbers except 2 and 3. Thus f is continuous at x for $x \neq 2$ and $x \neq 3$.

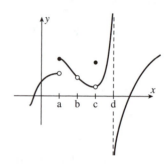

The figure at the right shows three situations where a function f is not continuous:

f has a **jump discontinuity** at $x = a$. The one-sided limits exist but are different numbers.

f has a **removable discontinuity** at $x = b$ and at $x = c$. The limits at b and c exist but $f(b)$ does not exist and while $f(c)$ exists, it is not $\lim\limits_{x \to c} f(x)$. The discontinuities are removable because redefining $f(b)$ and $f(c)$ would make f continuous.

f has an **infinite discontinuity** at $x = d$ because at least one of the one-sided limits is infinite.

Continuity from the right and left are defined in terms of one-sided limits:

A function f is **continuous from the right at a number a** if $\lim\limits_{x \to a^+} f(x) = f(a)$ and f is **continuous from the left at a** if $\lim\limits_{x \to a^-} f(x) = f(a)$.

For the figure above, f is right continuous at a but not left continuous.

For an interval $[a, b]$, f is **continuous on $[a, b]$** if f is right continuous at a, continuous at every point in (a, b), and left continuous at b.

1)

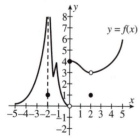

Determine the points where the function f above is not continuous (is discontinuous).

<Look for break points in the graph.>

Discontinuous at

$x = -2$, because $\lim\limits_{x \to -2} f(x)$ does not exist.

$x = 0$, because $\lim\limits_{x \to 0} f(x)$ does not exist.

$x = 2$, because $f(2) = 1$, but $\lim\limits_{x \to 2} f(x) = 3$.

2) For the graph in problem 1 determine whether f is continuous on:

    a) $[-4, -3]$

    b) $[.5, 1.5]$

    c) $[-1, 1]$

    d) $[-2, -1]$

    e) $[0, 1]$

    f) $(-2, -1]$

<Check whether there are any break points for each x in the given interval.>

Yes.

Yes.

No, not continuous at 0.

No, not continuous at -2.

Yes.

Yes.

3) Is $f(x) = \dfrac{(x+1)(x-2)}{x+1}$ continuous for all real numbers?

No. f is not defined at $x = -1$. f is continuous at every real number except -1.

4) Where is $f(x) = \dfrac{x}{(x-1)(x+2)}$ continuous?

<First recognize what type of function is given. Know where that type of function is continuous.>

Since f is rational, f is continuous everywhere it is defined: f is continuous for all x except $x = 1$, $x = -2$.

5) Find $\lim\limits_{x \to 0} \ln(x^2 + 1)$.

<First decide what happens to $x^2 + 1$ as x approaches 0.>

First note that $\lim\limits_{x \to 0} (x^2 + 1) = 1$. Therefore, the expression $\ln(x^2 + 1)$ is approaching $\ln(1) = 0$. Thus, $\lim\limits_{x \to 0} \ln(x^2 + 1) = 0$.

6) Is $f(x) = 4x^2 + 10x + 1$ continuous at $x = \sqrt{2}$?

<First recognize what type of function is given. Know where that type of function is continuous.>

Yes. Since f is a polynomial, it is continuous everywhere.

7) Where is $f(x) = \sqrt{\dfrac{1-x}{x}}$ continuous?

$(0, 1]$, f is algebraic and this is the domain of f.

8) Where is $f(x) = \tan\left(\dfrac{\pi}{2}x\right)$ discontinuous?

$f(x)$ is not continuous where it is not defined: at $x = \ldots, -5, -3, -1, 1, 3, 5, \ldots$.

f is continuous everywhere except the odd integers.

Page 125

B. The Intermediate Value Theorem:

If f is continuous on $[a, b]$ and N is between $f(a)$ and $f(b)$, then there is at least one c between a and b so that $f(c) = N$.

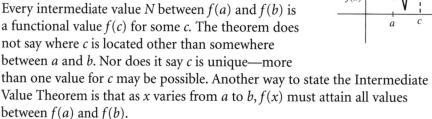

Every intermediate value N between $f(a)$ and $f(b)$ is a functional value $f(c)$ for some c. The theorem does not say where c is located other than somewhere between a and b. Nor does it say c is unique—more than one value for c may be possible. Another way to state the Intermediate Value Theorem is that as x varies from a to b, $f(x)$ must attain all values between $f(a)$ and $f(b)$.

Example: Let $f(x) = x + \dfrac{1}{x}$. The function is continuous on $\left[0.5, 3\right]$. We see that

$f(0.5) = 0.5 + \dfrac{1}{0.5} = 2.5$ and likewise,

$f(3) = 3.333$. Thus, the function f takes on all values between 2.5 and 3.333 as x ranges from 0.5 to 3.

The Intermediate Value Theorem does not apply to f on the interval $[-2, 2]$ because f is not continuous at $x = 0$. The number 1, for example, is an intermediate value between $f(-2) = -2.5$ and $f(2) = 2.5$, but there is no point c between -2 and 2 for which $f(c) = 1$.

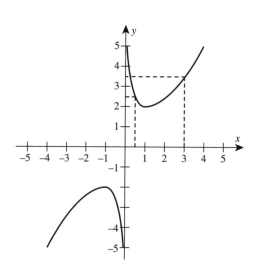

9) Mark all the possible values for c for which $f(c) = N$ in the Intermediate Value Theorem.

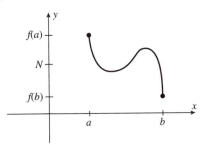

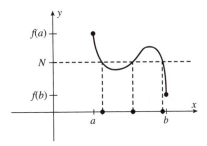

There are 3 possible values for c.

10) Does $f(x) = x^2 - x - 3$ have a root (a value w such that $f(w) = 0$) in $[1, 3]$?

<Use the Intermediate Value Theorem with $a = 1$ and $b = 3$.>

Yes. Since f is a polynomial, f is continuous on $[1, 3]$. The number 0 is between $f(1) = -3$ and $f(3) = 3$. By the Intermediate Value Theorem, there exists w between 1 and 3 such that $f(w) = 0$.

11) Is the hypothesis of the Intermediate Value Theorem satisfied for

$$f(x) = \begin{cases} \sin \frac{\pi}{x} & x \neq 0 \\ 0 & x = 0 \end{cases}$$

with domain $[-2, 2]$?

No. $f(x)$ is not continuous at 0, hence not continuous on $[-2, 2]$.

12) For $f(x) = x^2 - 3x$, is there a number $t \in [-1, 5]$ such that $f(t) = -2$?

The answer is "yes," but we cannot use the Intermediate Value Theorem to support the answer. $f(-1) = 4$ and $f(5) = 10$, but -2 is not between 4 and 10. Solving $x^2 - 3x = -2$ gives $x = 1$, $x = 2$, both in $[-1, 5]$.

13) Find an interval $[a, b]$ for which $f(x) = x^5 - 10x^2 + 6x - 1$ has a root.

<Find an interval for which 0 is between $f(a)$ and $f(b)$.>

Since f is a polynomial, f is continuous everywhere. We simply need to find an a with $f(a) < 0$ and b with $f(b) > 0$. $f(0) = -1$ and $f(3) = 170$, so $a = 0$ and $b = 3$ will do. By the Intermediate Value Theorem f has a root somewhere in $[0, 3]$.

C. Technology Plus. Use a computer algebra system or a graphing calculator to solve.

T-1) Use a graphing calculator to sketch the graph of $y = \left| \dfrac{1}{1 - \cos \pi x} \right|$.
Where is the function discontinuous?

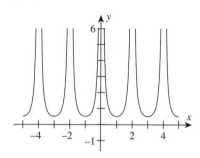

The function is discontinuous whenever x is an even integer.

T-2) Graph the function $f(x) = \dfrac{1}{5 \sin x + \cos x}$ for $-5 \le x \le 5$. For how many points in $[-5, 5]$ is f not continuous?

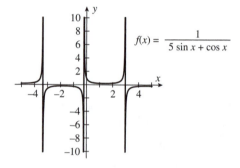

$$f(x) = \dfrac{1}{5 \sin x + \cos x}$$

There are three values for x where $f(x)$ is not continuous. By zooming in, we see that they are approximately -3.339, -0.197, and 2.944.

T-3) Sketch a graph of $f(x) = \cos\left(\dfrac{1}{x}\right)$.
What is $\lim\limits_{x \to 0} f(x)$? Is f continuous at $x = 0$?

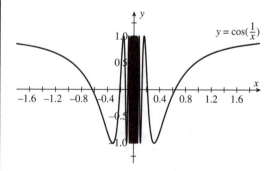

$y = \cos\left(\dfrac{1}{x}\right)$

$\lim\limits_{x \to 0} f(x)$ does not exist. f is not continuous at $x = 0$.

Section 2.6   Limits at Infinity; Horizontal Asymptotes

Limits at infinity describe the behavior of the graph of $y = f(x)$ as x gets larger and larger without bound. Just as $\lim_{x \to a} f(x) = \infty$ defined a vertical asymptote, $\lim_{x \to \infty} f(x) = L$ will define a horizontal asymptote.

Concepts to Master

A. Limits at infinity

B. Horizontal asymptote

C. Precise definition of $\lim_{x \to \infty} f(x) = L$

D. Technology Plus

Summary and Focus Questions

Page
130

A. $\lim_{x \to \infty} f(x) = L$ means that as x is assigned larger values without bound, the corresponding values of $f(x)$ approach L.

$\lim_{x \to \infty} f(x) = L$ is a description of what happens to the graph of f out to the "far right" side of the graph—the graph approaches the horizontal line $y = L$.

$\lim_{x \to -\infty} f(x) = L$ means that $f(x)$ approaches L as x takes more negative values without bound. On the left side of the graph of f the graph approaches the line $y = L$.

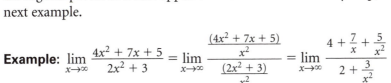

All of the usual limit theorems hold for limits at infinity. Thus, some of the techniques for evaluating ordinary limits may also be used for limits at infinity. In addition, for a limit involving a rational function, dividing the numerator and denominator by the highest power of x that appears in the denominator may help, as in the next example.

Example: $\lim_{x \to \infty} \dfrac{4x^2 + 7x + 5}{2x^2 + 3} = \lim_{x \to \infty} \dfrac{\frac{(4x^2 + 7x + 5)}{x^2}}{\frac{(2x^2 + 3)}{x^2}} = \lim_{x \to \infty} \dfrac{4 + \frac{7}{x} + \frac{5}{x^2}}{2 + \frac{3}{x^2}}$

$= \dfrac{4 + 0 + 0}{2 + 0} = 2.$

$\lim_{x \to \infty} f(x) = \infty$ means that as x increases without bound, then so do the corresponding $f(x)$ values.

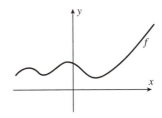

1) Evaluate $\lim\limits_{x\to\infty} f(x)$ and $\lim\limits_{x\to-\infty} f(x)$, where $y = f(x)$ has the given graph.

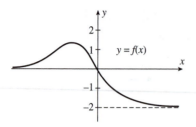

<Check the behavior of the graph to the far right and far left.>

$\lim\limits_{x\to\infty} f(x) = -2.$
$\lim\limits_{x\to-\infty} f(x) = 0.$

2) Evaluate

a) $\lim\limits_{x\to\infty} \dfrac{2}{x}$

$\lim\limits_{x\to\infty} \dfrac{2}{x} = 0.$ As x gets larger, $\dfrac{2}{x}$ gets

closer to 0.

b) $\lim\limits_{x\to\infty} \sin x$

Since the sine function oscillates between 1 and -1, the limit does not exist.

c) $\lim\limits_{x\to\infty} \dfrac{5x^3 + 7x + 1}{3x^3 + 2x^2 + 3}$

Divide numerator and denominator by x^3.

$$\lim\limits_{x\to\infty} \frac{5 + \dfrac{7}{x^2} + \dfrac{1}{x^3}}{3 + \dfrac{2}{x} + \dfrac{3}{x^3}} = \frac{5 + 0 + 0}{3 + 0 + 0} = \frac{5}{3}.$$

Limits of this type can also be "eye-balled" by observing that for very large positive x, $5x^3 + 7x + 1 \approx 5x^3$ and $3x^3 + 2x^2 + 3 \approx 3x^3$. Thus the limit is $\lim\limits_{x\to\infty} \dfrac{5x^3}{3x^3} = \dfrac{5}{3}.$

d) $\lim\limits_{x\to-\infty} \dfrac{2x + 5}{\sqrt{x^2 + 4}}$

<Divide numerator and denominator by $-x$.>

$$\lim\limits_{x\to-\infty} \frac{\dfrac{2x + 5}{-x}}{\dfrac{\sqrt{x^2 + 4}}{-x}} = \lim\limits_{x\to-\infty} \frac{\dfrac{2x + 5}{-x}}{\dfrac{\sqrt{x^2 + 4}}{\sqrt{x^2}}}$$

(Remember, since $x\to-\infty$, $x < 0$, which means $\sqrt{x^2} = -x$.)

$$= \lim\limits_{x\to-\infty} \frac{-2 - \dfrac{5}{x}}{\sqrt{1 + \dfrac{4}{x^2}}} = \frac{-2 - 0}{\sqrt{1 + 0}} = -2.$$

e) $\lim\limits_{x \to \infty} \dfrac{\cos x}{x}$

<Recognize the relative sizes of cos x and x. The values of cos x will be no greater than 1 while x increases without bound.>

Since $-1 \le \cos x \le 1$, $-\dfrac{1}{x} \le \dfrac{\cos x}{x} \le \dfrac{1}{x}$.

Both $\lim\limits_{x \to \infty} \dfrac{-1}{x} = 0$ and $\lim\limits_{x \to \infty} \dfrac{1}{x} = 0$.

Therefore, $\lim\limits_{x \to \infty} \dfrac{\cos x}{x} = 0$ by the Squeeze Theorem.

f) $\lim\limits_{x \to \infty} \dfrac{x^3 + 6x + 1}{2x^2 - 5x}$

<Recognize the function as a rational function.>

Dividing the numerator and denominator by x^2, we have $\lim\limits_{x \to \infty} \dfrac{x + \frac{6}{x} + \frac{1}{x^2}}{2 - \frac{5}{x}}$.

As x grows larger, the denominator approaches 2 but the numerator grows larger. Thus $\dfrac{x^3 + 6x + 1}{2x^2 - 5x}$ grows larger.

$\lim\limits_{x \to \infty} \dfrac{x^3 + 6x + 1}{2x^2 - 5x} = \infty$.

3) True, False: If all three limits exist, then
$\lim\limits_{x \to \infty} (f(x)g(x)) = \left(\lim\limits_{x \to \infty} f(x) \right)\left(\lim\limits_{x \to \infty} g(x) \right)$.

True.

4) $\lim\limits_{x \to -\infty} \left(\dfrac{1}{2} \right)^x =$ _____.

∞.

5) Find $\lim\limits_{x \to \infty} e^{-x}$.

As $x \to \infty$, $-x \to -\infty$. Thus $\lim\limits_{x \to \infty} e^{-x} = 0$.
Or let $z = -x$. Then $\lim\limits_{x \to \infty} e^{-x} = \lim\limits_{z \to -\infty} e^{z} = 0$.

6) Sometimes, Always, or Never:
If $a > 0$ and $a \ne 1$ then, $\lim\limits_{x \to \infty} a^x = \infty$.

Sometimes. True for $a > 1$. False for $0 < a < 1$.

7) Sketch a graph of $y = f(x)$ such that $\lim\limits_{x \to -\infty} f(x) = \infty$ and $\lim\limits_{x \to \infty} f(x) = -\infty$.

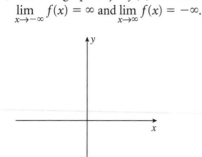

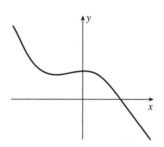

This graph is one of many possible answers. The behaviors at the ends are important.

Page 131

B. A **horizontal asymptote** for a function f is a line $y = L$ such that $\lim\limits_{x \to \infty} f(x) = L$, $\lim\limits_{x \to -\infty} f(x) = L$, or both. Here are the graphs of three examples:

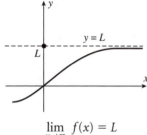

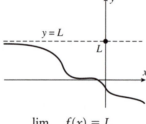

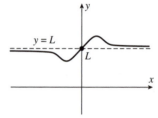

$$\lim\limits_{x \to \infty} f(x) = L \qquad\qquad \lim\limits_{x \to -\infty} f(x) = L \qquad\qquad \lim\limits_{x \to -\infty} f(x) = L \text{ and } \lim\limits_{x \to \infty} f(x) = L$$

A function may have at most two horizontal asymptotes, one in each direction.

8) Find the horizontal asymptotes for the graph:

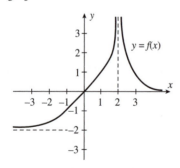

f has two horizontal asymptotes, $y = -2$ and $y = 0$. (Note: $x = 2$ is a vertical asymptote.)

9) Find the horizontal asymptotes for $f(x) = \dfrac{x}{|x| - 1}$.

There are two horizontal asymptotes $y = 1$, because $\lim\limits_{x \to \infty} \dfrac{x}{|x| - 1} = 1$ and $y = -1$, because $\lim\limits_{x \to -\infty} \dfrac{x}{|x| - 1} = -1$.

10) Find the horizontal asymptotes

of $f(x) = \dfrac{1 - \cos x}{x}$.

To find limits at infinity we use the Squeeze Theorem. Since $-1 \le \cos x \le 1$, then $-1 \le -\cos x \le 1$ and

$0 \le 1 - \cos x \le 2$. Thus $\dfrac{0}{x} \le \dfrac{1 - \cos x}{x} \le \dfrac{2}{x}$,

so $0 \le \dfrac{1 - \cos x}{x} \le \dfrac{2}{x}$ for $x > 0$. Since

$\lim\limits_{x \to \infty} \dfrac{2}{x} = 0$, $\lim\limits_{x \to \infty} \dfrac{1 - \cos x}{x} = 0$. In a similar

manner we can see that $\lim\limits_{x \to -\infty} \dfrac{1 - \cos x}{x} = 0$.

Thus, $y = 0$ is the only horizontal asymptote.

11) Find the asymptotes for $f(x) = \dfrac{x}{x^2 - 1}$.

<Evaluate limits at infinity to check for horizontal asymptotes. Check where the denominator is zero for vertical asymptotes.>

$\lim\limits_{x \to \infty} \dfrac{x}{x^2 - 1} = 0$ and $\lim\limits_{x \to -\infty} \dfrac{x}{x^2 - 1} = 0$, so

$y = 0$ is the only horizontal asymptote.

$\lim\limits_{x \to 1^+} \dfrac{x}{x^2 - 1} = \infty$ and $\lim\limits_{x \to 1^-} \dfrac{x}{x^2 - 1} = -\infty$ so

$x = 1$ is a vertical asymptote. Likewise,

$x = -1$ is a vertical asymptote because

$\lim\limits_{x \to -1^+} \dfrac{x}{x^2 - 1} = \infty$ and $\lim\limits_{x \to -1^-} \dfrac{x}{x^2 - 1} = -\infty$.

Page 137

C. The formal definition of $\lim\limits_{x \to \infty} f(x) = L$ is:

for each $\epsilon > 0$ there exists a number N such that if $x > N$ then $|f(x) - L| < \epsilon$.

Think of N as a large positive number. Then $f(x)$ will be near L (within ϵ) if we choose x large enough (greater than N).

12) $\lim\limits_{x \to \infty} f(x) = L$ means for all _____ there exists _____ such that if _____ then _____.

$\epsilon > 0$
N
$x > N$
$|f(x) - L| < \epsilon$

13) For $\epsilon = .01$, find a corresponding number N that fits the definition of $\lim\limits_{x \to \infty} \dfrac{x}{x+1} = 1$.

We work "backwards:"

$$|f(x) - L| < .01$$

$$\left|\frac{x}{x+1} - 1\right| < .01$$

$$\left|\frac{-1}{x+1}\right| < .01$$

$$\frac{1}{x+1} < .01$$

$$x + 1 > 100$$

$$x > 99.$$

For $N = 99$, if $x > N$ then $\left|\dfrac{x}{x+1} - 1\right| < .01$.

14) Give a definition of $\lim\limits_{x \to -\infty} f(x) = L$.

$\lim\limits_{x \to -\infty} f(x) = L$ means for all $\epsilon > 0$ there exists a number M such that if $x < M$ then $|f(x) - L| < \epsilon$.

D. Technology Plus. Use a computer algebra system or a graphing calculator to solve.

T-1) Use a calculator to make a table of values of $f(x) = \dfrac{x}{\sqrt{4x^2 + 900}}$ for large values of x.

Use the table to estimate $\lim\limits_{x \to \infty} f(x)$.

x	$f(x)$
100	0.4944681764
1,000	0.4999437595
10,000	0.4999994375
100,000	0.4999999944
1,000,000	0.4999999999
10,000,000	0.5000000000

From the table, we see that $\lim\limits_{x \to \infty} f(x) = \dfrac{1}{2}$.

Section 2.7   Derivatives and Rates of Change

This section returns us to the beginning of the chapter for another look at a special kind of limit—instantaneous rate of change. We will see that it is a limit of average rates of change.

Concepts to Master

A. Slope of the tangent line to the graph of a function

B. Rates of change; Instantaneous velocity

C. Definition of the derivative; Calculate and estimate the derivative of a function

D. Interpretations of the derivative

E. Technology Plus

Summary and Focus Questions

Page 143

A. For a function, f, the **tangent line to $y = f(x)$ at the point a** is the line through $(a, f(a))$ with slope

$$m = \lim_{x \to a} \frac{f(x) - f(a)}{x - a}.$$

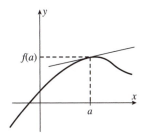

If m does not exist, the tangent line is not defined.

Another way to write the above limit is to first let $h = x - a$.

Then as $x \to a$ we have $h \to 0$. Since $h = x - a$, we can write $x = a + h$ and rewrite the definition of m:

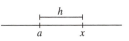

$$m = \lim_{h \to 0} \frac{f(a + h) - f(a)}{h}.$$

Either form for m may be used to calculate the slope of the tangent line to $y = f(x)$ at the point corresponding to $x = a$. The steps involved are the same as those described earlier.

For $m = \lim_{x \to a} \dfrac{f(x) - f(a)}{x - a}$:

   i) Find $f(x)$ and $f(a)$.

   ii) Calculate $\dfrac{f(x) - f(a)}{x - a}$ and simplify if possible when $x \neq a$.

   iii) Find the limit as $x \to a$ of the result of part ii).

For $m = \lim\limits_{h \to 0} \dfrac{f(a + h) - f(a)}{h}$:

  i) Find $f(a + h)$ and $f(a)$.

  ii) Calculate $\dfrac{f(a + h) - f(a)}{h}$ and simplify if possible when $h \neq 0$.

  iii) Find the limit as $h \to 0$ of the result of part ii).

Example: Find the slope of the tangent line to $y = x^2$ at the point $a = 5$.

Using $m = \lim\limits_{x \to a} \dfrac{f(x) - f(a)}{x - a}$:

  i) $f(5) = 5^2 = 25$.

  ii) $\dfrac{f(x) - f(5)}{x - 5} = \dfrac{x^2 - 25}{x - 5} = \dfrac{(x - 5)(x + 5)}{x - 5} = x + 5$, for $x \neq 5$.

  iii) $m = \lim\limits_{x \to 5} \dfrac{f(x) - f(5)}{x - 5} = \lim\limits_{x \to 5} (x + 5) = 10$.

Using $m = \lim\limits_{h \to 0} \dfrac{f(a + h) - f(a)}{h}$:

  i) $f(5 + h) = (5 + h)^2 = 25 + 10h + h^2$ and $f(5) = 25$.

  ii) $\dfrac{f(5 + h) - f(5)}{h} = \dfrac{(25 + 10h + h^2) - 25}{h} = \dfrac{10h + h^2}{h} = \dfrac{h(10 + h)}{h} = 10 + h$, for $h \neq 0$.

  iii) $m = \lim\limits_{h \to 0} \dfrac{f(5 + h) - f(5)}{h} = \lim\limits_{h \to 0} (10 + h) = 10$.

Notice that algebra is necessary in step ii) because direct substitution $x = a$ (or $h = 0$) in these limits always gives $\frac{0}{0}$.

1) Find the slope of the tangent line to $f(x) = x^2 + 2x$ at $x = 3$.

<First find $f(3)$ and simplify $\dfrac{f(x) - f(3)}{x - 3}$.>

  i) At $x = 3, f(3) = 3^2 + 2(3) = 15$.

  ii) $\dfrac{f(x) - f(a)}{x - a} = \dfrac{x^2 + 2x - 15}{x - 3} = \dfrac{(x - 3)(x + 5)}{x - 3}$
  $= x + 5$ for $x \neq 3$.

  iii) Thus, $\lim\limits_{x \to 3} \dfrac{f(x) - f(3)}{x - 3} = \lim\limits_{x \to 3} (x + 5) = 8$.

  The slope is 8.

2) Find $\lim\limits_{h \to 0} \dfrac{f(a + h) - f(a)}{h}$ where $f(x) = \dfrac{1}{x - 1}$ and $a = 2$.

<Construct the fraction $\dfrac{f(a + h) - f(a)}{h}$ and simplify it. Then evaluate the limit.>

  i) $f(2 + h) = \dfrac{1}{2 + h - 1} = \dfrac{1}{h + 1}$
  and $f(2) = \dfrac{1}{2 - 1} = \dfrac{1}{1}$.

2) *(continued)*

ii) $\dfrac{f(2+h)-f(2)}{h}$

$$= \dfrac{\frac{1}{h+1}-\frac{1}{1}}{h} = \dfrac{\frac{1}{h+1}-\frac{h+1}{h+1}}{h}$$

$$= \dfrac{\frac{1-(h+1)}{h+1}}{h} = \dfrac{\frac{-h}{h+1}}{h} = \dfrac{-1}{h+1}, \text{ for } h \neq 0.$$

iii) $\displaystyle\lim_{h\to 0} \dfrac{f(2+h)-f(2)}{h} = \lim_{h\to 0} \dfrac{-1}{h+1} = -1.$

3) For $G(x) = 4x + 3$

   a) find the slope of the tangent line to G at $a = 2$.

   i) $G(2) = 4(2) + 3 = 11.$

   ii) $\dfrac{G(x)-G(2)}{x-2} = \dfrac{(4x+3)-11}{x-2}$

$$= \dfrac{4x-8}{x-2} = \dfrac{4(x-2)}{x-2}$$

$$= 4, \text{ for } x \neq 2.$$

   iii) $\displaystyle\lim_{x\to 2} \dfrac{G(x)-G(2)}{x-2} = \lim_{x\to 2} 4 = 4.$

   b) what is the slope of the tangent line to G at any point a?

$G(x) = 4x + 3$ is a linear function, so the tangent line to G at $(a, G(a))$ will be the line itself, which has slope 4.

B. $\dfrac{f(x)-f(a)}{x-a}$ is the **average rate of change of $y = f(x)$ on the interval $[a, x]$.** Thus

Page 145

$\displaystyle\lim_{x\to a} \dfrac{f(x)-f(a)}{x-a}$ is the **instantaneous rate of change of $y = f(x)$ at $x = a$.**

In particular, if $y = f(x)$ is the position at time x of an object moving along an axis,

$\dfrac{f(x)-f(a)}{x-a}$ is the **average velocity over the time interval from a to x** and

$\displaystyle\lim_{x\to a} \dfrac{f(x)-f(a)}{x-a}$ is the **instantaneous velocity at $x = a$.**

4) Find the instantaneous velocity at time 3 seconds if a particle's position at time x is $f(x) = x^2 + 2x$ feet. See question 1.

8 ft/s. This question is asking for the same information as question 1. Here the rate of change is interpreted as velocity instead of slope.

5) a) A particle is moving along a scale in such a way that at time t, its position is $f(t) = 6t^2 - 4t + 1$ m. Find the average velocity over each of these intervals of time:

$[1, 4]$

<Compute $\dfrac{f(t) - f(a)}{t - a}$ for each.>

$$\frac{f(4) - f(1)}{4 - 1} = \frac{81 - 3}{3} = 26 \text{ m/s.}$$

$[1, 2]$

$$\frac{f(2) - f(1)}{2 - 1} = \frac{17 - 3}{1} = 14 \text{ m/s.}$$

$[1, 1.2]$

$$\frac{f(1.2) - f(1)}{1.2 - 1} = \frac{4.84 - 3}{0.2} = 9.2 \text{ m/s.}$$

$[1, 1.01]$

$$\frac{f(1.01) - f(1)}{1.01 - 1} = \frac{3.0806 - 3}{0.01} = 8.06 \text{ m/s.}$$

b) Find the instantaneous velocity of the particle at $t = 1$ second.

<Use the results from part **a)** to evaluate the limit as t approaches 1.>

From part **a)** it appears as though the answer is 8 ft/s. Let's find out:

$$\frac{f(t) - f(1)}{t - 1} = \frac{6t^2 - 4t + 1 - 3}{t - 1}$$
$$= \frac{6t^2 - 4t - 2}{t - 1}$$
$$= \frac{(6t + 2)(t - 1)}{t - 1}$$
$$= 6t + 2 \text{ for } t \neq 1.$$

Thus,

$$\lim_{t \to 1} \frac{f(t) - f(1)}{t - 1} = \lim_{t \to 1} (6t + 2) = 8 \text{ m/s.}$$

6) Fred is driving along a freeway. At 3:00 he passes mile marker 120 and at 5:00 he passes mile marker 250.

a) What was his average velocity during that time interval?

$$\frac{250 - 120}{5 - 3} = 65 \text{ mi/hr.}$$

b) When Fred passed mile marker 162, he saw that his speedometer read 54 mi/hr. What does 54 represent?

His instantaneous velocity as he passes mile marker 162.

Page
146

C. The **derivative of** $y = f(x)$ **at a point** a is

$$f'(a) = \lim_{h \to 0} \frac{f(a + h) - f(a)}{h}.$$

An equivalent way to write this is

$$f'(a) = \lim_{x \to a} \frac{f(x) - f(a)}{x - a}.$$

If the function is given algebraically, such as $f(x) = x^2 + 2x + 1$, then $f'(a)$ is calculated in the same manner as in part **A** of this section. If the function is given in table form, then the derivative is estimated after calculating several difference quotients $\dfrac{f(a + h) - f(a)}{h}$.

Example: Estimate the derivative at $a = 4$ for the function given at the right.

x	f(x)
0	1
2	5
4	8
6	11
8	13
10	16

First compute a table of difference quotients:

h	a + h	f(a + h)	f(a + h) − f(a)	$\frac{f(a+h) - f(a)}{h}$
−4	0	1	−7	$\frac{-7}{-4} = 1.75$
−2	2	5	−3	$\frac{-3}{-2} = 1.50$
2	6	11	3	$\frac{3}{2} = 1.50$
4	8	13	5	$\frac{5}{4} = 1.25$
6	10	16	8	$\frac{8}{6} = 1.33$

The table suggests that $f'(4)$ is somewhere between 1.25 and 1.75, and probably near 1.50 (since smaller values of h give difference quotients of 1.50). We estimate that $f'(4) \approx 1.50$.

7) True or False:

$$f'(a) = \lim_{h \to 0} \frac{f(h) - f(a)}{h}.$$

False.

8) Find $f'(3)$ for $f(x) = x^2 + 10x$.

<Use the three steps to find $f'(x)$.>

i) $f(3 + h) = (3 + h)^2 + 10(3 + h)$
$\qquad\qquad = 9 + 6h + h^2 + 30 + 10h$
$\qquad\qquad = 39 + 16h + h^2.$

$\quad f(3) = 3^2 + 10(3) = 39.$

ii) $\dfrac{f(3 + h) - f(3)}{h} = \dfrac{(39 + 16h + h^2) - 39}{h}$

$\qquad = \dfrac{16h + h^2}{h} = \dfrac{h(16 + h)}{h} = 16 + h,$ for $h \neq 0.$

iii) Finally, $f'(3) = \lim_{h \to 0} (16 + h) = 16.$

9) Find $f'(a)$ for $f(x) = 2x - x^2$.

i) $f(a + h) = 2(a + h) - (a + h)^2$
$$= 2a + 2h - a^2 - 2ah - h^2.$$
$$f(a) = 2a - a^2.$$

ii) $\dfrac{f(a + h) - f(a)}{h}$
$$= \frac{2a + 2h - a^2 - 2ah - h^2 - (2a - a^2)}{h}$$
$$= \frac{2h - 2ah - h^2}{h} = \frac{h(2 - 2a - h)}{h}$$
$$= 2 - 2a - h \text{ for } h \neq 0.$$

iii) $f'(a) = \lim_{h \to 0}(2 - 2a - h)$
$$= 2 - 2a - 0 = 2 - 2a.$$

10) The population $P(t)$ of the village of Rosebush is given by the table below. Estimate the instantaneous rate of change of the population for 1999.

t	P(t)
1996	841
1997	832
1998	821
1999	810
2000	801
2001	793

<Determine $P'(1999)$ by calculating several difference quotients.>

h	a + h	f(a + h)	f(a + h) − f(a)	$\dfrac{f(a+h)-f(a)}{h}$
−3	1996	841	31	−10.33
−2	1997	832	22	−11.00
−1	1998	821	11	−11.00
1	2000	801	−9	−9.00
2	2001	793	−17	−8.50

The difference quotients range from −8.50 to −11.00; we estimate $P'(1999) \approx -10.00$.

D. The slope of the tangent line (measured by $f'(a)$) is the same as the **instantaneous rate of change of $y = f(x)$ with respect to x at $x = a$.** The derivative measures how fast $f(x)$ is changing per unit change of x. Thus the derivative has many applications, such as velocity, metabolic rates, etc.

If $f'(a)$ exists, the tangent line to the graph of f at a is the line through $(a, f(a))$ with slope $f'(a)$. The tangent line has equation

$$y - f(a) = f'(a)(x - a).$$

If $f'(a)$ does not exist, then the tangent line might not exist, might be a vertical line, or might not be unique.

11) The slope of the tangent line to the graph of f at the point a is _____.

$f'(a)$, provided it exists.

12) True or False:
$f'(x)$ measures the average rate of change of $y = f(x)$ with respect to x.

False. $f'(x)$ measures instantaneous rate of change.

13) Given that $f'(x) = 3x^2 + 2$ for $f(x) = x^3 + 2x + 1$, find:

a) the slope of the tangent line to f at the point corresponding to $x = 2$.

The slope is $f'(2) = 3(2)^2 + 2 = 14$.

b) the equation of the tangent line to f at the point corresponding to $x = 2$.

*<Use part **a**) to find the slope. Also find the y-coordinate corresponding to x = 2.>*

The point of tangency has x-coordinate 2 and y-coordinate $f(2) = 2^3 + 2(2) + 1 = 13$. Hence the tangent line has equation

$$y - 13 = 14(x - 2) \text{ or } y = 14x - 15.$$

c) the instantaneous rate of change of $f(x)$ with respect to x at $x = 3$.

$f'(3) = 3(3)^2 + 2 = 29$.

d) If $f(x)$ represents the distance in feet of a particle from the origin at time x, find the instantaneous velocity at time $x = 4$ seconds.

$f'(4) = 3(4)^2 + 2 = 50$ ft/s.

14) An arrow is shot straight up into the air and falls back to earth after 8 seconds. Its height above the archer after t seconds is $H(t)$.

<Know that H'(t) is the instantaneous velocity of H at time t.>

a) What is the meaning of $H'(2)$?

$H'(2)$ is the instantaneous rate of change of the height of the arrow above the archer after 2 seconds. The rate of change of height (distance) is velocity, so $H'(2)$ is the instantaneous velocity of the arrow after 2 seconds.

b) What is $H'(9)$?

Since the arrow falls to earth after 8 seconds, it is not moving at time $t = 9$. Thus $H'(9) = 0$.

c) Is there a time a between 0 and 8 seconds for which $H'(a) = 0$?

Yes. $H'(a) = 0$ means the instantaneous velocity at time a is zero; that is, the arrow height does not change at time a.

At its highest point, before the arrow starts returning to earth, there is an instant when $H'(a) = 0$. Since the entire flight takes 8 seconds, you might guess (correctly) that $a = 4$ and $H'(4) = 0$.

15)

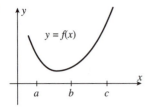

For the function *f* given above, which is largest: $f'(a), f'(b)$, or $f'(c)$?

<Use the fact that $f'(x)$ is the slope of the tangent line at $(x, f(x))$.>

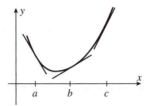

$f'(c)$ is the largest. $f'(a)$ is negative. Both $f'(b)$ and $f'(c)$ are positive, but the tangent line at $x = c$ is steeper than the tangent line at $x = b$.

E. Technology Plus. Use a computer algebra system or a graphing calculator to solve.

T-1) Let $f(x) = 2^x$ and $x = 1$.

a) Create the function

$$d(h) = \frac{f(1 + h) - f(1)}{h}.$$

Use a calculator to compute several values for $d(h)$ and estimate $f'(1)$.

h	1 + h	f(1 + h)	f(1 + h) − f(1)	$\frac{f(1+h)-f(x)}{h}$
1	2	4.00000	2.00000	2.00000
0.5	1.5	2.82843	0.82843	1.65685
0.1	1.1	2.14355	0.14355	1.43547
0.01	1.01	2.01391	0.01391	1.39111
0.001	1.001	2.00139	0.00139	1.38677
0.0001	1.0001	2.00014	0.00014	1.38634

We estimate $f'(1) \approx 1.386$.

b) Do the same as part a) for $f'(2)$.

h	2 + h	f(2 + h)	f(2 + h) − f(2)	$\frac{f(2+h)-f(2)}{h}$
1	3	8.00000	4.00000	4.00000
0.5	2.5	5.65685	1.65685	3.31371
0.1	2.1	4.28709	0.28709	2.87094
0.01	2.01	4.02782	0.02782	2.78222
0.001	2.001	4.00277	0.00277	2.77355
0.0001	2.0001	4.00028	0.00028	2.77268

We estimate $f'(2) \approx 2.772$.

Section 2.8   The Derivative as a Function

For a function *f*, by letting the number *a* vary, the values of the derivative $f'(a)$ will vary. Thus, f' is itself a function. The function f' will turn out to provide a lot of useful information about the function *f*. This section lists several other common notations for the derivative, each of which traditionally has been used in certain applications. We will also see some cases where $f'(a)$ does not exist.

Concepts to Master

A. Differentiability; Other differentiation notation

B. Relationship between continuity and differentiability

C. Relationship between the graphs of *f* and f'

D. Determining where a function is not differentiable

E. Second, third, and higher derivative; Acceleration and jerk

F. Technology Plus

Summary and Focus Questions

Page 154

A. By varying *x*, $f'(x)$ defines a function called **the derivative of *f* with respect to *x*:**

$$f'(x) = \lim_{h \to 0} \frac{f(x+h) - f(x)}{h}.$$

The function ***f* is differentiable at *x*** means that $f'(x)$ exists.

***f* is differentiable on an interval** if *f* is differentiable at every number in the interval.

All of these notations refer to the derivative of *f* at *x*:

$$f'(x), \quad y', \quad \frac{df}{dx}, \quad \frac{dy}{dx}, \quad \frac{d}{dx}f(x), \quad Df(x), \quad D_x f(x)$$

1) True or False:

$$f'(x) = \lim_{\Delta x \to 0} \frac{f(x + \Delta x) - f(x)}{\Delta x}.$$

True. This is the definition of $f'(x)$ using the symbol Δx instead of *h*.

2) Is $f(x) = x^3$ differentiable at $x = 2$?

Yes.

i) $f(2 + h) = (2 + h)^3 = 8 + 12h + 6h^2 + h^3$

ii) $\dfrac{f(2+h) - f(2)}{h} = \dfrac{8 + 12h + 6h^2 + h^3 - 8}{h}$

$$= 12 + 6h + h^3.$$

iii) $\lim\limits_{h \to 0} 12 + 6h + h^2 = 12.$

Since $f'(2)$ exists $(=12)$, *f* is differentiable at $x = 2$.

3) True or False:

For $y = f(x)$ the notation $f'(x)$ and $\frac{dx}{dy}$ have the same meaning.

False. $f'(x) = \frac{dy}{dx}$.

Page 158

B. If f is differentiable at a point a, then f is also continuous at a. In terms of the graph for f, this means that if you can draw a tangent line to the graph at the point $(a, f(a))$, then the graph must be unbroken at that point.

The converse is *false:* if a function f is continuous at a point b, it is not necessarily true that $f'(b)$ exists.

Example: The function graphed at the right is continuous at both a and b, differentiable at a, but not differentiable at b.

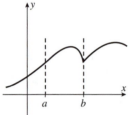

4) Can a function be continuous at $x = 3$ but not differentiable at $x = 3$?

Yes. $f(x) = |x - 3|$ is an example. f is continuous at $x = 3$, but $f'(3)$ does not exist.

5) Can a function be differentiable at $x = 3$ and not continuous at $x = 3$?

No. This can never happen.

6) True or False:

For the function f graphed below:

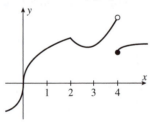

<For continuity look for an unbroken curve. For differentiability, look for a smooth curve at which you could draw a unique non-vertical tangent line.>

a) f is continuous at 0.

True.

b) f is differentiable at 0.

False. Tangent line is vertical and has no slope.

c) f is continuous at 2.

True.

d) f is differentiable at 2.

False. It is possible to draw multiple tangent lines at this point.

e) f is continuous at 3.

True.

f) f is differentiable at 3.

True.

g) f is continuous at 4.

False.

h) f is differentiable at 4.

False. This must be false because part **g)** is false.

i) f is differentiable on $[1, 3]$.

False.

j) f is differentiable on $[2.5, 3.5]$.

True.

Page 154

C. The graph of f' may be determined from the graph of f by remembering that $f'(x)$ is the slope of the tangent line at $(x, f(x))$:

   If c is the slope of the tangent line at $(x, f(x))$, then (x, c) is on the graph of $y = f'(x)$.

   Conversely, if (x, c) is on the graph of f', then $f'(x) = c$ and c is the slope of the tangent line to f at $(x, f(x))$.

In the first graph below of $y = f(x)$, we have labeled some values of the slopes of tangent lines. The second graph is the graph of $y = f'(x)$.

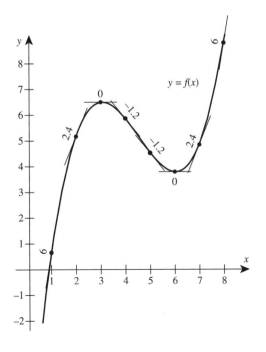

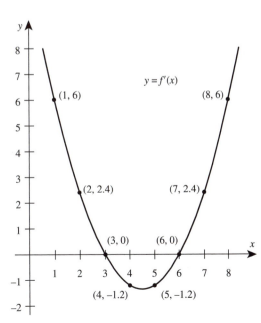

7) From the graph of *f* below, estimate $f'(1)$, $f'(2), f'(3)$, and $f'(4)$, and then sketch a graph of $y = f'(x)$.

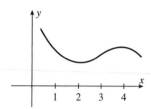

<Remember that $f'(x)$ is the slope of the tangent line.>

We estimate that $f'(1) = -2, f'(2) = 0$, $f'(3) = 1$, and $f'(4) = 0$. (Your estimates for $f'(1)$ and $f'(3)$ may be different.) For these estimates, the graph of f' is:

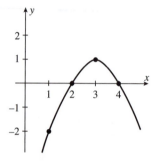

8) From the given graph of $y = f'(x)$, sketch a graph of $y = f(x)$.

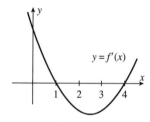

<f must have horizontal tangent lines at $x = 1$ and $x = 4$.>

One such graph of $y = f(x)$ is:

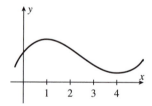

There are many other possible graphs, but all will have this shape.

9) Dan is on a ladder, 2.5 meters tall, above a trampoline which is 1 meter off the ground. He jumps onto the trampoline which propels him 3 meters into the air before he lands on the ground. Let $f(t)$ be his height after t seconds. Sketch a graph of each:

a) $y = f(t)$.

b) $y = f'(t)$.

2.5 m

1 m

<The important times are when Dan starts, when he reaches the trampoline, when the trampoline starts propelling him upward, when he reaches his highest point, and when he lands on the ground.>

When Dan is on the top of the ladder, $t = 0$ and $f(0) = 2.5$.

When Dan lands on the trampoline later, say at time $t = a$, $f(a) = 1$. For a very short time after that, he is sinking down below the trampoline rim. When the springs of the trampoline begin to propel Dan upward, the time is $t = b$, and $0 < f(b) < 1$.

Later ($t = c$), when Dan reaches his apex (3 meters high), $f(c) = 3$.

Finally, he lands on the ground at $t = d$, so $f(d) = 0$.

Between times 0 and b, Dan's velocity is negative because he is going down. Between times b and c, his velocity is positive because he is propelled upward. Between times c and d, he is headed back down so his velocity is negative.

a)

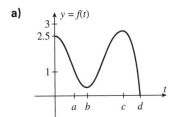

b)

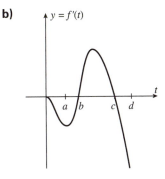

D. There are three common ways for a function to fail to be differentiable at a point.

i) The graph of the function has a corner or "kink" in it.

Example:

$$f(x) = \begin{cases} x^2 & x \leq 2 \\ (x-4)^2 & x > 2 \end{cases}.$$

f is not differentiable at $x = 2$.

Many tangent lines could be drawn at the point $(2, 4)$.

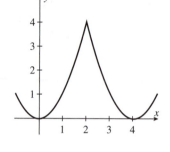

ii) The function is not continuous.

Example:

$$f(x) = \begin{cases} x^2 & x < 2 \\ 5 & x = 2 \\ 6-(x-2)^2 & x > 2 \end{cases}.$$

f is not differentiable at $x = 2$.

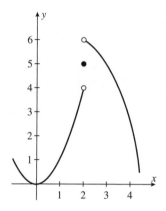

iii) The graph of the function has a tangent line, but it is a vertical line.

Example: $f(x) = \sqrt[3]{x - 2}$

f is not differentiable at $x = 2$.

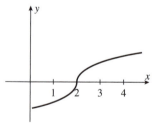

10) Where is $y = f(x)$ not differentiable?

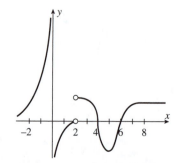

f is not differentiable at $x = 0$, $x = 2$, and $x = 4$.

Page 160

E. The **second derivative of** $y = f(x)$, denoted y'' or $f''(x)$, is the derivative of y'.
The **third derivative** is the derivative of the second derivative.
The **nth derivative of** $y = f(x)$ is denoted $y^{(n)}$.

Acceleration is the rate of change of velocity. (The gas pedal in a car is called the "accelerator" because when you push your foot on it the car speeds up—the velocity changes—the acceleration is positive.)

The second derivative is a measure of acceleration.

The third derivative measures the rate of change of acceleration and is called the **jerk**, indicating a change (think of it as sudden change) in acceleration.

Each of these are notations for the nth derivative of y with respect to x:

$$y^{(n)}, \quad f^{(n)}(x), \quad \frac{d^n y}{dx^n}$$

11) Find $f'(x)$ and $f''(x)$ for $f(x) = x^3 - 2x^2 + 3$.

$<f'(x)$ is found using the definition of the derivative. Once we have $f'(x)$ we use the definition again to find $f''(x)$.>

First find $f'(x)$.

i) $f(x + h) = (x + h)^3 - 2(x + h)^2 + 3$
$$= x^3 + 3x^2 h + 3xh^2 + h^3 - 2(x^2 + 2xh + h^2) + 3$$
$$= x^3 - 2x^2 + 3 + 3x^2 h + 3xh^2 + h^3 - 4xh - 2h^2$$
$$= (x^3 - 2x^2 + 3) + h(3x^2 + 3xh + h^2 - 4x - 2h).$$

ii) $\dfrac{f(x + h) - f(x)}{h}$
$$= \frac{[(x^3 - 2x^2 + 3) + h(3x^2 + 3xh + h^2 - 4x - 2h)] - (x^3 - 2x^2 + 3)}{h}$$
$$= \frac{h(3x^2 + 3xh + h^2 - 4x - 2h)}{h}$$
$$= 3x^2 + 3xh + h^2 - 4x - 2h, \text{ for } h \neq 0.$$

iii) $f'(x) = \lim\limits_{h \to 0} \dfrac{f(x + h) - f(x)}{h}$
$$= \lim\limits_{h \to 0} \left(3x^2 + 3xh + h^2 - 4x - 2h\right) = 3x^2 - 4x.$$

Now find $f''(x)$ by the same process, using $f'(x)$ as our function.

i) $f'(x + h) = 3(x + h)^2 - 4(x + h) = 3(x^2 + 2xh + h^2) - 4x - 4h$
$$= 3x^2 + 6xh + 3h^2 - 4x - 4h$$

ii) $\dfrac{f'(x + h) - f'(x)}{h} = \dfrac{\left[\left(3x^2 + 6xh + 3h^2 - 4x - 4h\right)\right] - \left(3x^2 - 4x\right)}{h}$
$$= \frac{6xh + 3h^2 - 4h}{h} = \frac{h(6x + 3h - 4)}{h} = 6x + 3h - 4, \text{ for } h \neq 0.$$

iii) $f''(x) = \lim\limits_{h \to 0} \dfrac{f'(x+h) - f'(x)}{h}$

$= \lim\limits_{h \to 0} (6x + 3h - 4x) = 6x - 4.$

12) A particle moves along an axis in such a way that at time x the particle is $x^3 - 2x^2 + 3$ meters from the origin. Find the velocity and acceleration at time $x = 3$. Is the particle moving left or right? Is the particle speeding up or slowing down? Use question 11.

The position function is $f(x) = x^3 - 2x^2 + 3$. The velocity is $f'(x) = 3x^2 - 4x$ and at $x = 2$, $f'(2) = 3(2)^2 - 4(2) = 12 - 8 = 4$ m/sec. The acceleration is $f''(x) = 6x - 4$ and at $x = 2, f''(2) = 6(2) - 4 = 12 - 4 = 8$ m/sec^2. Since the velocity at $x = 2$ is positive, the particle is moving to the right. Since the acceleration at $x = 2$ is positive, the particle is increasing in speed.

13) Use your best judgment to determine the graph of $y = f''(x)$ given this graph of $y = f(x)$.

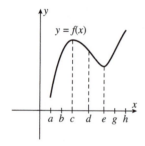

<$f'(x)$ is the slope of the tangent line to $y = f(x)$ and $f''(x)$ is the slope of the tangent line to $y = f'(x)$.>

From the slopes of tangents to $y = f(x)$ the graph of $y = f'(x)$ is:

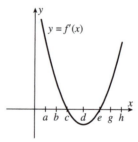

From the slopes of tangents to this graph of $y = f'(x)$ the graph of $y = f''(x)$ is:

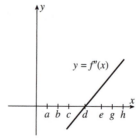

E. Technology Plus. Use a computer algebra system or a graphing calculator to solve.

T-1) Use a calculator to graph $f(x) = \left| 9 - x^2 \right|$ for $-5 \leq x \leq 5$. Where is f not differentiable?

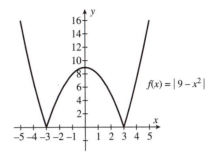

The function is not differentiable at $x = -3$ and $x = 3$.

Chapter 3 — Derivatives

Section 3.1 Derivatives of Polynomials and Exponential Functions

This section is fundamental to all of calculus. It shows you how to calculate the derivative for many types of functions using rules and formulas rather than calculating limits.

Concepts to Master

A. The derivatives of $f(x) = c$ and $f(x) = x^n$

B. The derivatives of sums and differences

C. Definition of e; derivative of e^x

Summary and Focus Questions

Page 174

A. The derivative of a constant function $f(x) = c$ is zero.
For example, if $f(x) = 7$, then $f'(x) = 0$ for all x.

(**General Power Rule**) For any real number n, the derivative of the function $f(x) = x^n$ is $f(x) = nx^{n-1}$.

For example, if $f(x) = x^4$, then $f'(x) = 4x^3$ for all x.
In symbols these rules are

$$\frac{d}{dx}(c) = 0 \qquad \text{and} \qquad \frac{d}{dx}(x^n) = nx^{n-1}.$$

1) True, False:

   a) For $f(x) = -5, f'(x) = 0$.

      True.

   b) For $f(x) = x^6, f'(x) = 6x$.

      False, $f'(x) = 6x^5$.

   c) For $f(x) = 2^x, f'(x) = x2^{x-1}$.

      False. The variable x must be in the base of the expression and the exponent must be a constant. We will have to wait a while before we find out that the derivative of 2^x is $(\ln 2)2^x$.

2) Find $f'(x)$ for each:

   a) $f(x) = 10$

                                           $f'(x) = 0.$

   b) $f(x) = x^{-3}$

                                 $f'(x) = -3x^{-4} = -\dfrac{3}{x^4}.$

   c) $f(x) = x$

                                 $f'(x) = 1.$ Since $f(x) = x = x^1$,
                                 $f'(x) = 1\, x^{1-1} = 1\, x^0 = 1 \cdot 1 = 1.$

B. The derivative of a sum or difference of functions, or the multiple of a function is found by putting together the derivatives of the components using these rules:

Page 177

Name	Rule	Example $H(x)$	$H'(x)$
Constant multiple	$[cf(x)]' = cf'(x)$	$10x^3$	$30x^2\,(= 10(3x^2))$
Sum Rule	$[f(x) + g(x)]' = f'(x) + g'(x)$	$x^3 + x^4$	$3x^2 + 4x^3$
Difference Rule	$[f(x) - g(x)]' = f'(x) - g'(x)$	$x^5 - x^2$	$5x^4 - 2x$

The derivative of a polynomial function is found by applying these rules, perhaps several times.

Example: To find the derivative of $f(x) = 9x^2 - 11x^3 + 7$ we first note that f is the sum or difference of three terms. We find the derivative of each:

$$\frac{d}{dx}(9x^2) = 9\frac{d}{dx}x^2 = 9(2x) = 18x.$$

$$\frac{d}{dx}(11x^3) = 11\frac{d}{dx}x^3 = 11(3x^2) = 33x^2.$$

$$\frac{d}{dx}(7) = 0.$$

Therefore, $f'(x) = 18x - 33x^2 + 0 = 18x - 33x^2.$

You should become very proficient at using these and other differentiation rules as they will be used throughout the text.

3) Find $f'(x)$ for each:

   a) $f(x) = 5x^4$

                                   $f'(x) = 20x^3.$

   b) $f(x) = x^4 + 4x^3$

                                   $f'(x) = 4x^3 + 12x^2.$

   c) $f(x) = 7x^3 + 6x^2 + 10x + 12$

                                   $f'(x) = 21x^2 + 12x + 10.$

d) $f(x) = x^{-7} - x^{-6}$

$f'(x) = -7x^{-8} + 6x^{-7}.$

e) $f(x) = x^{\pi}$

$f'(x) = \pi x^{\pi - 1}.$

f) $f(x) = 3x^3 - 6x^{-2}$

$f'(x) = 9x^2 + 12x^{-3}.$

g) $f(x) = (x + 1)^2$

For now we have no way to find the derivative except to first multiply out the expression:
$$f(x) = (x + 1)^2 = x^2 + 2x + 1.$$
Thus, $f'(x) = 2x + 2.$

(We will see another way to deal with these types of functions later in this chapter.)

4) Suppose the position of a particle along an axis at time t is $f(t) = t^2 + \frac{1}{t}$ meters. Find the velocity of the particle at time $t = 2$ seconds.

We can now use differentiation formulas to find velocities.
$$f(t) = t^2 + t^{-1}$$
$$f'(t) = 2t + (-1)t^{-2} = 2t - \frac{1}{t^2}$$
$$f'(2) = 2(2) - \frac{1}{2^2} = 3.75 \text{ m/s.}$$

5) For $f(x) = mx + b$, find $f'(x)$. What does this say about the tangent line to $f(x)$ at any point?

$f'(x) = m$, a constant. Therefore, the tangent line always has slope m. Since $f(x)$ has slope m, the tangent line at any point is the line itself.

Page 180

C. The number e is defined as the constant such that $\lim\limits_{h \to 0} \dfrac{e^h - 1}{h} = 1$.

The number e is an irrational number, approximately 2.71828.

In general, for $f(x) = a^x$, $f'(x) = f'(0) \cdot a^x$; that is,
$$(a^x)' = a^x \cdot \left(\lim_{h \to 0} \frac{a^h - 1}{h} \right).$$

Thus the number e was chosen in order to make this differentiation formula simple:

If $f(x) = e^x$, then $f'(x) = e^x$.

6) For what number(s) k is $\lim\limits_{h \to 0} \dfrac{k^h - 1}{h} = 1$?

$k = e$ is the only such number.

7) True or False:
If $f(x) = e^x$, then $f'(x) = f(x)$.

True. The derivative of e^x is e^x.

8) Find $f'(x)$ for:
  a) $f(x) = 3e^x$

$f'(x) = 3e^x$.

  b) $f(x) = e^3$

$f'(x) = 0$. (Remember e is a constant and therefore e^3 is a constant. The derivative of a constant is zero.)

  c) $f(x) = x^2 - 6x - e^x$

$f'(x) = 2x - 6 - e^x$

  d) $f(x) = \dfrac{e^x}{10}$

$<Recognize\ f\ as\ f(x) = \dfrac{1}{10}e^x.>$

$f'(x) = \dfrac{1}{10}e^x = \dfrac{e^x}{10}$.

  e) $f(x) = ex + e^x$

$f'(x) = e + e^x$.

9) For what a is $(a^x)' = a^x$?

Only for $a = e$.

10) Find the equation of the tangent line to the graph of $f(x) = x^2 + e^x$ at the point corresponding to $x = 3$.

For $x = 3$, $f(3) = 3^2 + e^3 = 9 + e^3$.
$f'(x) = 2x + e^x$.
$f'(3) = 2(3) + e^3 = 6 + e^3$.
The equation of the tangent line is
$y - (9 + e^3) = (6 + e^3)(x - 3)$
$y = (6 + e^3)x - 2e^3 - 9$.

Section 3.2   The Product and Quotient Rules

This section contains the rules for finding derivatives for functions formed from other functions by multiplication or division.

Concepts to Master

The derivatives of products and quotients

Summary and Focus Questions

The derivative of a product or quotient of functions is found by putting together the derivatives of the components using these rules. The rules are not as simple as the sum and difference rules.

Page 184

Name	Rule	Example $H(x)$	$H'(x)$
Product Rule	$[f(x) \cdot g(x)]' = f(x)g'(x) + f'(x)g(x)$	$x^3(x + 8)$	$x^3(1) + (x + 8)(3x^2)$
Quotient Rule	$\left[\dfrac{f(x)}{g(x)}\right]' = \dfrac{g(x)f'(x) - f(x)g'(x)}{(g(x))^2}$	$\dfrac{x^4}{x^2 + 1}$	$\dfrac{(x^2 + 1)(4x^3) - x^4(2x)}{(x^2 + 1)^2}$

To use the Quotient Rule for, say $f(x) = \dfrac{4x^2}{x^2 + 3}$, some people will

Draw a fraction line ($f'(x) = $ _____)

Insert the denominator squared below the line $\left(f'(x) = \dfrac{}{(x + 3)^2} \right)$

Insert the denominator above the line $\left(\dfrac{(x + 3)}{(x + 3)^2} \right)$

Complete the derivative $f'(x) = \dfrac{(x + 3)(8x) - (4x^2)(2x)}{(x + 3)^2}$.

1) Find $f'(x)$ for each:

a) $f(x) = 5x^4(x + 1)$

$f'(x) = 20x^3(x + 1) + 5x^4(1)$
$= 25x^4 + 20x^3.$

b) $f(x) = (x^2 + x)(3x + 1)$

$f'(x) = (x^2 + x)(3) + (3x + 1)(2x + 1)$
$= 9x^2 + 8x + 1.$

c) $f(x) = \dfrac{x^3}{x^2 + 10}$

$f'(x) = \dfrac{(x^2 + 10)(3x^2) - (x^3)(2x)}{(x^2 + 10)^2} = \dfrac{x^4 + 30x^2}{(x^2 + 10)^2}.$

d) $f(x) = \dfrac{1 + x^{-1}}{2 - x^{-2}}$

$f'(x) = \dfrac{(2 - x^{-2})(-x^{-2}) - (1 + x^{-1})(2x^{-3})}{(2 - x^{-2})^2}.$

e) $f(x) = \dfrac{\sqrt{x} - 1}{\sqrt{x} + 1}$

$f(x) = \dfrac{x^{1/2} - 1}{x^{1/2} + 1}$, so $f'(x) =$

$\dfrac{(x^{1/2} + 1)\left(\frac{1}{2}x^{-1/2}\right) - (x^{1/2} - 1)\left(\frac{1}{2}x^{-1/2}\right)}{(x^{1/2} + 1)^2}$

$= \dfrac{2\left(\frac{1}{2}x^{-1/2}\right)}{(x^{1/2} + 1)^2} = \dfrac{1}{\sqrt{x}(\sqrt{x} + 1)^2}.$

f) $f(x) = x^2 e^x$

$f'(x) = x^2(e^x) + e^x(2x) = (x^2 + 2x)e^x.$

2) Find $f'(x)$ for $f(x) = \dfrac{1}{x^2}$ two ways:

a) by the Power Rule.

$f(x) = \dfrac{1}{x^2} = x^{-2}.$

Thus, $f'(x) = -2x^{-3} = \dfrac{-2}{x^3}.$

b) by considering $\dfrac{1}{x^2}$ as a quotient.

$f'(x) = \dfrac{x^2(0) - 1(2x)}{(x^2)^2} = \dfrac{-2x}{x^4} = \dfrac{-2}{x^3}.$

3) Find $f'(x)$ for $f(x) = 4x^3$ using the product rule.

$f'(x) = (0)(x^3) + (4)(3x^2) = 12x^2.$

Note: This problem demonstrates that the constant multiple rule is a special case of the product rule. For problems like this one it is much easier to use the constant multiple rule.

4) True, False.

$\dfrac{d(f \cdot g)}{dx} = \dfrac{df}{dx} \cdot \dfrac{dg}{dx}$

False. If $f(x) = x^3 \cdot x^4$, then $f(x) = x^7$ so $f'(x) = 7x^6.$
$f'(x)$ is not $(3x^2)(4x^3) = 12x^5.$

5) Find the equation of the tangent line to

$f(x) = \dfrac{4x}{x + 1}$ at $x = 2.$

$f(x) = \dfrac{4x}{x + 1}$, so $f(2) = \dfrac{4(2)}{2 + 1} = \dfrac{8}{3}.$

$f'(x) = \dfrac{(x + 1)(4) - (4x)(1)}{(x + 1)^2} = \dfrac{4}{(x + 1)^2},$

so $f'(2) = \dfrac{4}{(2 + 1)^2} = \dfrac{4}{9}.$

The equation is $y - \dfrac{8}{3} = \dfrac{4}{9}(x - 2).$

Section 3.3   Derivatives of Trigonometric Functions

This section contains formulas for derivatives of another type of function—trigonometric functions. We are continuing to build up a base of derivative rules that will be useful throughout calculus. Here again, after the basic rules are mastered, you need to use them to find derivatives of more complex functions.

Concepts to Master

A. Limits involving trigonometric functions

B. Derivatives of the six trigonometric functions

C. Technology Plus

Summary and Focus Questions

Page 192

A. Two very important limits involving sine and cosine are:

$$\lim_{x \to 0} \frac{\sin x}{x} = 1 \ \text{ and } \ \lim_{x \to 0} \frac{\cos x - 1}{x} = 0$$

where the angle x is measured in radians.

You may use these two basic limits and the limit laws to find limits of more complex functions.

Example: Evaluate $\lim_{t \to 0} \frac{\sin 3t}{4t}$.

Since this resembles $\frac{\sin x}{x}$, we choose $x = 3t$. Then as $t \to 0$, $x \to 0$ and

$$\lim_{t \to 0} \frac{\sin 3t}{4t} = \lim_{t \to 0} \left(\frac{3}{4}\right)\frac{\sin 3t}{3t} = \frac{3}{4} \lim_{x \to 0} \frac{\sin x}{x} = \frac{3}{4}(1) = \frac{3}{4}.$$

1) $\lim_{x \to 0} \frac{1 - \cos x}{x} = $ _____.

$$\lim_{x \to 0} \frac{1 - \cos x}{x} = \lim_{x \to 0} -\frac{\cos x - 1}{x} = -0 = 0.$$

2) Find $\lim_{x \to 0} \frac{1}{x \cot x} = $ _____.

$$\frac{1}{x \cot x} = \frac{1}{x \frac{\cos x}{\sin x}} = \frac{\sin x}{x \cos x} = \frac{\sin x}{x} \cdot \frac{1}{\cos x}.$$

Then $\lim_{x \to 0} \frac{\sin x}{x} \cdot \frac{1}{\cos x} = 1 \cdot \frac{1}{1} = 1.$

3) Find $\lim\limits_{x \to 0} \dfrac{x^2}{2 \sin x}$.

$$\lim_{x \to 0} \frac{x^2}{2 \sin x} = \lim_{x \to 0} \frac{x}{2} \cdot \frac{x}{\sin x}$$

$$= \lim_{x \to 0} \frac{x}{2} \cdot \lim_{x \to 0} \frac{x}{\sin x}$$

$$= \lim_{x \to 0} \frac{x}{2} \cdot \lim_{x \to 0} \frac{1}{\frac{\sin x}{x}}$$

$$= (0)\left(\frac{1}{1}\right) = 0.$$

B. These six differentiation formulas *must* be memorized.

Page
194

$$\frac{d}{dx} \sin x = \cos x \qquad\qquad \frac{d}{dx} \cos x = -\sin x$$

$$\frac{d}{dx} \tan x = \sec^2 x \qquad\qquad \frac{d}{dx} \cot x = -\csc^2 x$$

$$\frac{d}{dx} \sec x = (\sec x)(\tan x) \qquad\qquad \frac{d}{dx} \csc x = -(\csc x)(\cot x)$$

4) Find y' for each:

    a) $y = \sin x - \cos x$

$$y' = \cos x - (-\sin x)$$
$$= \cos x + \sin x.$$

    b) $y = \dfrac{\tan x}{x + 1}$

$$y' = \frac{(x + 1)(\sec^2 x) - (\tan x)(1)}{(x + 1)^2}$$

$$= \frac{x \sec^2 x + \sec^2 x - \tan x}{(x + 1)^2}.$$

    c) $y = \sin \dfrac{\pi}{4}$

$y' = 0$, since y is a constant.

    d) $y = x^3 \sin x$

$$y' = x^3(\cos x) + (\sin x)3x^2$$
$$= x^3 \cos x + 3x^2 \sin x.$$

    e) $y = x^2 + 2x \cos x$

$$y' = 2x + 2x(-\sin x) + 2 \cos x$$
$$= 2(x - x \sin x + \cos x).$$

    f) $y = \dfrac{x}{\sec x + 1}$

$$y' = \frac{(\sec x + 1)1 - x(\sec x \tan x)}{(\sec x + 1)^2}$$

$$= \frac{\sec x + 1 - x \sec x \tan x}{(\sec x + 1)^2}.$$

    g) $y = \dfrac{x}{\cot x}$

<You could evaluate this using the quotient rule but the problem is probably easier if you first rewrite y with a trigonometric identity and then use the product rule.>

First rewrite as $y = x \tan x$.
$$y' = x \sec^2 x + \tan x.$$

5) Find where the graph of
$f(x) = \sqrt{3} \sin x + 3 \cos x$ has a horizontal
tangent.

<*If the tangent line is horizontal, then its slope
($f'(x)$) is zero. Set $f'(x) = 0$ and solve for x.*>

Since $f'(x) = \sqrt{3} \cos x - 3 \sin x$, set
$f'(x) = \sqrt{3} \cos x - 3 \sin x = 0$.
$\sqrt{3} \cos x = 3 \sin x$
$\dfrac{\sin x}{\cos x} = \dfrac{\sqrt{3}}{3}$
$\tan x = \dfrac{\sqrt{3}}{3}$.

There are horizontal tangents at
$x = \dfrac{\pi}{6} + k\pi$, where k is an integer.

6) A car repeatedly speeds up and slows down as
it drives away from you, so that at x seconds it
is $f(x) = x + \sin x$ meters from you.

   a) Draw a graph of $f(x)$ for $0 \le x \le 3\pi$.

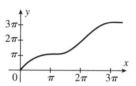

   b) What is the velocity of the car at
2 seconds?

$f'(x) = 1 + \cos x$.
$f'(2) = 1 + \cos 2 \approx 1 + (-.416)$
        $= 0.584$ m/s.

   c) What is the velocity of the car at
π seconds?

$f'(\pi) = 1 + \cos \pi = 1 + (-1) = 0$ m/s.
The car is motionless for an instant at
$x = \pi$ seconds.

C. Technology Plus. Use a computer algebra system or a graphing calculator to solve.

T-1) a) Use a computer or calculator to graph
$f(x) = x - \sin x$, for $-20 \le x \le 20$.

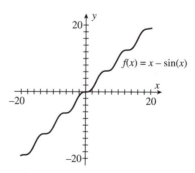

   b) Compute and graph $f'(x)$.

$f'(x) = 1 - \cos x$.

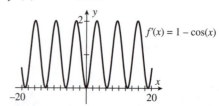

Section 3.4   The Chain Rule

The Chain Rule is used to find the derivative of a composition of functions. A good understanding of the rule is essential. The key steps to correctly using the Chain Rule are recognizing the function to be differentiated is a composite and determining the components of that composite.

Concepts to Master

A. Chain Rule for computing the derivative of a composition of functions

B. The derivatives of $e^{u(x)}$ and a^x

Summary and Focus Questions

Page 198

A. If $y = f(u)$ and $u = g(x)$ (and thus $y = (f \circ g)(x)$) then the **Chain Rule** for computing the derivative of y with respect to x is:

$$\frac{dy}{dx} = \frac{dy}{du} \cdot \frac{du}{dx}$$

or equivalently

$$[f(g(x))]' = f'(g(x)) \cdot g'(x).$$

The name "Chain" Rule comes from the fact that the terms in the product $\frac{dy}{du} \cdot \frac{du}{dx}$ are linked by the common symbol du.

Example: To find the derivative of $k(x) = (x^2 + 6x + 1)^4$, we first note that $k = (f \circ g)$, where $f(x) = x^4$ and $g(x) = x^2 + 6x + 1$. Since $f'(x) = 4x^3$ and $g'(x) = 2x + 6$,

$$k'(x) = \underbrace{4(x^2 + 6x + 1)^3}_{f'(g(x))} \cdot \underbrace{(2x + 6)}_{g'(x)}.$$

Recognizing components that may be easily differentiated is an important first step in correctly using the Chain Rule. Composite functions will often occur as expressions raised to a power, or as some function (such as sine or the exponential) of an expression. Here are several examples to help you recognize the composition and how to apply the Chain Rule.

Example: $y = (x + 3)^4$

$y = f(u) = u^4$ (the "outer" function)

$u = g(x) = x + 3$ (the "inner" function).

By the Chain Rule, $y' = 4(x + 3)^3 \cdot 1 = 4(x + 3)^3$.

Example: $y = \cos x^2$

$y = f(u) = \cos u$

$u = g(x) = x^2$.

By the Chain Rule, $y' = (-\sin x^2) \cdot 2x = -2x \sin x^2$.

Example: $y = \sqrt{1 + \frac{1}{x}}$

$y = f(u) = \sqrt{u}$

$u = g(x) = 1 + \frac{1}{x} = 1 + x^{-1}$.

By the Chain Rule, $y' = \frac{1}{2}\left(1 + \frac{1}{x}\right)^{-1/2} \cdot (0 + (-1)x^{-2}) = \frac{-1}{2x^2\sqrt{1 + \frac{1}{x}}}$.

Sometimes, more than two functions are involved, as in:

Example: $y = (3 + (x^3 - 2x)^5)^8$

$y = f(u) = u^8$ (the "outer" function)

$u = g(v) = 3 + v^5$ (the "middle" function)

$v = h(x) = x^3 - 2x$ (the "inner" function)

Here $y = (f \circ g \circ h)(x)$.

By the Chain Rule (applied twice), $\dfrac{dy}{dx} = \underbrace{8(3 + (x^3 - 2x)^5)^7}_{\frac{dy}{du}} \underbrace{(5(x^3 - 2x)^4)}_{\frac{du}{dv}} \underbrace{(3x^2 - 2)}_{\frac{dv}{dx}}$.

Example: $y = \tan^3 \sqrt{x}$

$y = f(u) = u^3$

$u = g(v) = \tan(v)$

$v = h(x) = \sqrt{x}$

By the Chain Rule (applied twice),

$$y' = \underbrace{\left(3 \tan^2 \sqrt{x}\right)}_{\frac{dy}{du}} \underbrace{\left(\sec^2\sqrt{x}\right)}_{\frac{du}{dv}} \underbrace{\left(\frac{1}{2}x^{-1/2}\right)}_{\frac{dv}{dx}} = \frac{3}{2}x^{-1/2} \tan^2 \sqrt{x} \, \sec^2\sqrt{x}.$$

1) Suppose $h(x) = f(g(x)), f'(7) = 3$, $g(4) = 7$, and $g'(4) = 5$. Find $h'(4)$.

$h'(4) = f'(g(4)) \cdot g'(4) = f'(7) \cdot 5$
$= 3 \cdot 5 = 15$.

2) Find $\frac{dy}{dx}$ for:

a) $y = \sqrt{x^3 + 6x}$

Since $y = (x^3 + 6x)^{1/2}$,

$$y' = \tfrac{1}{2}(x^3 + 6x)^{-1/2}(3x^2 + 6) = \frac{3x^2 + 6}{2\sqrt{x^3 + 6x}}.$$

b) $y = \sec x^2$

$$y' = (\sec x^2)(\tan x^2)(2x).$$

c) $y = \sec^2 x$

Since $y = (\sec x)^2$,

$$y' = 2(\sec x)^1(\sec x \tan x) = 2\sec^2 x \tan x.$$

d) $y = \cos^3 x^2$

Since $y = [\cos x^2]^3$,

$$y' = 3[\cos x^2]^2((-\sin x^2)(2x))$$
$$= -6x(\cos^2 x^2)\sin x^2.$$

e) $y = \sin 2x \cos 3x$

y is a product of $\sin 2x$ and $\cos 3x$, so

$$y' = (\sin 2x)[-\sin 3x \cdot 3]$$
$$+ (\cos 3x)[\cos 2x \cdot 2]$$
$$= -3 \sin 2x \sin 3x + 2 \cos 3x \cos 2x.$$

f) $y = \sin(\sec x)$

$$y' = \cos(\sec x)(\sec x \tan x).$$

g) $y = \tan(x^2 + 1)^4$

$$y' = \sec^2(x^2 + 1)^4(4(x^2 + 1)^3(2x))$$
$$= 8x(x^2 + 1)^3 \sec^2(x^2 + 1)^4.$$

h) $y = \dfrac{1}{(x^2 - 1)^4}$

Since $y = (x^2 - 1)^{-4}$,

$$y' = -4(x^2 - 1)^{-5}(2x) = \frac{-8x}{(x^2 - 1)^5}.$$

i) $y = x^3 \sqrt{x^2 + 1}$

Since $y = x^3(x^2 + 1)^{1/2}$,

$$y' = x^3\left(\tfrac{1}{2}(x^2 + 1)^{-1/2} \cdot 2x\right)$$
$$+ (x^2 + 1)^{1/2} \cdot 3x^2$$
$$= \frac{x^4}{\sqrt{x^2 + 1}} + 3x^2 \sqrt{x^2 + 1}$$
$$= \frac{4x^4 + 3x^2}{\sqrt{x^2 + 1}}.$$

3) Does the following make sense?

$$\frac{dh}{dt} = \frac{dh}{dx}\frac{dx}{du}\frac{du}{dv}\frac{dv}{dt}$$

Yes. There are four "links" in this use of the Chain Rule.

4) The temperature at noon each day x at the top of Mt. Arthur is approximately

$$50 + 30 \sin \frac{\pi(x - 200)}{180}$$

degrees Centigrade. (Assume a 360 day year.) Find the rate of change of the daily temperature for January 15.

<Find the derivative with respect to x and substitute $x = 15$.>

$$T(x) = 50 + 30 \sin \frac{\pi(x - 200)}{180}$$

$$T'(x) = 30 \cos \frac{\pi(x - 200)}{180} \cdot \frac{\pi}{180}$$

At $x = 15$, $T'(15) = 30 \cos \frac{\pi(-185)}{180} \cdot \frac{\pi}{180}$

$$\approx -.52 \text{ C°/day.}$$

On January 15, the noon temperature is decreasing about a half a degree per day.

Page 203

B. The derivative of the exponential function may be generalized using the Chain Rule:

$$\frac{d}{dx}(e^{u(x)}) = e^{u(x)}u'(x).$$

Example: If $f(x) = e^{3x^2}$, then $f'(x) = e^{3x^2}(6x)$.

If $f(x) = a^x$, then $f'(x) = a^x \ln a$.

Example: If $f(x) = 16^x$, then $f'(x) = 16^x (\ln 16)$.

5) Find $f'(x)$ for each:

a) $f(x) = 4^x$

$$f'(x) = 4^x \ln 4.$$

b) $f(x) = 3^{-x}$

$$f'(x) = 3^{-x} (\ln 3)(-1) = -3^{-x} \ln 3.$$

c) $f(x) = 2^x(x^2 + 1)$

$$f'(x) = 2^x (2x) + (x^2 + 1)(2^x \ln 2)$$
$$= 2^x(2x + (\ln 2)(x^2 + 1)).$$

d) $f(x) = e^{\sqrt{x}}$

$$f(x) = e^{x^{1/2}}$$
$$f'(x) = e^{x^{1/2}} \left(\frac{1}{2}x^{-1/2}\right) = \frac{e^{\sqrt{x}}}{2\sqrt{x}}.$$

e) $f(x) = x^2 \cos e^x$

$$f'(x) = x^2((-\sin e^x)e^x) + \cos e^x(2x)$$
$$= -x^2 e^x \sin e^x + 2x \cos e^x.$$

Section 3.5 Implicit Differentiation

This section shows you how to find the derivative of a function when you do not have an explicit rule for the function. If a function is expressed as a solution to an equation, you may differentiate the equation and determine the derivative from the result. This section also shows you how to find the derivatives of the inverse trigonometric functions.

Concepts to Master

A. Define a function implicitly

B. Find $\frac{dy}{dx}$ where the function y of x is given implicitly

C. Derivatives involving inverse trigonometric functions

Summary and Focus Questions

Page 209

A. If y is a function of x defined by an equation not of the form $y = f(x)$, we say y as a function of x is defined **implicitly** by the equation.

Example: The equation $x + y = 1$ defines y as a function of x but not in the form of $y = f(x)$. If we solve for y the function $y = 1 - x$ is the explicit description of the function defined implicitly by $x + y = 1$.

Example: $x + \sqrt{y} = 1$ defines y as a function of x. If we solve for y (and remember that y must be nonnegative) we get $\sqrt{y} = 1 - x$; $y = (1-x)^2$ with domain $(-\infty, 1]$.

The graph for an equation is not necessarily the graph of a function. However, a portion of the equation's graph may be the graph of a function. In this case, that function is defined implicitly by the equation. Some equations can implicity define more than one function.

Example: The equation $x = y^2$ represents a parabola opening to the right, which is not the graph of a function. The top half of the parabola is the graph of a function while the bottom half is a different function.

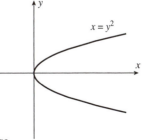

If we solve for y in $x = y^2$, $\sqrt{x} = \sqrt{y^2} = |y|$. Therefore, $y = \sqrt{x}$ (the top half of the parabola) or $y = -\sqrt{x}$ (the bottom half). Each of these functions is defined implicity by $x = y^2$. Their graphs are given in the figures on the next page.

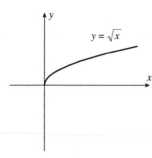

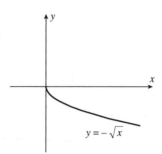

1) Sketch a graph of $\frac{x^2}{9} - \frac{y^2}{16} = 1$ and determine explicit forms for the functions that it defines.

<Recognize the graph of the equation is a hyperbola.>

The graph of the equation is a hyperbola opening left and right with asymptotes $y = \pm\frac{4}{3}x$.

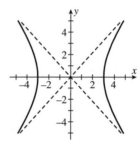

Solve for y:

$$-\frac{y^2}{16} = 1 - \frac{x^2}{9}$$
$$y^2 = \frac{16}{9}x^2 - 16$$

$$y = \sqrt{\frac{16}{9}x^2 - 16}, \text{ or } y = -\sqrt{\frac{16}{9}x^2 - 16}.$$

The graphs of these functions, each with domain $(-\infty, -3] \cup [3, \infty)$ are the top half and bottom half, respectively, of the hyperbola.

2) The graph below is $x^{2/3} + y^{2/3} = 1$, which implicitly defines two functions y of x. Find the functions explicitly.

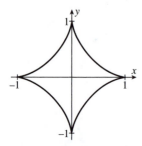

<Rewrite the equation in the form $y = \ldots$. Remember that when working with even roots and powers, be careful about introducing multiple solutions.>

Solve for y:
$$x^{2/3} + y^{2/3} = 1$$
$$y^{2/3} = 1 - x^{2/3}$$
$$y^2 = (1 - x^{2/3})^3$$
$$y = (1 - x^{2/3})^{3/2} \text{ (the top half) and}$$
$$y = -(1 - x^{2/3})^{3/2} \text{ (the bottom half).}$$

Page
210

B. Given an equation that defines y as an implicit function of x, $\dfrac{dy}{dx}$ may be found by **implicit differentiation**:

Step 1) Take the derivative with respect to x of both sides of the equation. (This will almost always use the Chain Rule.)

Step 2) Solve the result for $\dfrac{dy}{dx}$.

Example: Find the derivative of the function defined implicity by the equation $x^2 = y^3$. The derivative of the left side with respect to x is simply $2x$. The derivative of the right side is $3y^2y'$ by the Chain Rule. Remember that you are determining the derivative with respect to x and the expression y^3 is a composite—the outer function is the cube function and the inner function is y. Now solve for y':

$$2x = 3y^2y'$$
$$y' = \frac{2x}{3y^2}.$$

3) Find $\dfrac{dy}{dx}$ implicitly:

a) $3x^2 + y^5 = 10$

Step 1) Differentiate with respect to x:
$$6x + 5y^4y' = 0$$
(Since y is a function of x, the Chain Rule inserts the factor y'.)

Step 2) Solve for y':
$$5y^4y' = -6x$$
$$y' = -\frac{6x}{5y^4}.$$

b) $x^2y^3 = 2x + 1$

Step 1) Differentiate with respect to x (using the Product Rule because both x^2 and y^3 are functions of x):
$$x^2(3y^2y') + y^3(2x) = 2.$$

Step 2) Solve for y': $3x^2y^2y' = 2 - 2xy^3$
$$y' = \frac{2 - 2xy^3}{3x^2y^2}.$$

c) $3x^2 - 5xy + y^2 = 10$

Step 1) Differentiate:
$$6x - (5x(y') + y \cdot 5) + 2y \cdot y' = 0.$$

Step 2) Solve for y': $-5xy' + 2yy' = 5y - 6x$
$$(2y - 5x)y' = 5y - 6x$$
$$y' = \frac{5y - 6x}{2y - 5x}.$$

4) Find the slope of the tangent line to the curve defined by $x^2 + 2xy - y^2 = 41$ at the point $(5, 2)$.

The slope is $\dfrac{dy}{dx}$ at $(5, 2)$. First find $\dfrac{dy}{dx}$ implicitly:

Step 1) $2x + (2xy' + y \cdot 2) - 2yy' = 0$.

Step 2) $2y'(x - y) = -2(x + y)$
$$y' = -\frac{x + y}{x - y} = \frac{x + y}{y - x}.$$

Now use $x = 5$, $y = 2$:
$$\frac{dy}{dx} = \frac{5 + 2}{2 - 5} = \frac{7}{-3} = -\frac{7}{3}.$$

5) For $x^2 y = 1$ find y' both explicitly and implicitly.

Explicitly:

From $x^2 y = 1$, $y = x^{-2}$ so $y' = -2x^{-3}$.

Implicitly:

Step 1) $x^2 y' + 2xy = 0$.

Step 2) $x^2 y' = -2xy$
$$y' = -\frac{2y}{x}.$$

The two expressions for y' are the same because
$$\frac{-2y}{x} = \frac{-2x^{-2}}{x} = -2x^{-3} \text{ (remember } y = x^{-2}\text{)}.$$

Page 213

C. Recall from Chapter 1, section 6, that none of the six trigonometric functions has an inverse function. However, with a suitably restricted domain, each does. For example, the sine function has an inverse function when sine is given domain $[-1, 1]$. The inverse, $\sin^{-1}$, is a function from $\left[-\dfrac{\pi}{2}, \dfrac{\pi}{2}\right]$ to $[-1, 1]$ and $\sin^{-1} a = b$ if and only if $\sin b = a$.

Here are the derivatives of the inverse trigonometric functions.

$$\frac{d}{dx} \sin^{-1} x = \frac{1}{\sqrt{1 - x^2}} \qquad \frac{d}{dx} \cos^{-1} x = \frac{-1}{\sqrt{1 - x^2}}$$

$$\frac{d}{dx} \tan^{-1} x = \frac{1}{1 + x^2} \qquad \frac{d}{dx} \cot^{-1} x = \frac{-1}{1 + x^2}$$

$$\frac{d}{dx} \sec^{-1} x = \frac{1}{x\sqrt{x^2 - 1}} \qquad \frac{d}{dx} \csc^{-1} x = \frac{-1}{x\sqrt{x^2 - 1}}$$

6) Find $f'(x)$ for each:

   a) $f(x) = \sin^{-1}(x^3)$

   $$f'(x) = \frac{1}{\sqrt{1-(x^3)^2}}\,(3x^2) = \frac{3x^2}{\sqrt{1-x^6}}.$$

   b) $f(x) = (\tan^{-1}x)^3$

   $$f'(x) = 3(\tan^{-1}x)^2 \cdot \frac{1}{1+x^2} = \frac{3(\tan^{-1}x)^2}{1+x^2}.$$

   c) $f(x) = x\cos^{-1}x$

   $$f'(x) = x\left(\frac{-1}{\sqrt{1-x^2}}\right) + (\cos^{-1}x)(1)$$
   $$= \frac{-x}{\sqrt{1-x^2}} + \cos^{-1}x.$$

   d) $f(x) = e^{\csc^{-1}x^2}$

   $$f'(x) = e^{\csc^{-1}x^2}\left(\frac{-1}{x^2\sqrt{(x^2)^2-1}}\right)(2x)$$
   $$= \frac{-2e^{\csc^{-1}x^2}}{x\sqrt{x^4-1}}.$$

7) Find y' for $y = \sin^{-1}x + \cos^{-1}x$. What can you conclude about $\sin^{-1}x + \cos^{-1}x$?

   $$y' = \frac{1}{\sqrt{1-x^2}} + \frac{-1}{\sqrt{1-x^2}} = 0.$$ Thus y is a constant. To determine the constant, choose

   $x = 0$:

   $$y = \sin^{-1}0 + \cos^{-1}0 = 0 + \frac{\pi}{2} = \frac{\pi}{2}.$$

   Therefore $\sin^{-1}x + \cos^{-1}x = \frac{\pi}{2}$

   for all $-1 \le x \le 1$.

8) Find the equation of the tangent line to

   $$f(x) = \sin^{-1}\left(\frac{x}{2}\right) \text{ at } x = \sqrt{2}.$$

   $$f(\sqrt{2}) = \sin^{-1}\left(\frac{\sqrt{2}}{2}\right) = y \text{ if and only if}$$
   $$\sin y = \frac{\sqrt{2}}{2}. \text{ Thus } y = \frac{\pi}{4}.$$

   $$f'(x) = \frac{1}{\sqrt{1-\left(\frac{x}{2}\right)^2}} \cdot \frac{1}{2}$$

   $$= \frac{1}{\sqrt{4-x^2}}.$$

   $$f'(\sqrt{2}) = \frac{1}{\sqrt{4-(\sqrt{2})^2}} = \frac{1}{\sqrt{2}} = \frac{\sqrt{2}}{2}.$$

   The equation of the tangent line is

   $$y - \frac{\pi}{4} = \frac{\sqrt{2}}{2}(x - \sqrt{2}).$$

Section 3.6   Derivatives of Logarithmic Functions

This section completes the development of the derivative of logarithmic functions and provides a technique (logarithmic differentiation) for finding the derivative of certain complicated functions.

Concepts to Master

A. Derivatives of $\ln x$ and $\log_a x$
B. Logarithmic differentiation
C. The number e as a limiting value

Summary and Focus Questions

A. If $f(x) = \ln x$, then $f'(x) = \frac{1}{x}$. That is, $\frac{d}{dx} \ln x = \frac{1}{x}$.

Page 218

Also, $\frac{d}{dx} (\ln |x|) = \frac{1}{x}$.

By the Chain Rule, $\frac{d}{dx} (\ln u(x)) = \frac{1}{u(x)} \cdot u'(x) = \frac{u'(x)}{u(x)}$.

For a base a other than e, if $f(x) = \log_a x$, then $f'(x) = \frac{1}{x \ln a}$.

Examples: a) For $f(x) = \ln x$, $f'(x) = \frac{1}{x}$.

b) For $g(x) = \ln(x^2 - 1)$, $g'(x) = \frac{1}{x^2 - 1}(2x) = \frac{2x}{x^2 - 1}$.

c) For $h(x) = x^3 + \log_4 x$, $h'(x) = 3x^2 + \frac{1}{x(\ln 4)}$.

1) Find $f'(x)$ for each:

a) $f(x) = \ln x^2$

There are two ways to find $f'(x)$:
First, use properties of ln:
$f(x) = 2 \ln x$, so $f'(x) = 2\left(\frac{1}{x}\right) = \frac{2}{x}$.
Second, use the Chain Rule:
$f'(x) = \frac{1}{x^2}(2x) = \frac{2}{x}$.

b) $f(x) = \ln \cos x$

$f'(x) = \frac{1}{\cos x}(-\sin x) = -\tan x$.

c) $f(x) = x^2 \ln x$

$f'(x) = x^2\left(\frac{1}{x}\right) + \ln x\,(2x) = x + 2x \ln x$.

d) $f(x) = \log_2(x^2 + 1)$

$f'(x) = \frac{1}{(\ln 2)(x^2 + 1)} \cdot (2x)$

$= \frac{2x}{(\ln 2)(x^2 + 1)}$.

e) $f(x) = \ln \sqrt{\dfrac{x^2+1}{2x^3}}$ $(x>0)$

<Simplify f first using properties of logarithms.>

$$f(x) = \tfrac{1}{2}\ln\left(\dfrac{x^2+1}{2x^3}\right)$$
$$= \tfrac{1}{2}(\ln(x^2+1) - \ln 2 - 3\ln x).$$
$$f'(x) = \tfrac{1}{2}\left(\dfrac{2x}{x^2+1} - \dfrac{3}{x}\right)$$
$$= \dfrac{x}{x^2+1} - \dfrac{3}{2x}.$$

B. The derivative of a function $y = f(x)$ may be found using **logarithmic differentiation**—that is,

Page 220

  i) Take ln of both sides of $y = f(x)$: $\ln y = \ln(f(x))$.

  ii) Simplify $\ln(f(x))$ using properties of logarithms.

  iii) Differentiate both sides implicitly.

  iv) Solve for y': $y' = y\dfrac{d}{dx}\ln(f(x))$.

Example: Find y', where $y = x^{2/x}$.

  i) $\ln y = \ln x^{2/x}$

  ii) $\ln y = \dfrac{2}{x}\ln x$

  iii) $\dfrac{1}{y}\cdot y' = \dfrac{2}{x}\left(\dfrac{1}{x}\right) + \ln x\left(-\dfrac{2}{x^2}\right) = \dfrac{2}{x^2}(1-\ln x)$

  iv) $y' = y\left(\dfrac{2}{x^2}(1-\ln x)\right) = x^{2/x}\left(\dfrac{2}{x^2}(1-\ln x)\right) = 2x^{(-2+2/x)}(1-\ln x).$

2) Find y' for each using logarithmic differentiation.

  a) $y = \left(\dfrac{10x^3}{\sqrt{x+1}}\right)^4$

  i) $\ln y = \ln\left(\dfrac{10x^3}{\sqrt{x+1}}\right)^4.$

  ii) $\ln y = 4(\ln 10x^3 - \ln\sqrt{x+1})$
$$= 4\left(\ln 10 + 3\ln x - \tfrac{1}{2}\ln(x+1)\right).$$

  iii) $\dfrac{d}{dx}\ln y = 4\left(0 + \dfrac{3}{x} - \left(\dfrac{1}{2}\right)\dfrac{1}{x+1}(1)\right)$
$$\dfrac{1}{y}y' = 4\left(\dfrac{3}{x} - \dfrac{1}{2(x+1)}\right).$$

  iv) $y' = \left(\dfrac{10x^3}{\sqrt{x+1}}\right)^4\left[4\left(\dfrac{3}{x} - \dfrac{1}{2(x+1)}\right)\right].$

b) $y = 6^x$

i) $\ln y = \ln 6^x$.

ii) $\ln y = x \ln 6$.

iii) $\dfrac{d}{dx} \ln y = \ln 6$ (a constant).

$\dfrac{1}{y} y' = \ln 6$.

iv) $y' = y \ln 6 = 6^x \ln 6$.

Page 222

C. The number e may be written as

$$\lim_{x \to 0^+} (1 + x)^{1/x} = e.$$

If we let $t = \dfrac{1}{x}$, then as $x \to 0^+$, $t \to \infty$. Therefore the limit above may be rewritten as

$$\lim_{t \to \infty} \left(1 + \frac{1}{t}\right)^t = e.$$

3) Find each limit:

a) $\displaystyle\lim_{h \to 0^+} (1 + h)^{1/h}$

e.

b) $\displaystyle\lim_{x \to \infty} \left(1 + \frac{1}{x}\right)^{3x}$

<Rewrite so that the 3 is "outside" the limit.>

$$\lim_{x \to \infty} \left(1 + \frac{1}{x}\right)^{3x} = \lim_{x \to \infty} \left[\left(1 + \frac{1}{x}\right)^{x}\right]^3$$
$$= \left[\lim_{x \to \infty} \left(1 + \frac{1}{x}\right)^{x}\right]^3 = e^3.$$

c) $\displaystyle\lim_{x \to 1^+} x^{1/(x-1)}$

<Rewrite as a limit where the variable approaches zero. As $x \to 1^+$, $(x - 1) \to 0^+$.>

Let $t = x - 1$, hence $x = 1 + t$.

As $x \to 1^+$, $t \to 0^+$.

Thus $\displaystyle\lim_{x \to 1^+} x^{1/(x-1)} = \lim_{t \to 0^+} (1 + t)^{1/t} = e$.

4) We have three different ways to describe the number e. What are they?

First, e is the unique number a for which the slope of the tangent line to $y = a^x$ at $(0, 1)$ is 1.

Second, e is the unique number for which
$$\lim_{h \to 0^+} \frac{e^h - 1}{h} = 1.$$

Third, $e = \displaystyle\lim_{x \to 0^+} (1 + x)^{1/x}$, or equivalently
$$e = \lim_{x \to \infty} \left(1 + \frac{1}{x}\right)^x.$$

Section 3.7   Rates of Change in the Natural and Social Sciences

Rates of change play an important role in many fields, including the natural sciences (biology, chemistry, ...) and social sciences (economics, political science, ...).

Concepts to Master

Interpretation of a derivative as an instantaneous rate of change of a given quantity

Page 224

Summary and Focus Questions

Suppose $f(x)$ represents some quantity that varies depending on the value of the variable x, such as the amount of gasoline in a car's tank at time x, the total cost to produce x items, or the pressure within an inflating helium balloon after x seconds. Then $f'(x)$ measures the instantaneous rate of change of the quantity f with respect to x.

For example, if $f(x)$ represents the distance a car has traveled on a highway, then $f'(x)$ measures the instantaneous rate of change of that distance, that is, $f'(x)$ measures the instantaneous velocity at time x. The speedometer in your car measures instantaneous velocity.

---

1) Water is flowing out of a tank in such a fashion that after t minutes there are $10000 - 10t - t^3$ gallons of water remaining in the tank. How fast is the water flowing after 2 minutes?

Let $V(t) = 10000 - 10t - t^3$ be the volume at time t. The question asks for the rate of change of V after 2 minutes.

$V'(t) = -10 - 3t^2$. At $t = 2$,

$V'(2) = -10 - 3(2)^2 = -22$ gal/min.

Note: The answer is negative, indicating that the volume of water is decreasing.

2) A rocket is $16t + t^3$ meters from its launch pad t seconds after liftoff. What is its velocity after 3 seconds?

$d(t) = 16t + t^3$.

$d'(t) = 16 + 3t^2$.

$d'(3) = 16 + 3(3)^2 = 43$ m/s.

3) A particle is moving along an axis so that at time t its position is
$$f(t) = t^3 - 6t^2 + 6 \text{ meters.}$$

    a) What is its velocity at time t?

$f'(t) = 3t^2 - 12t.$

    b) What is the velocity at 3 seconds?

$f'(3) = 3(3)^2 - 12(3) = -9$ m/s.

    c) Is the particle moving left or right at 3 seconds?

Left, since $f'(3) = -9$ is negative.

    d) At what time(s) is the particle (instantaneously) motionless?

<Motionless means zero velocity. Solve the equation $f'(t) = 0$ for t.>
$f'(t) = 0$
$3t^2 - 12t = 0$
$3t(t - 4) = 0$
At $t = 0$ and $t = 4$ the particle has zero velocity.

4) A stone is thrown upward from a 70 m cliff so that its height above ground is $f(t) = 70 + 3t - t^2$. What is the velocity of the stone as it hits the ground?

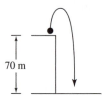

The time when the stone hits the ground is when $f(t) = 0$.
$70 + 3t - t^2 = 0$
$(10 - t)(7 + t) = 0$
$t = 10$ or $t = -7$ (Disregard $t = -7$.)
$f'(t) = 3 - 2t.$
$f'(10) = 3 - 20 = -17$ m/s.

5) The number of yellow perch in a bay in Lake Michigan was measured annually over a six year period. See the table. Estimate the rate of change of the perch population in the bay in 2008.

t	$P(t)$ (thousands of fish)
2005	420,000
2006	400,000
2007	370,000
2008	360,000
2009	330,000
2010	310,000

<Because the data is given in discrete values, average the slopes of the secant lines.>

The secant line for the period 2007–2008 has slope
$$\frac{360,000 - 370,000}{2008 - 2007} = \frac{-10,000}{1} = -10,000.$$

The secant line for the period 2008–2009 has slope
$$\frac{330,000 - 360,000}{2009 - 2008} = \frac{-30,000}{1} = -30,000.$$

We estimate $P'(2008)$ to be the average:
$$\frac{-10,000 + (-30,000)}{2} = \frac{-40,000}{2} = -20,000.$$

In 2008, the fish population was declining at a rate of about 20,000 fish per year.

Section 3.8   Exponential Growth and Decay

This section deals with rate of change problems where the rate of change of a quantity $y(t)$ is proportional to the quantity itself; that is, the rate of change for a quantity y is $y' = ky$ where k is a constant. For $k > 0$, the larger the value that y becomes, the greater the corresponding y' will be. Thus, as y increases, its rate of change also increases—this is exponential growth. The equation $y' = ky$ is an example of a **differential equation**—an equation that involves an unknown function y and its derivative. Much more about differential equations is in Chapter 9.

Concepts to Master

A. Solution of $\dfrac{dy}{dt} = ky$

B. Exponential growth and decay applications

C. Continuously compounded interest

Summary and Focus Questions

A. The solution to the differential equation $\dfrac{dy}{dt} = ky$ is $y(t) = y(0)e^{kt}$.

Page 237

1) If $y(0) = 3$ find the solution to the equation $\dfrac{dy}{dt} = 4y$.

Since $k = 4$ and $y(0) = 3$, the solution is $y(t) = 3e^{4t}$.

2) If a function $y(t)$ satisfies $\dfrac{y'}{y} = -2$ and the graph of the function crosses the y-axis at 5, find $y(t)$.

$\dfrac{y'}{y} = -2$ is $y' = -2y$ and has solution $y(t) = y(0)e^{-2t}$. Since the graph crosses the y axis at 5, $y(0) = 5$. Thus $y(t) = 5e^{-2t}$.

B. $\dfrac{dy}{dt} = ky$ is called the **law of natural growth** for $k > 0$ and the **law of natural decay** for $k < 0$. The number k is called the **relative growth constant.** The solution $y(t) = y(0)e^{kt}$ may be applied to problems of exponential growth or decay.

Page 237

Many growth and decay problems involving $y(t) = y(0)e^{kt}$ ask

"given $y = y_1$ at time t_1 and $y = y_2$ at t_2, find y at some third time t_3."

The procedure for solving $y' = ky$ depends on whether or not one of the times given is $t = 0$:

Case 1. If $t_1 = 0$, then we are given $y(0)$. Just solve $y_2 = y(0) \cdot e^{kt_2}$ for k.

Example: The half-life of radium is approximately 1600 years. What is the mass of a 10 mg sample of radium after 50 years?

At $t = 0$, $y = 10$ and at $t = 1600$, $y = 5$ (half the mass).

Thus, $y(t) = 10e^{kt}$

$$5 = 10e^{k(1600)}$$

$$\frac{1}{2} = e^{k(1600)}$$

$$k(1600) = \ln\frac{1}{2}$$

$$k = \frac{\ln 1/2}{1600}.$$

Thus, $y(t) = 10e^{\frac{\ln 1/2}{1600}t} = 10(e^{\ln 1/2})^{t/1600} = 10\left(\frac{1}{2}\right)^{t/1600}$.

Finally, at $t = 50$, $y(50) = 10\left(\frac{1}{2}\right)^{50/1600} \approx 9.79$ mg.

Case 2. If both t_1 and t_2 are nonzero, solve the equations $y_1 = y(0) \cdot e^{kt_1}$ and $y_2 = y(0) \cdot e^{kt_2}$ simultaneously for $y(0)$ and k:

i) Divide one equation by the other to eliminate the $y(0)$ term and solve for k.

ii) Solve for $y(0)$.

iii) Find $y(t_3)$ using $y(t) = y(0)e^{kt}$.

Example: A colony of bees moved its hive to a new location 6 years ago. Three years ago there were 400 bees in the new hive and today there are 600 bees. Assuming exponential population growth, what was the initial bee population and what will it be next year?

Using the initial year as $t = 0$, we are given $y(3) = 400$ and $y(6) = 600$ for the population model $y(t) = y(0)e^{kt}$ and asked to find $y(0)$ and $y(7)$.

i) In terms of the model we are given $400 = y(0)e^{3k}$ and $600 = y(0)e^{6k}$.

Divide the second equation by the first and simplify:

$$\frac{600}{400} = \frac{y(0)e^{6k}}{y(0)e^{3k}}$$

$$1.5 = e^{3k}$$

$$3k = \ln 1.5$$

$$k = \tfrac{1}{3}\ln 1.5.$$

ii) Use $400 = y(0)e^{3k}$ to solve for $y(0)$:

$$400 = y(0)e^{3\left(\frac{1}{3}\ln 1.5\right)} = y(0)e^{\ln 1.5} = 1.5y(0).$$

$$y(0) = \frac{400}{1.5} \approx 267, \text{the initial population.}$$

The model is $y(t) = 267e^{\frac{1}{3}\ln 1.5\, t} = 267(e^{\ln 1.5})^{t/3} = 267(1.5)^{t/3}$.

iii) Next year, at $t = 7$, the population is $y(7) = 267(1.5)^{7/3} \approx 688$.

3) A village had a population of 1000 in 1980 and 1200 in 1990. Assuming the population is experiencing natural growth, what was the population in 2010?

<*Identify* 1980 *as the year corresponding to* $t = 0$.>

Let 1980 be $t = 0$. Then $y(0) = 1000$.

At $t = 10$ (year 1990), $y = 1200$

so $1200 = 1000e^{k(10)}$

$1.2 = e^{10k}$

$10k = \ln 1.2$

$k = \frac{1}{10} \ln 1.2$.

Thus, $y = 1000e^{(1/10 \ln 1.2)t} = 1000(e^{\ln 1.2})^{t/10}$

$= 1000(1.2)^{t/10}$.

In the year 2010, $t = 30$, so

$y = 1000(1.2)^{30/10} = 1000(1.2)^3 \approx 1728$.

4) One hour after a bacteria culture was started there were 300 organisms and after 2 hours from the start there were 900. How many organisms will there be 4 hours after the start?

<*We are not given* $y(0)$, *so we must solve two equations for* k.>

At $t = 1, y = 300$, so $300 = y(0)e^{k \cdot 1}$.

At $t = 2, y = 900$, so $900 = y(0)e^{k \cdot 2}$

i) Dividing the second equation by the first

$\frac{900}{300} = \frac{y(0)e^{2k}}{y(0)e^k}$

$3 = e^k$

$k = \ln 3$.

ii) Now use $300 = y(0)e^{k \cdot 1}$ to solve for $y(0)$:

$300 = y(0)e^{\ln 3}$

$300 = y(0)3$

$y(0) = 100$.

The model is

$y = 100e^{\ln 3t} = 100(e^{\ln 3})^t = 100(3^t)$.

So $y = 100(3^t)$.

iii) At $t = 4, y = 100(3^4) = 8100$.

5) A sample of an isotope initially weighs 90 mg. After 100 days 60 mg remain. Assuming natural decay of the isotope, how many days will elapse until the sample mass is 20 mg?

At $t_1 = 0, y(0) = 90$.

At $t_2 = 100, y = 60$, so $60 = 90e^{k(100)}$.

$\frac{2}{3} = e^{100k}$; $100k = \ln \frac{2}{3}$; $k = \frac{1}{100} \ln \frac{2}{3}$

Thus $y = 90e^{(1/100 \ln 2/3)t} = 90(e^{\ln 2/3})^{t/100}$

$= 90\left(\frac{2}{3}\right)^{t/100}$.

When will $y = 20$?

$20 = 90\left(\frac{2}{3}\right)^{t/100}$

$\frac{2}{9} = \left(\frac{2}{3}\right)^{t/100}$

$\ln\left(\frac{2}{9}\right) = \ln\left(\frac{2}{3}\right)^{t/100} = \frac{t}{100} \ln\left(\frac{2}{3}\right)$

$t = \frac{100 \ln \frac{2}{9}}{\ln \frac{2}{3}} \approx 371$ days.

Page 241

C. If an initial amount A_0 is invested at an annual rate of r compounded n times per year, then the amount of money accumulated after t years is

$$A(t) = A_0\left(1 + \frac{r}{n}\right)^{nt}.$$

If, instead, interest is **compounded continuously** (take the limit as $n \to \infty$) then

$$A(t) = A_0 e^{rt}.$$

6) Suppose $1000 is invested at 4%.

    a) Find the value of the investment after 5 years if interest is compounded quarterly.

        $A(0) = 1000$, $r = .04$, and $n = 4$.

        $A(5) = 1000\left(1 + \frac{.04}{4}\right)^{4(5)} \approx \$1220.19.$

    b) Find the value after 5 years if interest is compounded continuously.

        $A(5) = 1000e^{.04(5)} \approx \$1221.40.$

    c) What is the differential equation that models compounding continuously at 4%; include an initial amount of $1000.

        $\frac{dA}{dt} = .04A$ with initial value of $A(0) = 1000.$

7) Eric invests some money in a fund that is paying 5% annual interest compounded continuously. How long before Eric's amount will be double his initial investment?

    Let M be the amount Eric invests and $A(t)$ be the amount accumulated after t years. Then $A(t) = Me^{.05t}$. The question asks for what value of t is $A(t) = 2M$?

$$2M = Me^{.05t}$$
$$2 = e^{.05t}$$
$$.05t = \ln 2$$
$$t = \frac{\ln 2}{.05} \approx 13.86 \text{ years.}$$

Section 3.9   **Related Rates**

Suppose you pour milk from a 2 liter jug into a tall glass and a bubble is on top of the milk in the glass. The faster that you pour the milk from the jug, the faster the bubble rises in the glass. If you pour the milk slowly the bubble rises slowly. This example illustrates the notion of related rates—the rate of change of one quantity (the volume of milk in the jug) is related to the rate of change of another quantity (the height of the bubble in the glass).

This section deals with relating rates of change. If two quantities $a(t)$ and $b(t)$ are related by an equation, then by implicit differentiation their derivatives, $a'(t)$ and $b'(t)$, are also related. Thus, if we know the value of $a'(t)$ we can determine the value of $b'(t)$.

Concepts to Master

Solve related rates problems

Summary and Focus Questions

Page 244

A related rate problem is an application in which one or more rates of change are given and you are asked to find the rate of change of some other (related) quantity. The procedure to solve a related rate problem is:

   i) Illustrate the problem with a picture, if possible.

   ii) Identify and label with constants all fixed quantities and with variables all quantities that vary with time. Identify each rate of change given in the problem as a rate of change of one of these variables. The unknown rate will also be the rate of change of one of these variables.

   iii) Find an equation that relates the constants, the variables whose rates are given, and the variable whose rate is unknown.

   iv) Differentiate the equation in **iii)** with respect to time (usually by implicit differentiation).

   v) Substitute all known quantities in the result from **iv)** and solve for the unknown rate.

Step **iii)** is often the hardest and will call upon your skills in remembering relationships from geometry, trigonometry, and other branches of mathematics. Remember not to substitute known quantities until after differentiating in Step **iv)**.

Example: A ladder 13 ft tall is leaning against the side of a building. If the bottom of the ladder slides away from the wall at a rate of 2 ft/sec, how fast is the top of the ladder sliding down the wall when the bottom of the ladder is 5 feet from the wall?

i)

ladder 13 ft. | wall

y

x

ii) Let y = height of wall.

Let x = distance from the wall to the bottom of the ladder.

$\frac{dx}{dt}$ is known to be 2 ft/sec. $\frac{dy}{dt}$ is unknown.

iii) Relate y, x, and 13: $x^2 + y^2 = 13^2$.

iv) Use implicit differentiation, where y and x are functions of time t:

$$2x\frac{dx}{dt} + 2y\frac{dy}{dt} = 0$$

$$2y\frac{dy}{dt} = -2x\frac{dx}{dt}$$

$$\frac{dy}{dt} = \frac{-x}{y}\frac{dx}{dt}.$$

v) Substitute known values:

$\frac{dx}{dt} = 2$ and $x = 5$ are given. The value of y may be calculated when $x = 5$ using

$y^2 = 13^2 - 5^2 = 169 - 25 = 144$

$y = 12$.

Therefore $\frac{dy}{dt} = \frac{-5}{12}(2) = \frac{-10}{12} = \frac{-5}{6}$ ft/sec.

1) An observer, 300 meters from the launch pad of a rocket, watches it ascend vertically at 60 m/s. Find the rate of change of the distance between the rocket and the observer when the rocket is 400 meters high.

i)

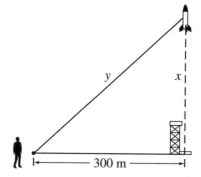

y x

300 m

ii) Let y = distance between rocket and observer. Let x = distance between rocket and pad.

$\frac{dx}{dt}$ is known (60). $\frac{dy}{dt}$ is unknown.

1) *(continued)*

iii) Relate y, x, and 300: $y^2 = 300^2 + x^2$.

iv) Use implicit differentiation, remembering y and x are functions of time t:

$$2y\frac{dy}{dt} = 0 + 2x\frac{dx}{dt}$$
$$y\frac{dy}{dt} = x\frac{dx}{dt}.$$

v) Substitute known values:

$\frac{dx}{dt} = 60$, $x = 400$. The value of y when

$x = 400$ is calculated from:

$$y^2 = 300^2 + (400)^2,$$

$$y = 500.$$

From $y\frac{dy}{dt} = x\frac{dx}{dt}$

$$500\frac{dy}{dt} = 400(60).$$

Therefore $\frac{dy}{dt} = \frac{400(60)}{500} = 48$ m/s.

2) A spherical ball has its diameter increasing at 2 cm/s. How fast is the volume changing when the radius is 10 inches?

i)

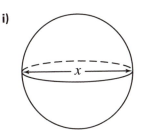

ii) Let x be the diameter and V be the volume. We are given $\frac{dx}{dt} = 2$.

iii) Relate V and x using the formula for the volume of a sphere:

$$V = \frac{4}{3}\pi\left(\frac{x}{2}\right)^3 = \frac{\pi}{6}x^3.$$

iv) Differentiate:

$$\frac{dV}{dt} = \frac{3\pi}{6}x^2\frac{dx}{dt} = \frac{\pi x^2}{2}\frac{dx}{dt}.$$

v) When the radius is 10, $x = 20$.

$$\frac{dV}{dt} = \frac{\pi(20)^2}{2}(2) = 400\pi \text{ cm}^3/\text{s}.$$

3) A light is on top of a 12 ft vertical pole. A 6 ft tall woman walks away from the pole base at 4 ft/s. How fast is the angle made by the woman's head, the light, and the pole base changing 2 seconds after she starts walking?

i)

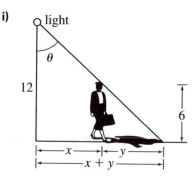

ii) Let θ be the desired angle. Let x be the distance between the pole base and the woman. Let y be the distance from the woman to the tip of her shadow.

iii) We are given $\frac{dx}{dt}$ $(= 4)$ and we must find $\frac{d\theta}{dt}$. Thus, we must relate x and θ. We can relate $x + y$ and θ so we first determine y in terms of x. By similar triangles, $\frac{x + y}{12} = \frac{y}{6}$. Thus,

$6x + 6y = 12y$

$6x = 6y$, or $x = y$.

Thus, $x + y = x + x = 2x$.

The relationship between x and θ is:

$\tan \theta = \frac{2x}{12}$

$6 \tan \theta = x$.

iv) Differentiate the relation:

$6 \sec^2 \theta \cdot \frac{d\theta}{dt} = \frac{dx}{dt}$.

v) After 2 seconds, $x = 2(4) = 8$ so $2x = 16$ ft.

$\sec \theta = \frac{d}{12}$, so $\sec^2 \theta = \frac{d^2}{144}$.

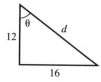

From the figure, $12^2 + 16^2 = d^2$, so $d^2 = 400$ and $\sec^2 \theta = \frac{400}{144}$. Finally, substituting into the result from **iv)**:

$6\left(\frac{400}{144}\right)\frac{d\theta}{dt} = 4$

$\frac{d\theta}{dt} = \frac{4(144)}{6(400)} = 0.24$ radians/s.

4)

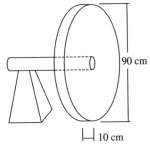

90 cm

10 cm

A 10-cm-thick mill wheel is initially 90 cm in diameter and is wearing away (uniformly) at 55 cm³/hr. How fast is the diameter changing after 20 hours?

i)

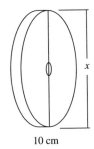

x

10 cm

ii) Let V be the volume and x be the diameter; both V and x are functions of time.

$\frac{dV}{dt}$ is known (-55), $\frac{dx}{dt}$ is unknown.

iii) From the formula for the volume of a cylinder: $V = 10\pi\left(\frac{x}{2}\right)^2$, so $V = 2.5\pi x^2$.

iv) Differentiate with respect to time:
$\frac{dV}{dt} = 5\pi x \cdot \frac{dx}{dt}$.

v) 20 hours later the amount of material worn away is $(55)(20) = 1100$ cm³.
The volume after 20 hours is
$2.5\pi(90)^2 - 1100 \approx 62517.25$ cm³.
Thus the diameter x after 20 hours is the solution to $62517.25 = 2.5\pi x^2$.

$x = \sqrt{\frac{62517.25}{2.5\pi}} \approx 89.2$ cm.

Finally, from $\frac{dV}{dt} = 5\pi x \frac{dx}{dt}$

$-55 = 5\pi(89.2)\frac{dx}{dt}$.

$\frac{dx}{dt} = \frac{-55}{5\pi(89.2)} \approx -0.039$ cm/hr.

Section 3.10   Linear Approximations and Differentials

The tangent line to a function $y = f(x)$ at a point a may be used to estimate a value for $f(x)$ when x is near a. Because the tangent line is a linear function, computing its values are easy, often much easier than computing $f(x)$ values. This section shows how to find such approximations.

Concepts to Master

A. Linear approximation of a function value

B. Differentials; Approximations with differentials

C. Technology Plus

Summary and Focus Questions

Page 250

A. Let $L(x)$ be the equation of the tangent line to $y = f(x)$ at $x = a$:

$$L(x) = f(a) + f'(a)(x - a).$$

$L(x)$ is called a **linearization** of f at a.

Both $L(x)$ and $f(x)$ have the same first derivative at $x = a$ so they both "head in the same direction" at $x = a$. For x near a, $L(x)$ is near $f(x)$ and is easy to compute. Therefore, $L(x)$ may be used as a linear approximation to $f(x)$: $L(x) \approx f(x)$.

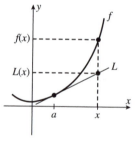

Example: Let $f(x) = \sqrt{x}$. Find the linear approximation to $f(x)$ at $x = 4$.

$L(x) = f(4) + f'(4)(x - 4)$.

$f(4) = 2$.

$f'(x) = \dfrac{1}{2}x^{-\frac{1}{2}} = \dfrac{1}{2\sqrt{x}}$, so $f'(4) = \dfrac{1}{2\sqrt{4}} = \dfrac{1}{4}$.

$L(x) = 2 + \dfrac{1}{4}(x - 4)$.

We can use the above example to provide a good estimate for $\sqrt{4.02}$.

Since 4.02 is close to $a = 4$, $L(4.02) = 2 + \dfrac{1}{4}(4.02 - 4) + 1 = 2.005$ is a

good approximation to $f(4.02) = \sqrt{4.02}$. Note to 5 decimals $\sqrt{4.02} = 2.00499$.

1) Find the linear approximation to
$f(x) = 5x^3 + 6x$ at $x = 2$.

<What you are asked to do is find the equation of the tangent line.>
$f(2) = 5(2)^3 + 6(2) = 52$.
$f'(x) = 15x^2 + 6$, so
$f'(2) = 15(2)^2 + 6 = 66$.
Thus, $L(x) = 52 + 66(x - 2)$.

2) Approximate $f(1.98)$ for the function in question 1.

We use $f(1.98) \approx L(1.98)$.
$L(1.98) = 52 + 66(1.98 - 2) = 50.68$.
Note: $f(1.98) = 50.69196$, so $L(1.98)$ is rather close.

3) Use a calculator to find $\sqrt{66}$, then approximate $\sqrt{66}$ using a linear approximation.

By calculator $\sqrt{66} \approx 8.1240384$.
Choose $f(x) = \sqrt{x}$ and $a = 64$ (64 is near 66 and $f(64) = 8$ is easy to calculate).
Then $f(64) = 8, f'(x) = \dfrac{1}{2\sqrt{x}}$,
and $f'(64) = \dfrac{1}{2\sqrt{64}} = \dfrac{1}{16}$.
The linear approximation is:
$L(x) = 8 + \dfrac{1}{16}(x - 64)$.
At $x = 66$,
$L(66) = 8 + \dfrac{1}{16}(66 - 64) = 8.125$.
Although $x = 66$ and $a = 64$ are a full 2 units apart, $L(66)$ and $f(66)$ are less than 0.001 units apart.

4) Find the linear approximation of
$f(x) = \sin x$ at $a = 0$. Use it to estimate
$\sin\left(\dfrac{\pi}{15}\right)$.

$f(x) = \sin x$
$f(0) = \sin 0 = 0$.
$f'(x) = \cos x$
$f'(0) = \cos 0 = 1$.
So $L(x) = f(0) + f'(0)(x - 0)$
$\qquad = 0 + 1(x - 0) = x$.
Thus, $\sin x \approx x$, for x near 0. (This is a common approximation used in physics and other sciences.)
For $x = \dfrac{\pi}{15}$, $\sin\left(\dfrac{\pi}{15}\right) \approx \dfrac{\pi}{15}$. (Using a calculator, $\sin\dfrac{\pi}{15} \approx .2079$ and $\dfrac{\pi}{15} \approx .2094$.)

B. Let $y = f(x)$ be a differentiable function. The differential dx is an independent variable (may take on any value). The **differential** dy is defined as $dy = f'(x)\,dx$.
Note that dy is a function of the two variables x and dx.

If we let $dx = \Delta x$, then for small values of dx, the change in the function (Δy) is approximately the same as the change in the tangent line dy:

$$dy \approx \Delta y, \text{ when } dx \text{ is small.}$$

This is handy since dy may be easier to calculate than Δy.

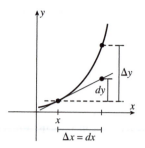

dy may be thought of as the **error** in calculating a value for y if an error of dx is made in estimating x. $\dfrac{dy}{y}$ is the **relative error.**

Example: The sides of a square field are measured and found to be 50 m with a possible error of .02 m in the measurement. We calculate the area to be $50^2 = 2500$ m^2. Estimate the maximum error and relative error in this calculation.

Let $x =$ the side of the field and A be the area.
We are given $\Delta x = dx = .02$ when $x = 50$.
$A = x^2$ and $A'(x) = 2x$, so $dA = A'(x)dx = 2x\,dx$.
We estimate the maximum error ΔA with dA:

$$\Delta A \approx dA = 2(50)(.02) = 2 \text{ m}^2.$$

An estimate of the relative error with $\dfrac{dA}{A}$ is

$$\frac{dA}{A} = \frac{2 \text{ m}^2}{2500 \text{ m}^2} = .0008 = .08\%$$

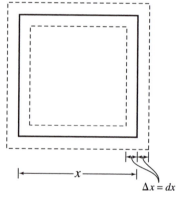

5) True or False:

   a) $\Delta x = dx$

True.

   b) $\Delta y = dy$

False. (dy is an approximation of Δy.)

6) Compute dy and Δy for $f(x) = x^2 + 3x$ at $x = 2$ with $\Delta x = dx = 0.1$.

$f(x) = x^2 + 3x$.
$f'(x) = 2x + 3$.
At $x = 2, f'(2) = 2(2) + 3 = 7$.
Thus, $dy = f'(x)\,dx = f'(2)(0.1)$
$$= 7(0.1) = 0.7.$$
At $x = 2, y = f(2) = 2^2 + 3(2) = 10$.
At $x = 2 + \Delta x = 2 + 0.1 = 2.1$,
$y = f(2.1) = (2.1)^2 + 3(2.1) = 10.71$.
Thus, $\Delta y = f(2.1) - f(2)$
$$= 10.71 - 10 = 0.71.$$
Note that $dy = 0.7$ and $\Delta y = 0.71$ are quite close but dy is easier to calculate.

7) A circle has a radius of 20 cm with a possible measurement error of 0.02 cm. Use differentials to estimate the maximum error and relative error for the area of the circle.

Let x be the radius of circle and A be the area of circle. We are given $\Delta x = dx = 0.02$ for $x = 20$.

The error is ΔA which we estimate with dA:
$A = \pi x^2$, so
$dA = 2\pi x\, dx = 2\pi(20) \cdot (0.02) = 0.8\pi.$
The relative error is
$$\frac{dA}{A} = \frac{0.8\pi}{\pi(20)^2} \approx .002 = .2\%.$$

C. Technology Plus. Use a computer algebra system or a graphing calculator to solve.

T-1) Let $f(x) = \sqrt{1 + x}$.

  a) Find the equation of the tangent line at $x = 3$.

$f(x) = (1 + x)^{1/2}; f(3) = (1 + 3)^{1/2} = 2$

$f'(x) = \frac{1}{2}(1 + x)^{-1/2} = \frac{1}{2\sqrt{1 + x}}$

$f'(3) = \frac{1}{2\sqrt{1 + 3}} = \frac{1}{4}.$

The tangent line is $y = 2 + \frac{1}{4}(x - 3)$.

  b) Graph $f(x)$ and the tangent line on the same screen. Use a window of $[-1, 6]$ by $[-1, 6]$.

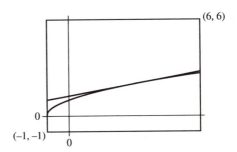

  c) Use a calculator and the equation for the tangent line from part **a)** to complete this table of values for x near 3:

x	f(x)	tan line	difference
2.50			
2.80			
2.90			
2.95			
2.99			

x	f(x)	tan line	difference
2.50	1.87083	1.87500	−0.00417
2.80	1.94936	1.95000	−0.00064
2.90	1.97484	1.97500	−0.00016
2.95	1.98746	1.98750	−0.00004
2.99	1.99750	1.99750	0.00000

Section 3.11 Hyperbolic Functions

Recall that the trigonometric functions (sin, cos, tan, ...) are sometimes called "circular functions" because some of their properties refer to the unit circle $x^2 + y^2 = 1$. For example, the identity $\cos^2 x + \sin^2 x = 1$ may be remembered by noting that $(\cos x, \sin x)$ is a point on the unit circle and therefore satisfies the equation $x^2 + y^2 = 1$.

In this section we shall see certain combinations of e^x and e^{-x} give rise to functions called **hyperbolic functions** (sinh, cosh, tanh, ...) because some of their properties refer to the hyperbola $x^2 - y^2 = 1$. For example, we can remember identities such as $\cosh^2 x - \sinh^2 x = 1$ by noting that the point $(\cosh x, \sinh x)$ is a point on the hyperbola.

Concepts to Master

A. Define the hyperbolic trigonometric functions

B. Derivatives involving hyperbolic functions

C. Definition and derivatives of inverse hyperbolic functions

Summary and Focus Questions

Page 257

A. The **hyperbolic sine and cosine functions** are:

$$\sinh x = \frac{e^x - e^{-x}}{2} \qquad\qquad \cosh x = \frac{e^x + e^{-x}}{2}$$

The other **hyperbolic trigonometric functions** are:

$$\tanh x = \frac{\sinh x}{\cosh x} \qquad\qquad \coth x = \frac{\cosh x}{\sinh x}$$

$$\operatorname{sech} x = \frac{1}{\cosh x} \qquad\qquad \operatorname{csch} x = \frac{1}{\sinh x}$$

Identities for these functions include:

$$\sinh(-x) = \sinh x \qquad\qquad \cosh(-x) = \cosh x$$
$$\cosh^2 x - \sinh^2 x = 1 \qquad\quad 1 - \tanh^2 x = \operatorname{sech}^2 x$$
$$\sinh(x + y) = \sinh x \cosh y + \cosh x \sinh y$$
$$\cosh(x + y) = \cosh x \cosh y + \sinh x \sinh y$$

1) Evaluate sinh 2.

$$\sinh 2 = \frac{e^2 - e^{-2}}{2} = \frac{e^4 - 1}{2e^2}.$$

2) Write coth x in terms of powers of e.

$$\coth x = \frac{\cosh x}{\sinh x} = \frac{\dfrac{e^x + e^{-x}}{2}}{\dfrac{e^x - e^{-x}}{2}} = \frac{e^x + e^{-x}}{e^x - e^{-x}}.$$

3) Verify $\sinh 2x = 2 \sinh x \cosh x$.

$$2 \sinh x \cosh x = 2\frac{e^x - e^{-x}}{2} \cdot \frac{e^x + e^{-x}}{2}$$
$$= \frac{(e^x - e^{-x})(e^x + e^{-x})}{2}$$
$$= \frac{e^{2x} - e^{-2x}}{2} = \sinh 2x.$$

4) What is the domain and range of $y = \cosh x$?

$\cosh x = \dfrac{e^x + e^{-x}}{2}$ is defined for all real x and is always greater than or equal to one (since e^x and e^{-x} are reciprocals). Thus the domain is all reals and range is $[1, \infty)$.

5) Evaluate $\tanh(\ln 5)$.

$$\tanh(\ln 5) = \frac{\sinh(\ln 5)}{\cosh(\ln 5)}$$
$$= \frac{\frac{e^{\ln 5} - e^{-\ln 5}}{2}}{\frac{e^{\ln 5} + e^{-\ln 5}}{2}} = \frac{5 - \frac{1}{5}}{5 + \frac{1}{5}} = \frac{12}{13}.$$

B. Here are derivatives for the hyperbolic functions:

$$\frac{d}{dx} \sinh x = \cosh x \qquad \frac{d}{dx} \tanh x = \operatorname{sech}^2 x \qquad \frac{d}{dx} \operatorname{sech} x = -\operatorname{sech} x \tanh x$$

$$\frac{d}{dx} \cosh x = \sinh x \qquad \frac{d}{dx} \coth x = -\operatorname{csch}^2 x \qquad \frac{d}{dx} \operatorname{csch} x = -\operatorname{csch} x \coth x$$

Page 259

6) Find $f'(x)$ for each:

   a) $f(x) = \cosh x$

$f'(x) = \sinh x$. (Remember that, unlike the cosine function, there is no minus sign in the derivative of cosh.)

   b) $f(x) = \tanh\left(\frac{x}{2}\right)$

$f'(x) = \operatorname{sech}^2\left(\frac{x}{2}\right)\left(\frac{1}{2}\right) = \frac{1}{2}\operatorname{sech}^2\left(\frac{x}{2}\right).$

   c) $f(x) = \sqrt{\cosh x}$

$f'(x) = \frac{1}{2}(\cosh)^{-1/2}(\sinh x) = \dfrac{\sinh x}{2\sqrt{\cosh x}}.$

   d) $f(x) = \sinh x + \cosh x$

$f'(x) = \cosh x + \sinh x$. Thus $f'(x) = f(x)$. If this seems familiar it should since $\sinh x + \cosh x = e^x$ and $(e^x)' = e^x$.

C. Like the trigonometric functions, we restrict the domains of some of the hyperbolic functions so that the inverse is a function. Since the hyperbolic functions are defined in terms of e^x, each inverse may be explicitly written in terms of natural logarithms.

Page 260

The inverses of the hyperbolic functions are:

Function	Domain	Range	Explicit Form
$\sinh^{-1} x$	all reals	all reals	$\ln(x + \sqrt{x^2 + 1})$
$\cosh^{-1} x$	$[1, \infty)$	$[0, \infty)$	$\ln(x + \sqrt{x^2 - 1})$
$\tanh^{-1} x$	$(-1, 1)$	all reals	$\frac{1}{2} \ln\left(\frac{1 + x}{1 - x}\right)$
$\coth^{-1} x$	$(-\infty, -1) \cup (1, \infty)$	all reals except 0	$\frac{1}{2} \ln\left(\frac{1 + x}{x - 1}\right)$
$\text{sech}^{-1} x$	$(0, 1]$	$[0, \infty)$	$\ln\left(\frac{1}{x} + \frac{\sqrt{1 - x^2}}{x}\right)$
$\text{csch}^{-1} x$	all reals except 0	all reals except 0	$\ln\left(\frac{1}{x} + \frac{\sqrt{1 + x^2}}{|x|}\right)$

The derivatives of the inverse hyperbolic functions are:

$$\frac{d}{dx}\left(\sinh^{-1} x\right) = -\frac{1}{\sqrt{1 + x^2}} \qquad \frac{d}{dx}\left(\text{csch}^{-1} x\right) = \frac{-1}{|x|\sqrt{x^2 + 1}}$$

$$\frac{d}{dx}\left(\cosh^{-1} x\right) = \frac{1}{\sqrt{x^2 - 1}} \qquad \frac{d}{dx}\left(\text{sech}^{-1} x\right) = \frac{-1}{x\sqrt{1 - x^2}}$$

$$\frac{d}{dx}\left(\tanh^{-1} x\right) = \frac{1}{1 - x^2} \qquad \frac{d}{dx}\left(\coth^{-1} x\right) = \frac{1}{1 - x^2}$$

7) Evaluate $\tanh^{-1} \frac{1}{2}$.

<Use the definition of $y = \tanh^{-1} x$.>

$\tanh^{-1} \frac{1}{2} = y$ means $\tanh y = \frac{1}{2}$.

$\dfrac{e^y - e^{-y}}{e^y + e^{-y}} = \dfrac{1}{2}$

$2e^y - 2e^{-y} = e^y + e^{-y}$

$e^y = 3e^{-y}$

$e^{2y} = 3$

$2y = \ln 3$

$y = \frac{1}{2} \ln 3.$

Alternately, we could use the explicit form for $\tanh^{-1}$:

$$\tanh^{-1} \frac{1}{2} = \frac{1}{2} \ln\left(\frac{1 + \frac{1}{2}}{1 - \frac{1}{2}}\right) = \frac{1}{2} \ln\left(\frac{\frac{3}{2}}{\frac{1}{2}}\right)$$

$$= \frac{1}{2} \ln 3.$$

8) Show that the derivative of $\tanh^{-1} x$ is always positive.

<Show that $f'(x) > 0$ for each x in its domain.>

$$f'(x) = \frac{1}{1 - x^2}$$

Since the domain is $(-1, 1)$ we have

$-1 < x < 1$, thus $1 - x^2 > 0$.

Therefore, $\dfrac{1}{1 - x^2} > 0$.

9) Verify the derivative $\dfrac{d}{dx}\left(\coth^{-1} x\right) = \dfrac{1}{1 - x^2}$, using the explicit form for $\coth^{-1} x$.

$\coth^{-1} x = \dfrac{1}{2} \ln\left(\dfrac{1 + x}{x - 1}\right)$.

Therefore, $\dfrac{d}{dx}\left(\coth^{-1} x\right)$

$$= \frac{1}{2}\left(\frac{1}{\frac{1+x}{x-1}}\right)\frac{(x-1)(1) - (1+x)(1)}{(x-1)^2}$$

$$= \frac{1}{2}\left(\frac{x-1}{1+x}\right)\frac{-2}{(x-1)^2} = \frac{-1}{(1+x)(x-1)}$$

$$= \frac{1}{1 - x^2}.$$

10) Find $f'(x)$ for:

a) $f(x) = \sinh^{-1} x^2$.

$$f'(x) = \frac{1}{\sqrt{1 + (x^2)^2}}(2x) = \frac{2x}{\sqrt{1 + x^4}}.$$

b) $f(x) = \tanh^{-1}\sqrt{1 - x^2}$.

$$f'(x) = \frac{1}{1 - (\sqrt{1 - x^2})^2}\frac{1}{2}(1 - x^2)^{-1/2}(-2x)$$

$$= \frac{1}{x^2}\frac{1}{2}\frac{1}{\sqrt{1 - x^2}}(-2x)$$

$$= \frac{-1}{x\sqrt{1 - x^2}}.$$

Chapter 4 — Applications of Differentiation

Section 4.1 Maximum and Minimum Values

This section gives you some tools for finding the largest and smallest values of $f(x)$ for a given function f, where x is from a set D. The largest and smallest values of $f(x)$ for all x in D are called absolute extrema. Those points where $f(x)$ is largest or smallest in a region around the point x are called relative extrema. Extreme values are important in applications (such as maximum profit, minimum force, etc.) and helpful in drawing graphs.

Concepts to Master

A. Absolute maxima, minima, and extrema; The Extreme Value Theorem

B. Relative (or local) maxima, minima, and extrema

C. Critical Numbers; Fermat's Theorem about local extrema

D. The Closed Interval Method for finding extrema for a continuous function on a closed interval

E. Technology Plus

Summary and Focus Questions

Page 274

A. Let f be a function with domain D. f has an **absolute maximum** at c (and $f(c)$ is the **maximum value**) means $f(x) \leq f(c)$ for all $x \in D$.

Thus, a highest point on the graph of f occurs at $(c, f(c))$.

f has an **absolute minimum** at d means $f(x) \geq f(d)$ for all $x \in D$, so $(d, f(d))$ is a lowest point on the graph.

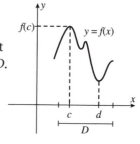

The **extreme values** of f are the maximum of f (if there is one) and the minimum of f (if there is one).

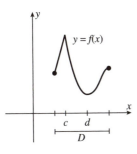 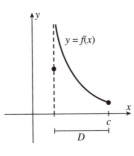 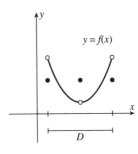

From left to right, the three graphs above have two, one, and no extreme values on *D*. Note that the functions in the middle and on the right are not continuous on *D*.

The Extreme Value Theorem: The extreme values must always exist for a continuous function whose domain is a closed interval.

Example: Let $y = f(x) = x^2 + 1$ with domain $[-1, 2]$. Because *f* is a polynomial *f* is continuous on $[-1, 2]$. By the Extreme Value Theorem, the extreme values of *f* exist. The graph of *f* at the right is a parabola with absolute minimum at $x = 0$ and absolute maximum at $x = 2$. The maximum value is 5 and the minimum value is 1.

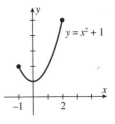

1) If $f(x) \geq f(c)$ for all *x* in the domain of the function *f*, then *f* has an _____ at *c*.

absolute minimum

2) Sketch a graph of a continuous function *f* with domain $[0, 5]$ such that

    a) the absolute maximum is at $x = 4$ and the absolute minimum is at $x = 1$.

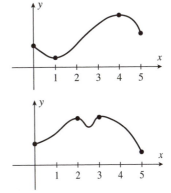

    b) the absolute maximum is at $x = 2$ and $x = 3$ and the absolute minimum is at $x = 5$.

    c) the absolute maximum is at $x = 3$ but there is no absolute minimum.

Not possible. By the Extreme Value Theorem, the absolute minimum must exist.

3) Where does the absolute minimum value occur for $f(x) = 9 - x^2$ with domain $[-2, 2]$?

$f(x) = 9 - x^2$ with domain $[-2, 2]$ has a minimum value of 5 which occurs at both $x = 2$ and $x = -2$.

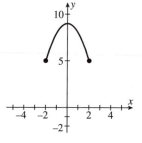

4) Does the Extreme Value Theorem guarantee that the extreme values exist for the function graphed?

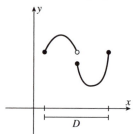

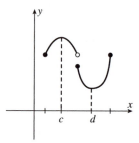

No. The function is not continuous on D. For this function, the extreme values exist, (at c and d) but their existence is not guaranteed by the Extreme Value Theorem.

Page 274

B. If there is some open interval I containing c such that $f(c) \geq f(x)$ for all $x \in I$, then f has a **local** (or **relative**) **maximum at c**. In case $f(c) \leq f(x)$ for all $x \in I$, we say f has a **local** (or **relative**) **minimum at c**.

The following is the graph of a function with domain $[a, b]$ with four local maxima at $x = c, p, r, t$ and three local minima at $x = d, q, s$.

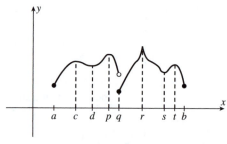

Note: There is a local minimum at $x = q$ but f is not continuous at q.

5) True or False:
If f has domain $[a, b]$, $c \in (a, b)$, and f has an absolute minimum at $x = c$, then f has a local minimum at $x = c$.

True.

6) How many local maxima and minima does the following function with domain $[a, b]$ have?

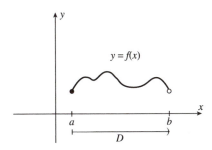

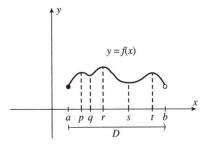

f has 3 local maxima (at $x = p, r$, and t) and 2 local minima (at $x = q$ and s). Because a is an endpoint there is no local minimum at $x = a$.

7) Sketch a graph of $f(x) = 2x + \dfrac{1}{x^2}$ on $[0.5, 3]$

and find the local and absolute extrema.

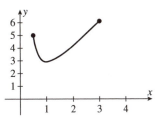

From the graph, f has a local minimum at $x = 1$. The absolute maximum occurs at $x = 3$ and is approximately 6.111. The absolute minimum is 3, which occurs at $x = 1$.

Page 276

C. A **critical number** c is a number in the domain of f for which either $f'(c) = 0$ or $f'(c)$ does not exist.

The method to find where $f'(x) = 0$ depends greatly on the nature of this equation. One way is to use a computer algebra system or calculator to approximate the solutions. If solving without using technology, factoring may help, as in this next example.

Example: To find the critical points of $f(x) = 3x^4 + 20x^3 - 36x^2 + 7$, we first note that f is a polynomial. Therefore, $f'(x)$ exists everywhere.

$f'(x) = 12x^3 + 60x^2 - 72x = 12x(x - 1)(x + 6)$.
$f'(x) = 0$ at $x = 0$, $x = 1$, $x = -6$.

Therefore, f has three critical numbers, at $x = 0, 1,$ and -6.

The method to find where $f'(x)$ does not exist also depends on the nature of $f'(x)$. For algebraic functions this may be where a denominator is zero or where an even root of a negative number occurs. Sometimes knowing the shape of the graph helps.

Examples:

Find all points in the domain of f at which $f'(x)$ does not exist.

a) $f(x) = 9\sqrt{x} + x^{3/2}$.

Here the domain of f is $[0, \infty]$, $f'(x) = \dfrac{9}{2\sqrt{x}} + \dfrac{3\sqrt{x}}{2}$, and $f'(x)$ does not exist for $x = 0$.

b) $f(x) = \sqrt{4 - x}$.

The domain of f is $(-\infty, 4]$ and $f'(x) = \dfrac{-1}{2\sqrt{4 - x}}$ does not exist at $x = 4$.

c) $f(x) = |x - 2| + 1$.

The graph of f is at the right. The derivative does not exist at $x = 2$.

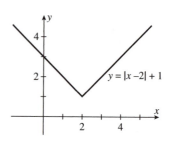

Fermat's Theorem: If f has a local maximum or minimum at c and $f'(c)$ exists, then $f'(c) = 0$.

In other words, if $f(c)$ is a local extremum for f, then c is a critical number of f. The converse is false—a critical number does not have to be a point where a local maximum or local minimum occurs.

8) True or False:
If $f'(c) = 0$, then c is a local maximum or local minimum.

False. For example, $f(x) = x^3$ with $c = 0$ has $f'(c) = 0$ but $c = 0$ is not a local extremum. The graph of f is below.

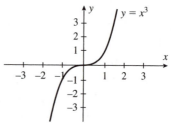

9) True or False:
If c is a local maximum or local minimum, then $f'(c) = 0$.

False. This statement sounds like Fermat's Theorem, but it does not have the hypothesis that $f'(c)$ exists. For the function $f(x) = |x - 2| + 1$ in the example above, f has a local (and absolute) minimum at $c = 2$ but $f'(2)$ does not exist.

10) Find the critical numbers for:

    a) $f(x) = \sqrt[5]{x}$.

$f(x) = x^{1/5}$.

$f'(x) = \frac{1}{5}x^{-4/5} = \frac{1}{5\sqrt[5]{x^4}}$. This does not exist for $x = 0$. $f'(x)$ is never 0, so the only critical number is $x = 0$.

    b) $f(x) = x + \frac{1}{x}$.

$f(x) = x + x^{-1}$.

$f'(x) = 1 + (-1)x^{-2} = 1 - \frac{1}{x^2}$.

$f'(x) = 0$ at $x = 1, x = -1$.

$f'(x)$ does not exist at $x = 0$ (because 0 is not in the domain of f). The critical numbers are $x = 1$ and $x = -1$.

Page 278

D. An important result: The extreme values of a continuous function f on a closed interval $[a, b]$ always exist—they occur either at a, at b, or at a critical number of f in (a, b). Thus, to find the absolute maximum and minimum values of f on a closed interval $[a, b]$, use the **Closed Interval Method:**

    i) Find $f(x)$ for all critical numbers x in (a, b).

    ii) Find $f(a)$ and $f(b)$.

    iii) Choose the largest and smallest values from the results of Steps i) and ii).

11) Find the extreme values of $f(x) = \sqrt{10x - x^2}$ on $[2, 10]$.

<Check the endpoints and critical numbers.>

Since f is continuous on $[2, 10]$, the extreme values occur at 2, 10, or some critical number between 2 and 10.

First, find the critical numbers.

$f'(x) = \frac{1}{2}(10x - x^2)^{-1/2}(10 - 2x)$

$\quad\quad = \frac{5 - x}{\sqrt{10x - x^2}}$.

On $[2, 10]$, $f'(x) = 0$ at $x = 5$.

$f'(x)$ does not exist at $x = 10$. Don't worry about $x = 0$ since 0 is not in the specified domain $[2, 10]$.

Thus 5 is the only critical number between 2 and 10. Compute function values:

x	2	10	5
f(x)	4	0	5

The absolute minimum is 0 and occurs at $x = 10$. The absolute maximum is 5 and occurs at $x = 5$.

12) Do the extreme values of
$f(x) = 6 + 12x - x^2$ on $(4, 10]$ exist?

<Note that the endpoint 4 is not in the domain. Check just the endpoint 10 and the critical numbers.>

f is a polynomial and, therefore, continuous on $(4, 10]$. However, $(4, 10]$ is not a closed interval so we cannot immediately conclude the extreme values of *f* exist. From the graph of *f* below (it is a portion of a parabola) we see that the extreme values do exist. The absolute maximum is 42 at $x = 6$ and the absolute minimum is 26 at $x = 10$.

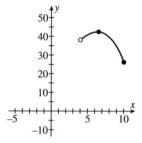

13) Find the extreme values of
$f(x) = 2x + \sin x$ on $[0, 2\pi]$.

First, determine the critical numbers:
$f'(x) = 2 + \cos x = 0$
$\cos x = -2$ has no solution. Thus, $f'(x)$ is never zero. Also, $f'(x)$ exists for all x. Therefore, there are no critical numbers. The extreme values occur at the endpoints.
$f(0) = 2(0) + \sin 0 = 0$.
$f(2\pi) = 2(2\pi) + \sin 2\pi = 4\pi$.
The absolute minimum is 0 and the absolute maximum is 4π.

14) Sketch a function *f* on $[-2, 2]$ for which extrema do not exist.

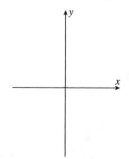

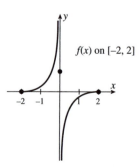

f(x) on [–2, 2]

There are many other possible answers. Note that *f* cannot be continuous on $[-2, 2]$.

E. Technology Plus. Use a computer algebra system or a graphing calculator to solve.

T-1) Let $f(x) = x^4 - 3x^3 + 5x$. Using a graph of the function estimate:

   a) the critical points of f.

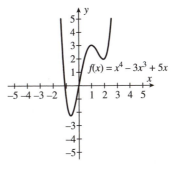

There are three critical points; they are (approximately) -0.656, 1.000, and 1.906.

   b) the absolute minimum and maximum of f on $[0, 2]$.

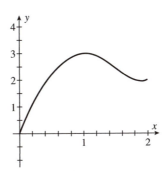

By using a window of $[0, 2]$ by $[-1, 4]$ we see that the absolute minimum occurs at $x = 0$ and is $f(0) = 0$ and the absolute maximum occurs at $x = 1$ and is $f(1) = 3$.

Section 4.2 The Mean Value Theorem

The Mean Value Theorem is extremely important because it provides an equation relating both functional values $f(x)$ and values of the derivative $f'(x)$. There are very few theorems that do this. Rolle's Theorem is a special case of the Mean Value Theorem.

Concepts to Master

A. Rolle's Theorem; Mean Value Theorem

B. Relationship between functions f and g when $f' = g'$

C. Technology Plus

Summary and Focus Questions

Page
284

A. Rolle's Theorem:
   If the function f
   i) is continuous on $[a, b]$,
   ii) is differentiable on (a, b), and
   iii) has $f(a) = f(b)$,
   then there is at least one number $c \in (a, b)$
   such that $f'(c) = 0$.

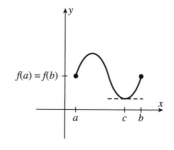

Rolle's Theorem says that if the three hypotheses are true, then there is at least one horizontal tangent between the endpoints of the graph.

The Mean Value Theorem is similar to Rolle's Theorem except we do not assume that $f(a)$ equals $f(b)$.

Mean Value Theorem:
If the function f
   i) is continuous on $[a, b]$,
   ii) is differentiable on (a, b),
   then there is at least one $c \in (a, b)$ such that
   $$f'(c) = \frac{f(b) - f(a)}{b - a}.$$

The Mean Value Theorem is an "existence theorem"—it does not say where in the interval (a, b) that c must be. It simply says that there must be at least one such c somewhere in the interval.

The number $\dfrac{f(b) - f(a)}{b - a}$ is the average rate of change (mean value) of the function f over the interval $[a, b]$. It is the slope of the secant line joining the endpoints of the graph. Therefore, the Mean Value Theorem says there is at least one point on the graph somewhere between the endpoints where the tangent line has the same slope as the line joining the endpoints.

The Mean Value Theorem is important because

$$f'(c) = \frac{f(b) - f(a)}{b - a}$$

is an equation involving both f and f'.

1) For $f(x) = 1 - x^2$ on $[-2, 1]$, do the hypotheses and conclusion of Rolle's Theorem hold?

<Check to see if each of the three hypotheses hold.>

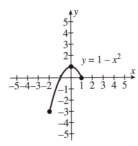

f is continuous on $[-2, 1]$ and differentiable on $(-2, 1)$ because f is a polynomial. Since $f(-2) = -3$ and $f(1) = 0$ are unequal, the third hypothesis fails. Nevertheless, the conclusion holds because $0 \in (-2, 1)$ and $f'(0) = 0$.

2) Find all numbers c that satisfy the conclusion of the Mean Value Theorem for $f(x) = x^3 - 3x^2 + x$ on $[0, 3]$.

<Find the mean value, M and solve the equation $f'(x) = M$.>

The mean value on $[0, 3]$ is

$$\frac{f(3) - f(0)}{3 - 0} = \frac{3 - 0}{3 - 0} = 1.$$

$f'(x) = 3x^2 - 6x + 1.$

$3x^2 - 6x + 1 = 1$

$3x^2 - 6x = 0$

$3x(x - 2) = 0.$

Thus, $f'(x) = 1$ at $x = 0, x = 2$. Since 0 is not in the open interval $(0, 3)$ $c = 2$ is the only point satisfying the Mean Value Theorem.

3) Do the hypotheses and conclusion of the Mean Value Theorem hold for
$$f(x) = 3x - \frac{x^3}{3} \text{ on } [-1, 4]?$$

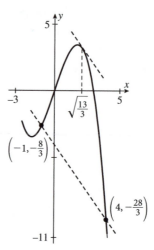

The hypotheses hold because f is a polynomial and therefore continuous and differentiable everywhere. Since the hypotheses are true, the conclusion must be true as well.

$$f(4) = 3(4) - \frac{4^3}{3} = -\frac{28}{3}.$$

$$f(-1) = 3(-1) - \frac{(-1)^3}{3} = -\frac{8}{3}.$$

$$\frac{f(4) - f(-1)}{4 - (-1)} = \frac{-\frac{28}{3} - \frac{-8}{3}}{5} = -\frac{4}{3} \text{ is the mean value.}$$

$$f'(x) = 3 - x^2 = -\frac{4}{3}.$$

$$x^2 = \frac{13}{3}, \text{ so } x = \sqrt{\frac{13}{3}}.$$

$$\left(\text{Note that } -\sqrt{\frac{13}{3}} \text{ is not in } [-1, 4].\right)$$

4) Mark on the *x*-axis the point(s) c in the conclusion of the Mean Value Theorem for the function below.

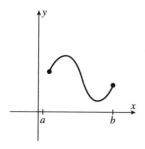

<Draw the secant line between the endpoints and find those points where the tangent line has the same slope as this secant line. >

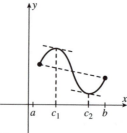

There are two points (c_1 and c_2) that satisfy the conclusion of the Mean Value Theorem.

5) David is driving on an interstate highway which has a speed limit of 55 mi/hr. At 2 p.m. he is at milepost 110 and at 5 p.m. he is at milepost 290. Is this enough evidence to prove that David is guilty of speeding?

Yes. Let $f(t)$ be his position at time t. $f(2) = 110$ and $f(5) = 290$. The mean value on $[2, 5]$ is $\dfrac{f(5) - f(2)}{5 - 2} = \dfrac{290 - 110}{5 - 2} = 60$. By the Mean Value Theorem there is a time $c \in (2, 5)$ such that $f'(c) = 60$. So at least once (at time c) David's instantaneous velocity ($f'(c)$) was 60 mi/hr.

6) Suppose $f'(x)$ exists and $f'(x) \leq 2$ for all x. If $f(1) = 8$ what is the largest possible value that $f(5)$ could be?

Since $f'(x)$ exists everywhere we can use the Mean Value Theorem for f on $[1, 5]$. For some $c \in (1, 5)$,

$$f'(c) = \frac{f(5) - f(1)}{5 - 1} = \frac{f(5) - 8}{4}.$$

Since $f'(c) \leq 2$

$$\frac{f(5) - 8}{4} \leq 2$$

$$f(5) - 8 \leq 8$$

$$f(5) \leq 16.$$

So $f(5)$ can be at most 16.

Page 288

B. If $f'(x) = g'(x)$ for all $x \in (a, b)$, then $f(x) = g(x) + c$ on (a, b), where c is a constant. In other words, two functions that have the same derivative will differ by a constant value.

7) True or False:

a) If $f(x) = g(x)$ for all x, then $f'(x) = g'(x)$.

True.

b) If $f'(x) = g'(x)$, then $f(x) = g(x)$.

False. For example, $f(x) = x^2$ and $g(x) = x^2 + 6$ have the same derivative, $2x$.

8) If $f'(x) = 3x^2$ then what is $f(x)$? (Hint: If $g(x) = x^3$ then $g'(x) = 3x^2$.)

<Think of the differentiation rules in reverse—what function should you start with so that its derivative is $3x^2$? >

We know that $\dfrac{d}{dx}(x^3) = 3x^2$, so $f(x)$ has to

have the form $f(x) = x^3 + c$, where c is some constant.

Here are three of many valid solutions:

$x^3 + 1$
$x^3 - 10$
$x^3 + \pi$.

C. Technology Plus. Use a computer algebra system or a graphing calculator to solve.

T-1) Let $f(x) = x^3 - x^2 + 2$.

    a) On the same screen, draw all three of these:

$y = f(x)$

the secant line joining $(0, f(0))$ and $(3, f(3))$

the line tangent to f at $c = 2$.

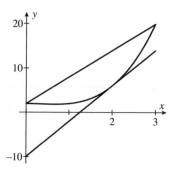

The line joining the endpoints has slope

$$\frac{f(3) - f(0)}{3 - 0} = \frac{20 - 2}{3 - 0} = 6 \text{ and equation}$$

$y = 6x + 2$.

$f'(x) = 3x^2 - 2x$ so $f'(2) = 8$. $f(2) = 6$.
The tangent line at $c = 2$ has equation
$y - 6 = 8(x - 2)$, or $y = 8x - 10$.

    b) Is $c = 2$ a value that satisfies the conclusion of the Mean Value Theorem?

No. Since $f'(2) = 8$, the tangent at $c = 2$ is too steep; it would need to have a slope of 6.

Section 4.3   How Derivatives Affect the Shape of a Graph

This section shows how values for f' determine how fast the graph of f increases or decreases, and it shows how values for f'' determine how fast the graph of f is curving.

Concepts to Master

A. The Increasing/Decreasing Test

B. The First Derivative Test

C. Concave upward and downward; Concavity Test

D. Point of inflection; Second Derivative Test

E. Technology Plus

Summary and Focus Questions

Page 290

A. Recall that a function f is increasing on an interval I means that the graph of f rises as you move from left to right. Therefore, the tangent lines to f must also rise as you move from left to right and hence their slopes (the derivative) are positive. The Increasing/Decreasing Test describes this very important relationship between f' and the graph of f.

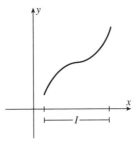

Increasing/Decreasing Test:
If f is differentiable on an interval I

a) $f'(x) > 0$ for all $x \in I$ implies f is increasing on I.

b) $f'(x) < 0$ for all $x \in I$ implies f is decreasing on I.

Example: For $f(x) = (x - 4)^2, f'(x) = 2(x - 4) = 2x - 8$. $2x - 8 > 0$ for $x > 4$ and $2x - 8 < 0$ for $x < 4$. Thus the graph of f is increasing for $x > 4$ and decreasing for $x < 4$. The graph of f is the parabola through $(4, 0)$ given at the right.

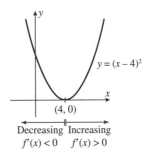

1) Is $f(x) = x^2 + 6x$ increasing on $(-1, 2)$?

<Determine where $f'(x)$ is positive and negative in the interval $(-1, 2)$. >

Yes. $f'(x) = 2x + 6$. $2x + 6 > 0$ when $2x > -6$ or $x > -3$. Thus, for all $x \in (-1, 2), f'(x) > 0$. Therefore, f is increasing on $(-1, 2)$.

2) Where is $f(x) = x^3 - 3x^2$ increasing and where is it decreasing?

<Determine where $f'(x)$ is positive and where it is negative. >

$f'(x) = 3x^2 - 6x = 3x(x - 2) = 0$ at $x = 0$, $x = 2$. At other values of $x, f'(x)$ will be either positive or negative. We make a table of the intervals where the factors $3x$ and $x - 2$ are positive or negative.

x	$3x$	$x - 2$	$3x(x - 2)$	f
$(-\infty, 0)$	−	−	+	increasing
$(0, 2)$	+	−	−	decreasing
$(2, \infty)$	+	+	+	increasing

f is increasing on $(-\infty, 0)$ and $(2, \infty)$.
f is decreasing on $(0, 2)$.

3) Sketch a graph of a differentiable function having all these properties:
  i) $f(0) = 1, f(2) = 3, f(5) = 0$,
  ii) $f'(x) > 0$ for
    $x \in (0, 2)$ and $x \in (5, \infty)$,
  iii) $f'(x) < 0$ for
    $x \in (-\infty, 0)$ and $x \in (2, 5)$.

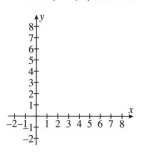

Property **ii)** tells us that f is increasing on $(0, 2)$ and $(5, \infty)$. Property **iii)** tells us that f is decreasing on $(-\infty, 0)$ and $(2, 5)$.

One such graph is:

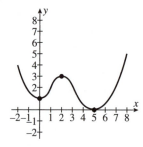

4) The graph of f' is given below. Where is f increasing? decreasing?

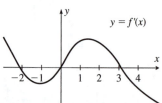

<Determine where $f'(x)$ is positive and negative. >

$f'(x) > 0$ for $x < -2$ and $0 < x < 3$.
$f'(x) < 0$ for $-2 < x < 0$ and $x > 3$.
Therefore, f is increasing on $(-\infty, -2)$ and on $(0, 3)$, and f is decreasing on $(-2, 0)$, and on $(3, \infty)$.

Page 291

B. For a local maximum, such as the point c in the graph pictured at the right, the curve to the left of c must rise up to $(c, f(c))$ and thus $f'(x)$ is positive to the left of c. Likewise, the curve to the right of c must fall downward from $(c, f(c))$ and thus $f'(x)$ is negative to the right of c.

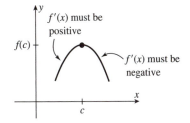

The First Derivative Test tells us when a critical number is a relative extremum.

The First Derivative Test

a) If $f'(x)$ is positive in (a, c) and negative in (c, b), then f has a local maximum at c.

b) If $f'(x)$ is negative in (a, c) and positive in (c, b), then f has a local minimum at c.

c) If $f'(x)$ has the same sign (either positive or negative) in both (a, c) and (c, b), then $f(c)$ is not a local extremum.

4) Suppose 8 is a critical number for a function f with $f'(x) > 0$ for $x \in (8, 9)$ and $f'(x) < 0$ for $x \in (7, 8)$. Then 8 is:

A. a local maximum

B. a local minimum

C. not a local extremum

B. (Use the First Derivative Test.)

5) The critical numbers for a differentiable function f are $x = 1, 3, 4, 6, 8$. What can you conclude about the local extrema of f from the following?

Interval	Sign of $f'(x)$
$(-\infty, 1)$	positive
$(1, 3)$	negative
$(3, 4)$	negative
$(4, 6)$	positive
$(6, 8)$	positive
$(8, \infty)$	negative

Interval	Sign of $f'(x)$	f
$(-\infty, 1)$	positive	increasing
$(1, 3)$	negative	decreasing
$(3, 4)$	negative	decreasing
$(4, 6)$	positive	increasing
$(6, 8)$	positive	increasing
$(8, \infty)$	negative	decreasing

f has a local maximum at $x = 1$ and $x = 8$. f has a local minimum at $x = 4$. There are no local extrema at $x = 3$ or $x = 6$.

6) Is $x = 5$ a local minimum for $f(x) = 10x - x^2 + 1$?

No. Since $f'(x) = 10 - 2x$, 5 is a critical point. But $f'(x) > 0$ for $x < 5$ and $f'(x) < 0$ for $x > 5$. f has a local *maximum* at $x = 5$.

7) Find where the local maximum and minimum values occur for $f(x) = 4x^3 - x^4$.

$f'(x) = 12x^2 - 4x^3 = 4x^2(3 - x)$

$f'(x) = 0$ at $x = 0, x = 3$, so there are two critical numbers. Since $f'(x) \geq 0$ for $x < 3$ and $f'(x) \leq 0$ for $x > 3$, 3 is a local maximum. No extreme value occurs at $x = 0$.

8) Find where the local maximum and minimum values occur for $f(x) = x + \sin x$.

$f'(x) = 1 + \cos x$. Since $-1 \leq \cos x \leq 1$, $f'(x) \geq 0$ for all x. Therefore, f has no local extreme values.

<When $f'(x)$ is positive, the function is increasing. When $f'(x)$ is negative, the function is decreasing.>

9) Sketch the graph of a function f with the following properties:

Vertical asymptote at $x = 2$
$f'(x) < 0$ for $x < 0$
$f'(x) > 0$ for $0 < x < 2$
$f'(x) > 0$ for $2 < x < 3$
$f'(x) < 0$ for $x > 3$

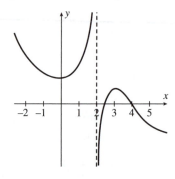

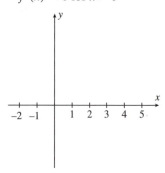

Page 293

C. Concavity is a term used to describe how a curve is bending.

A function f is **concave upward on an interval I** if f lies above all tangent lines to f in I. f is **concave downward on I** if the graph of f is below all tangent lines.

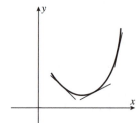

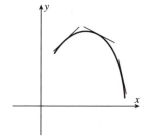

concave upward   concave downward

To help remember the definition of concavity, remember that the letter "U" in the word "UP" is concave up.

The test for concavity uses the second derivative:
Concavity Test: If f is twice differentiable on an interval I (meaning $f''(x)$ exists for all $x \in I$) then:

a) If $f''(x) > 0$ for all $x \in I$, f is concave *upward* on I.

b) If $f''(x) < 0$ for all $x \in I$, f is concave *downward* on I.

Example: For $f(x) = 1 - x^3$, $f''(x) = -6x$. Thus $f''(x) > 0$ for $x < 0$ and $f''(x) < 0$ for $x > 0$.

x	$f''(x)$	Concavity
$(-\infty, 0)$	+	Up
0	0	
$(0, \infty)$	−	Down

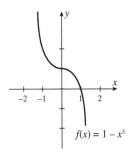

$f(x) = 1 - x^3$

The graph is concave up to the left of $x = 0$ and concave down to the right of $x = 0$. See the graph at the right.

10) Use the graph below to answer true or false to each.

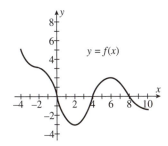

$y = f(x)$

a) $f''(x) > 0$ for $x \in (2, 4)$.   True.

b) $f''(x) < 0$ for $x \in (-4, -2)$.   False.

c) $f''(6) = 0$.   False. $f''(6)$ is negative.

d) $f''(2) > 0$.   True.

e) f is concave upward on $(0, 2)$.   True.

f) $f''(x) < 0$ for $x \in (4, 8)$.   True.

g) Both $f'(3)$ and $f''(3)$ are positive.   True.

11) What can be said about $f'(x)$ and $f''(x)$ for each?

a)

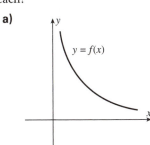

$y = f(x)$

$f'(x) < 0$ and $f''(x) > 0$.

b)

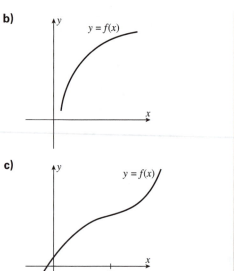

$f'(x) > 0$ and $f''(x) < 0$.

c)

$f'(x) \geq 0$ for all x,
$f''(x) > 0$ for $x > 2$,
$f''(x) < 0$ for $x < 2$.

Page 294

D. A **point of inflection** for f is a point on the graph of f where concavity changes from concave downward to concave upward or from concave upward to concave downward. Here are two examples for each case.

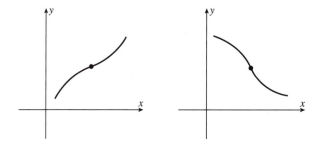

From concave downward to concave upward

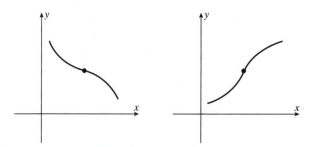

From concave upward to concave downward

Since an inflection point $(c, f(c))$ for a function f occurs where the graph of f changes concavity, $f''(c)$ will either be 0 or not exist. Therefore, we can determine inflection points in a manner similar to the way we find critical points.

Example: Find the points of inflection of

$$f(x) = \tfrac{1}{4}x^4 - \tfrac{13}{6}x^3 + 5x^2.$$

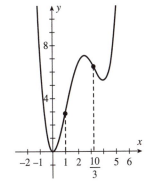

Because f is a polynomial, all critical and inflection points will occur where $f'(x) = 0$ and $f''(x) = 0$, respectively.

$$f'(x) = x^3 - \tfrac{13}{2}x^2 + 10x.$$

$$f''(x) = 3x^2 - 13x + 10 = (3x - 10)(x - 1).$$

$$f''(x) = 0 \text{ at } x = \tfrac{10}{3} \text{ and } x = 1.$$

The points of inflection are at $x = \tfrac{10}{3}$ and $x = 1$.
See the graph at the right.

In addition to finding inflection points, the second derivative also can help determine whether a critical point is a relative maximum or relative minimum.

The Second Derivative Test:

If f'' is continuous on (a, b), $c \in (a, b)$ and $f'(c) = 0$, then

a) if $f''(c) < 0$, f has a local maximum at c.

b) if $f''(c) > 0$, f has a local minimum at c.

If it should happen that both $f'(c) = 0$ and $f''(c) = 0$ then no conclusion can be made about an extremum at c.

Example: Find the relative maxima and minima of $f(x) = \tfrac{1}{4}x^4 - \tfrac{13}{6}x^3 + 5x^2$.

From the example above, $f'(x) = x^3 - \tfrac{13}{2}x^2 + 10x = x\left(x^2 - \tfrac{13}{2}x + 10\right) = x\left(x - \tfrac{5}{2}\right)(x - 4)$.

There are three critical points at $x = 0$, $x = \tfrac{5}{2}$, and $x = 4$.

$f''(x) = 3x^2 - 13x + 10$. By the Second Derivative Test:

$f''(0) = 3(0)^2 - 13(0) + 10 = 10 > 0$, so there is a relative minimum at $x = 0$.

$f''\left(\tfrac{5}{2}\right) = 3\left(\tfrac{5}{2}\right)^2 - 13\left(\tfrac{5}{2}\right) + 10 = -\tfrac{15}{4} < 0$, so there is a relative maximum at $x = \tfrac{5}{2}$.

$f''(4) = 3(4)^2 - 13(4) + 10 = 6 > 0$, so there is a relative minimum at $x = 4$.
These results agree with the graph above.

12) What are the points of inflection for the function in question 10)?

$x = -2, 0, 4, 8.$

13) Find the points of inflection for $f(x) = 8x^3 - x^4$.

$f'(x) = 24x^2 - 4x^3.$
$f''(x) = 48x - 12x^2 = 12x(4 - x) = 0$
at $x = 0, x = 4.$

x	f''(x)	Concavity
$(-\infty, 0)$	–	Down
0	0	
$(0, 4)$	+	Up
4	0	
$(4, \infty)$	–	Down

Since f is a polynomial the only points of inflection are at $x = 0$ and $x = 4.$

14) Find all local extrema of

a) $f(x) = x^3 + 6x^2 - 36x.$

$f'(x) = 3x^2 + 12x - 36 = 3(x - 2)(x + 6).$
The critical points are at $x = 2, -6.$
$f''(x) = 6x + 12.$
At $x = 2, f''(2) = 24 > 0$ so f has a local minimum at 2.
At $x = -6, f''(-6) = -24 < 0$ so f has a local maximum at $-6.$

b) $f(x) = x^2 + 2 \cos x.$

$f'(x) = 2x - 2 \sin x = 0$ when $x = \sin x.$
The only solution to this equation is $x = 0.$
$f''(x) = 2 - 2 \cos x$ but $f''(0) = 0$ so the Second Derivative Test fails to give additional information. However $f''(x) \geq 0,$ for all $x,$ means f is concave upward and thus f has a local minimum at 0.

15) Sketch a graph of a function f having all these properties:

   i) $f(-1) = 4, f(0) = 2, f(2) = 1, f(3) = 0$
   ii) $f'(x) < 0$ for $x < 3$ and
   iii) $f'(x) > 0$ for $x > 3.$
   iv) $f''(x) < 0$ for $0 < x < 2$ and
   v) $f''(x) > 0$ elsewhere.

Property ii) tells us f is decreasing on $(-\infty, 3)$ while property iii) tells us f is increasing on $(3, -\infty).$ Property iv) says f is concave down in $(0, 2)$ and property v) says f is concave up elsewhere.

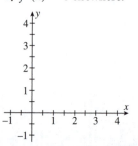

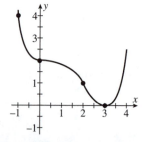

16) Let $f(x) = x^{4/3} + 8x^{1/3}$.

   a) Where f is increasing? decreasing?

$$f'(x) = \tfrac{4}{3}x^{1/3} + 8\left(\tfrac{1}{3}x^{-2/3}\right) = \frac{4x + 8}{3x^{2/3}}.$$

Since $x^{2/3}$ is positive for $x \neq 0$,

$f'(x) > 0$ when $4x + 8 > 0$. Thus $x > -2$.
f is increasing for $x > -2$.
Likewise, f is decreasing for $x < -2$.

   b) Find the local maxima and minima.

$f'(x) = 0$ when $4x + 8 = 0$. Thus $x = -2$.

$$f''(x) = \tfrac{4}{3}\left(\tfrac{1}{3}x^{-2/3}\right) + \tfrac{8}{3}\left(-\tfrac{2}{3}x^{-5/3}\right) = \frac{4x - 16}{9x^{5/3}}.$$

$f(-2) = (-2)^{4/3} + 8(-2)^{1/3} \approx -7.6$.

$$f''(-2) = \frac{4(-2) - 16}{9(-2)^{5/3}} \approx 0.84. \text{ Therefore,}$$

$(-2, -7.6)$ is a local minimum.
$f'(x)$ does not exist for $x = 0$.
$f(0) = 0$. Therefore, since f is increasing for $x > -2$, $(0, 0)$ is not a local extremum.

   c) Find where f is concave up, concave down, and has inflection points.

$$f''(x) = \frac{4x - 16}{9x^{5/3}} = 0 \text{ at } x = 4 \text{ and}$$

does not exist at $x = 0$.

$f(4) = 4^{4/3} + 8(4)^{1/3} \approx 19$.

Interval	f''	f is
$(-\infty, 0)$	positive	concave up
$(0, 4)$	negative	concave down
$(4, \infty)$	positive	concave up

Both $(0, 0)$ and $(4, 19)$ are inflection points.

   d) Sketch the graph of f.

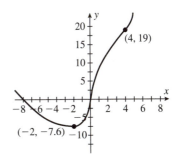

17) Orlando's car is stopped at a stop sign on a straight road. He then accelerates to 30 mi/hr and maintains this speed. He next sees a red traffic light ahead and slows down to 15 mi/hr as he approaches the light. The instant he has slowed to 15 mi/hr the light turns green and he resumes his cruising speed of 30 mi/hr. Sketch a graph of the position function $p(t)$ of Orlando's distance from the stop sign after t hours.

We use the following times t:

t	Position $p(t)$
0	At stop sign
a	Speed reaches 30 mi/hr
b	Sees red light. Starts to slow down.
c	Speed is 15 mi/hr and light turns green. Starts to speed up again.
d	Speed reaches 30 mi/hr again

Orlando continues to move forward at all times so $p'(t)$ is always positive. He increases his speed on the interval $(0, a)$ from zero to 30 mi/hr. On (b, c) he decreases speed from 30 down to 15 mi/hr and then on (c, d) the speed increases from 15 back up to 30 mi/hr. On (a, b) and (d, ∞) his speed is a constant 30 mi/hr and thus $p''(t) = 0$.

t	$p'(t)$	$p''(t)$	$p(t)$ is
$(0, a)$	+	+	concave up
a	30	0	
(a, b)	+	0	linear
b	30	0	
(b, c)	+	–	concave down
c	15	0	
(c, d)	+	+	concave up
d	30	0	
(d, ∞)	+	0	linear

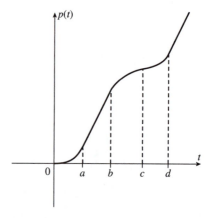

E. Technology Plus. Use a computer algebra system or a graphing calculator to solve.

T-1) Let $f(x) = x^4 - 3x^3 - x^2 + 3x$. Draw the graph of f'' on a calculator with window $[-2, 4]$ by $[-12, 6]$. Where is f concave up and concave down?

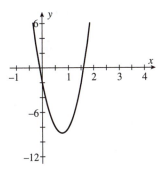

The function f'' crosses the x-axis at about -0.014 and 1.604. $f''(x) \geq 0$ on $(-\infty, -0.104] \cup [1.604, \infty)$, so f is concave upward there. On $[-0.104, 1.604]$ $f''(x) \leq 0$ and so f is concave downward there.

Section 4.4 Indeterminate Forms and l'Hospital's Rule

Sometimes applying limit theorems of Chapter 2 to a limit results in an indeterminable expression such as $\frac{0}{0}$. We have handled cases like this with clever factoring or rewriting the limit. This section describes solutions using derivatives.

Concepts to Master

A. Indeterminate forms; L'Hospital's Rule for $\frac{0}{0}$ and $\frac{\infty}{\infty}$

B. Other indeterminate forms: $0 \cdot \infty$, $\infty - \infty$, 0^0, 1^∞, and ∞^0

Summary and Focus Questions

Page 301

A. Indeterminate forms $\frac{0}{0}$ and $\frac{\infty}{\infty}$.

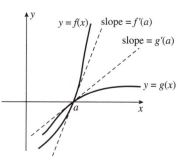

If $\lim_{x \to a} f(x) = 0$ and $\lim_{x \to a} g(x) = 0$, the quotient limit law does not apply to $\lim_{x \to a} \frac{f(x)}{g(x)}$ because the result is $\frac{0}{0}$. However, if f and g are differentiable and x is near a, then the ratio $\frac{f(x)}{g(x)}$ is approximately $\frac{f'(x)}{g'(x)}$, the ratio of the slopes of the tangent lines. See the figure.

L'Hospital's Rule for $\frac{0}{0}$.

If f and g are differentiable, $g'(x) \neq 0$ for x near a, and $\lim_{x \to a} f(x) = 0$ and $\lim_{x \to a} g(x) = 0$, then

$$\lim_{x \to a} \frac{f(x)}{g(x)} = \lim_{x \to a} \frac{f'(x)}{g'(x)}.$$

L'Hospital's Rule for $\frac{0}{0}$ says we may replace $\lim_{x \to a} \frac{f(x)}{g(x)}$ by $\lim_{x \to a} \frac{f'(x)}{g'(x)}$ and then try the limit laws for this limit.

Example: $\lim_{x \to 2} \frac{x^2 - 4}{3x - 6}$ has form $\frac{0}{0}$. Although it can be evaluated by factoring, canceling, and applying the limit theorems, l'Hospital's Rule is quicker:

$$\lim_{x \to 2} \frac{x^2 - 4}{3x - 6} = \lim_{x \to 2} \frac{2x}{3} = \frac{4}{3}.$$

L'Hospital's Rule for $\frac{\infty}{\infty}$.

If f and g are differentiable and $\lim\limits_{x\to a} f(x) = \infty$ and $\lim\limits_{x\to a} g(x) = \infty$, then

$$\lim_{x\to a} \frac{f(x)}{g(x)} = \lim_{x\to a} \frac{f'(x)}{g'(x)}.$$

Example: $\lim\limits_{x\to 0^+} \dfrac{-\ln x}{1/x^2}$ has form $\frac{\infty}{\infty}$. Therefore, $\lim\limits_{x\to 0^+} \dfrac{-\ln x}{1/x^2} = \lim\limits_{x\to 0^+} \dfrac{-x^{-1}}{-2x^{-3}} = \lim\limits_{x\to 0^+} \dfrac{x^2}{2} = 0.$

L'Hospital's Rule also applies to limits at infinity and one-sided limits.

Example: $\lim\limits_{x\to\infty} \dfrac{x^2+1}{3x^2-4}$ has form $\frac{\infty}{\infty}$. Therefore, $\lim\limits_{x\to\infty} \dfrac{x^2+1}{3x^2-4} = \lim\limits_{x\to\infty} \dfrac{2x}{6x} = \lim\limits_{x\to\infty} \dfrac{2}{6} = \dfrac{1}{3}.$

1) Evaluate each:

a) $\lim\limits_{x\to 3} \dfrac{x^2-7x+12}{x^2-9}$

This limit has form $\frac{0}{0}$.

$$\lim_{x\to 3} \frac{x^2-7x+12}{x^2-9} = \lim_{x\to 3} \frac{2x-7}{2x}$$

$$= -\frac{1}{6}.$$

b) $\lim\limits_{x\to 0} \dfrac{\sin x}{x}$ (a familiar limit)

This limit has form $\frac{0}{0}$.

$$\lim_{x\to 0} \frac{\sin x}{x} = \lim_{x\to 0} \frac{\cos x}{1}$$

$$= \frac{1}{1} = 1.$$

c) $\lim\limits_{x\to 0^+} \dfrac{-\ln x}{x+\frac{1}{x}}$

This limit has the form $\frac{\infty}{\infty}$.

$$\lim_{x\to 0^+} \frac{-\ln x}{x+\frac{1}{x}} = \lim_{x\to 0^+} \frac{-\frac{1}{x}}{1-\frac{1}{x^2}} =$$

$$\lim_{x\to 0^+} \frac{-x}{x^2-1} = \frac{-0}{0-1} = 0.$$

d) $\lim\limits_{x\to\infty} \dfrac{3x^2}{e^x}$

This limit has the form $\frac{\infty}{\infty}$.

$$\lim_{x\to\infty} \frac{3x^2}{e^x} = \lim_{x\to\infty} \frac{6x}{e^x}.$$ This limit also has

the form $\frac{\infty}{\infty}$ so we apply l'Hospital's Rule again.

$$\lim_{x\to\infty} \frac{6x}{e^x} = \lim_{x\to\infty} \frac{6}{e^x} = 0.$$

Page 305

B. There are five other indeterminate forms, each of which may be handled by reducing the problem to either the form $\frac{0}{0}$ or the form $\frac{\infty}{\infty}$.

Indeterminate Form	Example	Procedure
$\lim\limits_{x\to a}(f(x)\cdot g(x)) = 0\cdot\infty$	$\lim\limits_{x\to 0}(e^x - 1)\csc x$	Rewrite $0\cdot\infty$ as $\frac{f(x)}{1/g(x)}$ or $\frac{g(x)}{1/f(x)}$ to get $\frac{0}{0}$ or $\frac{\infty}{\infty}$.
$\lim\limits_{x\to a}(f(x) - g(x)) = \infty - \infty$	$\lim\limits_{x\to \pi/2^-}(\sec x - \tan x)$	Rewrite $f(x) - g(x)$ using algebra, identities, etc., to fit $\frac{0}{0}$ or $\frac{\infty}{\infty}$.
$\lim\limits_{x\to a} f(x)^{g(x)} = 0^0$	$\lim\limits_{x\to 0}(\sin x)^x$	i) Let $y = f(x)^{g(x)}$. Then $\ln y = g(x)(\ln f(x))$.
$\lim\limits_{x\to a} f(x)^{g(x)} = 1^\infty$	$\lim\limits_{x\to 0^+}(1+x)^{\cot x}$	ii) Evaluate $\lim\limits_{x\to a} g(x)(\ln f(x))$ using l'Hospital's Rule.
$\lim\limits_{x\to a} f(x)^{g(x)} = \infty^0$	$\lim\limits_{x\to 0^+}(-\ln x)^x$	iii) If $\lim\limits_{x\to a} g(x)\ln f(x) = L$, then $\lim\limits_{x\to a} f(x)^{g(x)}=e^L$.

Examples: Here are the solutions to the examples in the table.

a) $\lim\limits_{x\to 0}(e^x - 1)\csc x = \lim\limits_{x\to 0}\dfrac{e^x - 1}{\sin x}\left(\text{form } \dfrac{0}{0}\right) = \lim\limits_{x\to 0}\dfrac{e^x}{\cos x} = \lim\limits_{x\to 0}\dfrac{e^0}{\cos 0} = \dfrac{1}{1} = 1.$

b) $\lim\limits_{x\to \pi/2^-}(\sec x - \tan x) = \lim\limits_{x\to \pi/2^-}\left(\dfrac{1}{\cos x} - \dfrac{\sin x}{\cos x}\right)$

$\qquad\qquad = \lim\limits_{x\to \pi/2^-}\dfrac{1 - \sin x}{\cos x}\left(\text{form } \dfrac{0}{0}\right) = \lim\limits_{x\to \pi/2^-}\dfrac{-\cos x}{-\sin x} = 0.$

c) $\lim\limits_{x\to 0}(\sin x)^x$. Let $y = (\sin x)^x$. Then $\ln y = x\ln(\sin x) = \dfrac{\ln(\sin x)}{x^{-1}}$.

$\lim\limits_{x\to 0}\dfrac{\ln(\sin x)}{x^{-1}} = \lim\limits_{x\to 0}\dfrac{\dfrac{1}{\sin x}(\cos x)}{-1x^{-2}} = \lim\limits_{x\to 0}\dfrac{-x^2}{\tan x}.$ Since this limit is of the form $\dfrac{0}{0}$,

$\lim\limits_{x\to 0}\dfrac{-x^2}{\tan x} = \lim\limits_{x\to 0}\dfrac{-2x}{\sec^2 x} = \lim\limits_{x\to 0}(-2x)\cos^2 x = (-2)(0)(1)^2 = 0.$ Therefore,

$\lim\limits_{x\to 0}(\sin x)^x = e^0 = 1.$

d) $\lim\limits_{x\to 0^+}(1 + x)^{\cot x}$. Let $y = (1 + x)^{\cot x}$. Then $\ln y = \ln(1 + x)^{\cot x} = (\cot x)\ln(1 + x).$

$\lim\limits_{x\to 0^+}(\cot x)\ln(1 + x)\left(\text{form } \infty\cdot 0\right) = \lim\limits_{x\to 0^+}\dfrac{\ln(1 + x)}{\tan x}\left(\text{form } \dfrac{0}{0}\right) = \lim\limits_{x\to 0^+}\dfrac{\frac{1}{1+x}}{\sec^2 x} =$

$\lim\limits_{x\to 0^+}\dfrac{\cos^2 x}{1 + x} = \dfrac{1^2}{1 + 0} = 1.$ Therefore, $\lim\limits_{x\to 0^+}(1 + x)^{\cot x} = e^1 = e.$

e) $\lim\limits_{x\to 0^+}(-\ln x)^x$. Let $y = (-\ln x)^x$. Then $\ln y = x\ln(-\ln x) = \dfrac{\ln(-\ln x)}{x^{-1}}$.

$$\lim_{x\to 0^+}\frac{\ln(-\ln x)}{x^{-1}}\left(\text{form }\frac{0}{0}\right) = \lim_{x\to 0^+}\frac{\frac{1}{-\ln x}\left(\frac{-1}{x}\right)}{-1x^{-2}} = \lim_{x\to 0^+}\frac{-x}{\ln x} = 0.\ \text{Therefore,}$$

$$\lim_{x\to 0^+}(-\ln x)^x = e^0 = 1.$$

2) Is ∞^∞ an indeterminant form?

No. ∞^∞ would mean the base and exponent grow larger. A limit with this form is ∞.

3) Evaluate each:

a) $\lim\limits_{x\to 0^+} x\cot x$

This has form $0\cdot\infty$. Rewrite $x\cot x$ in the form $\dfrac{0}{0}$ as $x\cdot\dfrac{\cos x}{\sin x} = \dfrac{x}{\sin x}\cdot\cos x$.

$$\lim_{x\to 0^+} x\cot x = \lim_{x\to 0^+}\frac{x}{\sin x}\cdot\cos x$$

$$= \left(\lim_{x\to 0^+}\frac{x}{\sin x}\right)\cdot\left(\lim_{x\to 0^+}\cos x\right)\left(\text{form }\frac{0}{0}\right)$$

$$= \lim_{x\to 0^+}\frac{1}{\cos x}\cdot\lim_{x\to 0^+}\cos x = \frac{1}{1}\cdot 1 = 1.$$

b) $\lim\limits_{x\to\infty}\left(\dfrac{1}{x^2-1} - \dfrac{1}{x-1}\right)$

This limit is $0 - 0 = 0$. (Not an indeterminate form.)

c) $\lim\limits_{x\to 0^+}\left(\dfrac{1}{x} - \dfrac{1}{e^x-1}\right)$

<Apply l'Hospital's Rule twice.>

$$\lim_{x\to 0^+}\left(\frac{1}{x} - \frac{1}{e^x-1}\right)(\text{form }\infty-\infty)$$

$$= \lim_{x\to 0^+}\frac{e^x-1-x}{x(e^x-1)}\left(\text{form }\frac{0}{0}\right)$$

$$= \lim_{x\to 0^+}\frac{e^x-1}{xe^x+e^x-1}\left(\text{form }\frac{0}{0}\text{ again}\right)$$

$$= \lim_{x\to 0^+}\frac{e^x}{xe^x+2e^x} = \frac{1}{0+2} = \frac{1}{2}.$$

d) $\lim\limits_{x\to 0^+}\left(\dfrac{1}{\sin^2 x} - \dfrac{\cot x}{x}\right)$

<Apply l'Hospital's Rule twice.>

$$\lim_{x\to 0^+}\left(\frac{1}{\sin^2 x} - \frac{\cot x}{x}\right)$$

$$= \lim_{x\to 0^+}\frac{x-\cos x\sin x}{x\sin^2 x}\left(\text{form }\frac{0}{0}\right)$$

$$= \lim_{x\to 0^+}\frac{1-\cos^2 x+\sin^2 x}{\sin^2 x+2x\sin x\cos x}$$

$$= \lim_{x\to 0^+}\frac{2\sin^2 x}{\sin x(\sin x+2x\cos x)}$$

$$= \lim_{x\to 0^+}\frac{2\sin x}{\sin x+2x\cos x}\left(\text{form }\frac{0}{0}\right)$$

$$= \lim_{x\to 0^+}\frac{2\cos x}{3\cos x-2x\sin x} = \frac{2}{3}.$$

e) $\lim\limits_{x \to 0^+} (\csc x)^x$

This has form ∞^0.

Let $y = (\csc x)^x$.

$\ln y = x \ln(\csc x)$.

$\lim\limits_{x \to 0^+} x(\ln \csc x) \ (\text{form } 0 \cdot \infty)$

$= \lim\limits_{x \to 0^+} \dfrac{\ln \csc x}{x^{-1}} \left(\text{form } \dfrac{\infty}{\infty}\right)$

$= \lim\limits_{x \to 0^+} \dfrac{\frac{1}{\csc x} \cdot (-\csc x \cot x)}{(-1)x^{-2}}$

$= \lim\limits_{x \to 0^+} x^2 \cot x \ (\text{form } 0 \cdot \infty)$

$= \lim\limits_{x \to 0^+} \dfrac{x^2}{\tan x} \left(\text{form } \dfrac{0}{0}\right)$

$= \lim\limits_{x \to 0^+} \dfrac{2x}{\sec^2 x} = \dfrac{2(0)}{1} = 0.$

Thus, $\lim\limits_{x \to 0^+} (\csc x)^x = e^0 = 1$.

f) $\lim\limits_{x \to 0^+} x^x$

This has form 0^0.

Let $y = x^x$.

$\ln y = \ln x^x = x \ln x$.

$\lim\limits_{x \to 0^+} x \ln x = \lim\limits_{x \to 0^+} \dfrac{\ln x}{\frac{1}{x}} \left(\text{form } \dfrac{\infty}{\infty}\right)$

$= \lim\limits_{x \to 0^+} \dfrac{\frac{1}{x}}{\frac{-1}{x^2}}$

$= \lim\limits_{x \to 0^+} -x = 0.$

Thus, $\lim\limits_{x \to 0^+} x^x = e^0 = 1$.

Section 4.5   Summary of Curve Sketching

This section reviews and applies the curve sketching calculus concepts of the previous sections. It also introduces the notion of a slant asymptote of a function $y = f(x)$—a non-vertical line that the graph of the function approaches for large positive or negative values of x.

Concepts to Master

A. Curve sketching using information obtained through calculus concepts

B. Slant asymptotes

C. Technology Plus

Summary and Focus Questions

Page 310

A. You should develop a checklist of features to apply before drawing a graph. Features to consider in sketching a graph of $y = f(x)$ include:

A. Domain of f For what x is $f(x)$ defined?

B. x-intercept(s)   What are the solution(s) (if any) to $f(x) = 0$?
    y-intercept   What value (if any) is $f(0)$?

C. Symmetry about y-axis   Is $f(-x) = f(x)$?
    Symmetry about origin   Is $f(-x) = -f(x)$?
    Periodicity   Is there a number p such that $f(x + p) = f(x)$ for all x in the domain?

D. Horizontal asymptote(s)   Does $\lim\limits_{x \to \infty} f(x)$ exist?
         Does $\lim\limits_{x \to -\infty} f(x)$ exist?
    Vertical asymptote(s)   For what a is $\lim\limits_{x \to a^+} f(x) = \infty$ or $-\infty$?
         For what a is $\lim\limits_{x \to a^-} f(x) = \infty$ or $-\infty$?

E. Increasing   On what intervals is $f'(x) > 0$?
    Decreasing   On what intervals if $f'(x) < 0$?

F. Critical numbers   Where does $f'(x) = 0$ or not exist?
    Local extrema   Where are the local maxima or minima (if any)?
         (Use the First and Second Derivative Tests.)

G. Concave upward   On what intervals is $f''(x) > 0$?
    Concave downward   On what intervals is $f''(x) < 0$?
    Inflection points   Where does f change concavity?
         Where does $f''(x) = 0$ or not exist?

H. Sketch the graph.

1) Sketch a graph of $y = f(x)$ that has all these properties:

a) domain $= (-\infty, -1) \cup (1, \infty)$

b) $f(2) = 0$

c) $f(x) = -f(-x)$

d) $\lim\limits_{x \to \infty} f(x) = 4$ and $\lim\limits_{x \to 1+} f(x) = -\infty$

e) $f'(3) = 0, f'(5) = 0$

f) $f'(x) > 0$ on $(1, 3) \cup (5, \infty)$

g) $f'(x) < 0$ on $(3, 5)$

h) local maximum at 3

i) local minimum at 5

j) $f''(x) > 0$ on $(4, 6)$

k) $f''(x) < 0$ on $(1, 4) \cup (6, \infty)$

l) $f''(4) = 0, f''(6) = 0$

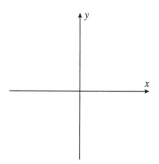

A. By property a), the domain is all reals except those in $[-1, 1]$.

B. By property b), 2 is x-intercept. There is no y-intercept because 0 is not in the domain.

C. The graph is symmetric about the origin by property c).

D. By property d), $y = 4$ is a horizontal asymptote and $x = 1$ is a vertical asymptote.

E. Properties f) and g) say that f is increasing on $(1, 3)$, increasing on $(5, \infty)$ and decreasing on $(3, 5)$.

F. By h) and i), 3 is a local maximum and 5 is a local minimum.

G. By j), f is a concave up on $(4, 6)$. By k), f is concave down on $(1, 4)$ and on $(4, 6)$. By j), k), and l), 4 and 6 are inflection points.

H. We draw the "right" half (for $x > 1$) of the graph and then use part C to draw the other half.

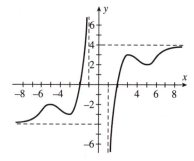

2) Sketch a graph of $f(x) = \dfrac{1}{x^2 - 4}$.

A. The domain is all x except where $x^2 - 4 = 0$; all x except $x = 2, x = -2$.

B. $\dfrac{1}{x^2 - 4}$ is never 0, so f has no x-intercepts.

At $x = 0, y = \dfrac{1}{0^2 - 4} = -\dfrac{1}{4}$. The y-intercept is $y = -\dfrac{1}{4}$.

C. $f(-x) = \dfrac{1}{(-x)^2 - 4} = \dfrac{1}{x^2 - 4} = f(x)$, so there is symmetry about the y-axis.

D. $\lim\limits_{x\to\infty} \dfrac{1}{x^2-4} = 0$ and $\lim\limits_{x\to-\infty} \dfrac{1}{x^2-4} = 0$.

$y = 0$ is the only horizontal asymptote.

$$\lim_{x\to 2^+} \frac{1}{x^2-4} = \infty, \lim_{x\to 2^-} \frac{1}{x^2-4} = -\infty,$$

$$\lim_{x\to -2^+} \frac{1}{x^2-4} = -\infty, \text{ and } \lim_{x\to -2^-} \frac{1}{x^2-4} = \infty.$$

Therefore, $x = 2$ and $x = -2$ are vertical asymptotes.

E. $f'(x) = (-1)(x^2-4)^{-2}(2x)$

$$= -2x(x^2-4)^{-2} = \frac{-2x}{(x^2-4)^2}.$$

For $x > 0$, $x \neq 2$, $f'(x) < 0$. For $x < 0$, $x \neq -2$, $f'(x) > 0$.

x	$f'(x)$	f
$(-\infty, -2)$	$+$	Increasing
-2	does not exist	
$(-2, 0)$	$+$	Increasing
0	0	
$(0, 2)$	$-$	Decreasing
2	does not exist	
$(2, \infty)$	$-$	Decreasing

F. $f'(x) = \dfrac{-2x}{(x^2-4)^2} = 0$ at $x = 0$.

$$f''(x) = (-2x)[-2(x^2-4)^{-3} \cdot 2x]$$
$$+ (x^2-4)^{-2}(-2)$$
$$= (x^2-4)^{-3}[8x^2 - 2(x^2-4)]$$
$$= \frac{6x^2+8}{(x^2-4)^3}.$$

$f''(0) = \dfrac{8}{(-4)^3} = -\dfrac{1}{8}$. Thus f has a local maximum at $x = 0$.

G. Since $f''(x) = \dfrac{6x^2+8}{(x^2-4)^3}$, the sign of $f''(x)$ is determined by $(x^2-4)^3$ and thus by $x^2 - 4$.

$x^2 - 4 > 0$ for $x > 2$ and $x < -2$ and $x^2 - 4 < 0$ for $-2 < x < 2$.

x	$f''(x)$	Concavity
$(-\infty, -2)$	$+$	Up
-2	does not exist	
$(-2, 0)$	$-$	Down
0	$-\dfrac{1}{8}$	
$(0, 2)$	$-$	Down
2	does not exist	
$(2, \infty)$	$+$	Up

H. Here is a sketch:

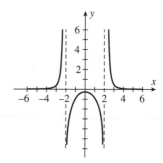

3) Sketch a graph of $f(x) = x - \ln x$.

A. The domain is $(0, \infty)$.

B. $x - \ln x$ is never 0, so there is no x-intercept.

0 is not in the domain, so there is no y-intercept.

C. Since $x > 0$ there is no symmetry.

D. $\lim_{x \to 0^+} (x - \ln x) = \infty$.

$\lim_{x \to \infty} (x - \ln x) = \infty$.

E. $f'(x) = 1 - \frac{1}{x}$.

For $0 < x < 1, f'(x) < 0$.
For $x > 1, f'(x) > 0$.

x	$f'(x)$	f
$(0, 1)$	–	Decreasing
1	0	
$(1, \infty)$	+	Increasing

F. $f'(x) = 1 - \frac{1}{x} = 0$ at $x = 1$.

$f''(x) = x^{-2}$, so $f''(1) = 1 > 0$.
f has a local minimum at $x = 1$.

G. Since $f''(x) > 0$, the graph is always concave upward.

H. Here is a sketch:

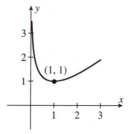

Page 315

B. The line $y = mx + b$ is a **slant asymptote** of $y = f(x)$ means that
$$\lim_{x \to \infty} [f(x) - (mx + b)] = 0 \text{ or } \lim_{x \to -\infty} [f(x) - (mx + b)] = 0.$$

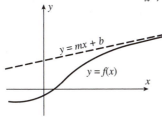

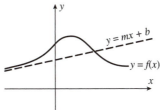

A slant asymptote is a nonvertical, nonhorizontal line that the graph of $f(x)$ approaches for large (positive or negative) values of x.

4) Find a slant asymptote for
$f(x) = \frac{x^2 + x + 1}{x}$.

$f(x) = \frac{x^2 + x + 1}{x} = x + 1 + \frac{1}{x}$.
The line $y = x + 1$ is a slant asymptote because

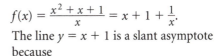

$$\lim_{x \to \infty} (f(x) - (x + 1))$$
$$= \lim_{x \to \infty} \left((x + 1 + \frac{1}{x}) - (x + 1) \right)$$
$$= \lim_{x \to \infty} \frac{1}{x} = 0.$$

5) Can a polynomial of degree greater than one have a slant asymptote?

No. If $f(x)$ is a polynomial whose degree is two or more then $f(x) - (mx + b)$ will also have degree two or more. Thus, $\lim_{x \to \infty} [f(x) - (mx + b)]$ will not be zero (it will be ∞ or $-\infty$).

6) Draw the graph of a function that has two parallel slant asymptotes.

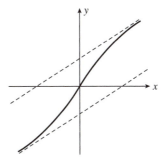

7) Show that $y = \sqrt{x^2 - 4}$ has two slant asymptotes.

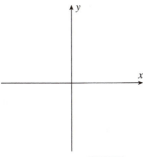

For large x, $\sqrt{x^2 - 4}$ is approximately $\sqrt{x^2} = x$, so we guess that $y = x$ is a slant asymptote.

$$\lim_{x \to \infty} (\sqrt{x^2 - 4} - x)$$
$$= \lim_{x \to \infty} (\sqrt{x^2 - 4} - x) \frac{(\sqrt{x^2 - 4} + x)}{(\sqrt{x^2 - 4} + x)}$$
$$= \lim_{x \to \infty} \frac{(x^2 - 4) - x^2}{(\sqrt{x^2 - 4} + x)}$$
$$= \lim_{x \to \infty} \frac{-4}{(\sqrt{x^2 - 4} + x)} = 0.$$

For large negative x,
$\sqrt{x^2 - 4} \approx \sqrt{x^2} = -x$. Thus, $y = -x$, is the other slant asymptote.

We note that $y = \sqrt{x^2 - 4}$ is the top half of the hyperbola $x^2 - y^2 = 4$ with asymptotes $y = x$ and $y = -x$.

A graph of $y = \sqrt{x^2 - 4}$ is given here:

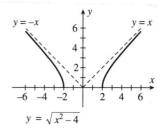

$$y = \sqrt{x^2 - 4}$$

8) Can a function have three slant asymptotes?

No. A function can have at most two slant asymptotes.

C. Technology Plus. Use a computer algebra system or a graphing calculator to solve.

T-1) Let $f(x) = \dfrac{1}{x^2} - \dfrac{1}{(x - 2)^2}$. Draw the graph of f on a calculator with window $[-2, 4]$ by $[-6, 6]$. Describe the graph.

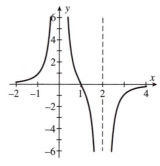

A. The domain is $(-\infty, 0) \cup (0, 2) \cup (2, \infty)$.

B. The x-intercept is 1; there is no y-intercept.

C. The graph is not symmetric about the y-axis or origin. (It is symmetric about the point $(1, 0)$.)

D. $y = 0$ is the horizontal asymptote. There are vertical asymptotes at $x = 0$ and $x = 2$.

E. Increasing on $(-\infty, 0) \cup (2, \infty)$; decreasing on $(0, 2)$.

F. The graph has no relative extrema.

G. $(1, 0)$ is an inflection point. The graph is concave up on $(-\infty, 0) \cup (0, 1)$ and concave down for $(1, 2] \cup [2, \infty)$.

Section 4.6   Graphing with Calculus *and* Calculators

This section shows you how some knowledge of calculus will help you use your calculator to draw more representative graphs.

Concepts to Master

A. Use calculus to enhance information from graphing calculator displays
B. Determine the graphs of families of functions

Summary and Focus Questions

Page 318

A. A graphing calculator can display an initial graph of a function. The concepts of calculus can then be used to refine it so that details such as relative extrema and concavity stand out.

1) Use a graphing calculator to graph $f(x) = 9x^4 - 76x^3 + 180x^2$. Then use calculus to refine it.

An initial graph from a calculator with window $[-1, 4]$ by $[-1, 330]$ will look something like:

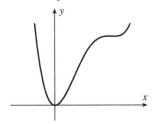

$$f'(x) = 36x^3 - 228x^2 + 360x$$
$$= 12x(3x^2 - 19x + 30)$$
$$= 12x(x - 3)(3x - 10) = 0$$
at $x = 0$, $x = 3$, $x = \frac{10}{3}$.
$$f''(x) = 108x^2 - 456x + 360.$$

x	f	f'	f''	Type
0	0	0	360	relative minimum
3	297	0	-36	relative maximum
$\frac{10}{3}$	296.3	0	40	relative minimum

The graph of f with its critical numbers 0, 3, and $\frac{10}{3}$ is:

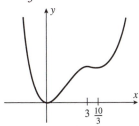

Zooming in, using a window of [2.8, 3.5] by [295, 300] we see

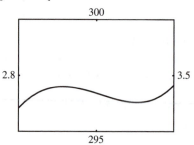

B. A **family of functions** is a collection of functions defined by a formula with one or more arbitrary constants. Varying the values of the constants results in changes in the graphs of the functions.

For example, the family described by $f(x) = x^2 + bx + 1$ has graphs which are parabolas passing through $(0, 1)$ with vertex (relative minimum) at $x = -\dfrac{b}{2}$.

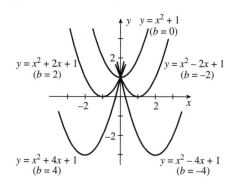

2) Describe the graphs of the family of functions $f(x) = x^3 - 3ax$.

Each function passes through $(0, 0)$. $f'(x) = 3(x^2 - a)$. For $a > 0$, f has a relative maximum at $x = -\sqrt{a}$ and relative minimum at $x = \sqrt{a}$. For $a \le 0$, f has no extrema.

$f''(x) = 6x$ so f is concave up for $x > 0$ and concave down for $x < 0$. Thus $x = 0$ is an inflection point. The family of graphs is:

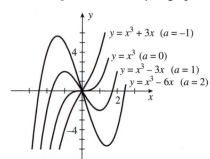

Section 4.7   Optimization Problems

An optimization problem seeks to find the largest (or smallest) value of a quantity (such as maximum revenue or minimum surface area) given certain limits or constraints. These problems arise in many areas of application. The method to follow that will help set up and solve them is based on the First Derivative Test.

Concepts to Master

A. Solve applied extrema problems
B. Technology Plus

Summary and Focus Questions

Page 325

A. An optimization problem can usually be expressed as "find the maximum (or minimum) value of some quantity Q under a certain set of given conditions."

For example, suppose we wish to find the dimensions of the largest rectangular field enclosed by 100 feet of fencing. The quantity to be maximized is the area, subject to the condition that the perimeter is 100.

A procedure to help solve problems of this type is:
   1. Determine known and unknown quantities.
   2. Draw a diagram (if helpful).
   3. Introduce notation identifying the quantity Q to be maximized (or minimized) and the other variable(s).
   4. Write Q as a function of the other variables. Also, express any relationships or conditions among the other variables with equations.
   5. Rewrite, if necessary, Q as a function of just one of the variables, using given relationships and determine the domain D of Q.
   6. Use the methods in Sections 4.1 and 4.3 to find the absolute extrema of Q on domain D.

In our example above, the steps in the solution are:
   1. The perimeter is 100. The area is unknown. The dimensions of the rectangle are unknown.

   2.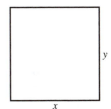

   3. Let A = area, x = width, y = height
   4. $A = xy$, $2x + 2y = 100$.
   5. From $2x + 2y = 100$, $y = 50 - x$.
      Thus, $A = x(50 - x) = 50x - x^2$. The domain is $[0, 50]$.

6. $A'(x) = 50 - 2x = 0$ at $x = 25$. Thus, $y = 50 - 25 = 25$. $A(0) = 0$, $A(50) = 0$, and $A(25) = 625$. So the largest area enclosed by 100 feet of fencing is a square 25 ft by 25 ft.

For steps 4 and 5 in the procedure you may need to recall facts and relationships from geometry, trigonometry, etc.

Some applications in economics use these concepts:

The **price** (or **demand**) **function** $d(x)$ is the price per item that must be charged to sell x items.

The (**total**) **revenue function** $R(x) = x\, d(x)$ is the revenue obtained by selling x items at $d(x)$ dollars each.

Marginal revenue is the derivative $R'(x)$ of total revenue.

The **profit function** is $P(x) = R(x) - C(x)$.

$P(x)$ is maximum when marginal revenue equals marginal costs.

1) A gardener wishes to create two equal-sized gardens by enclosing a rectangular area with 300 feet of fencing and fence it down the middle. What is the largest rectangular area that may be enclosed?

1. The amount of fence is known (100). The area enclosed is unknown. The dimensions of the rectangle are unknown.

2.

3. Label one side as x and the half side of the other as y (since that side is bisected). Let A be the area of the enclosed rectangle.

4. $A = x(2y) = 2xy$. The total amount of fence is $3x + 4y = 300$. Thus, $y = \dfrac{300 - 3x}{4}$.

5. Therefore,
$A = 2xy = 2x\left(\dfrac{300 - 3x}{4}\right) = 150x - \dfrac{3}{2}x^2$.
Since $3x + 4y = 300$ and both $x \geq 0$ and $y \geq 0$, $x \in [0, 100]$.

6. Maximize $A = 150x - \dfrac{3}{2}x^2$ on $[0, 100]$. Clearly 0 and 100 produce minimum values, so A is maximized when $A'(x) = 0$. $A'(x) = 150 - 3x = 0$ at $x = 50$. Then $y = \dfrac{300 - 3(50)}{4} = 37.5$. Thus, the maximum value of A is $2xy = 2(50)(37.5) = 3750 \text{ ft}^2$.

2) A cylindrical cup is to have a surface area of 12 square inches of aluminum. What is the largest possible volume of such a cup?

1. The surface area is known (12). The volume is unknown. The height and radius are unknown.

2.

3. Maximize the volume V of a cup whose circular base has radius r and height h.

4. $V = \pi r^2 h$. The total surface area (bottom plus side) is $12 = \pi r^2 + 2\pi rh$.

Thus, $h = \dfrac{12 - \pi r^2}{2\pi r}$.

5. $V = \pi r^2 \left(\dfrac{12 - \pi r^2}{2\pi r}\right)$.

$V = 6r - \dfrac{\pi}{2}r^3$, where $r \in \left[0, \sqrt{\dfrac{12}{\pi}}\right]$.

6. $V' = 6 - \dfrac{3\pi}{2}r^2 = 0$

$r^2 = \dfrac{4}{\pi}$

$r = \dfrac{2}{\sqrt{\pi}}$.

$V'' = -3\pi r$ so $r = \dfrac{2}{\sqrt{\pi}}$ is a maximum.

At $r = \dfrac{2}{\sqrt{\pi}}$ the maximum volume is

$V = 6\left(\dfrac{2}{\sqrt{\pi}}\right) - \dfrac{\pi}{2}\left(\dfrac{2}{\sqrt{\pi}}\right)^3 = \dfrac{8}{\sqrt{\pi}}$ in^3.

3) Find the dimensions of the largest rectangular peg that can be put into a round hole with diameter 10 cm.

1. The diameter of the hole is known (10). The length and width of a cross section of the peg are unknown. The area is unknown.

2.

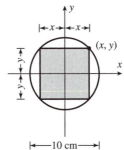

3. "Largest peg" means one with the largest cross-sectional area. The cross section of the peg in the hole is given in the diagram.

4. The quantity to be maximized is the area
$A = (2x)(2y) = 4xy$.

The equation for the perimeter of the hole is $x^2 + y^2 = 25$. Therefore, $y = \sqrt{25 - x^2}$.

5. $A(x) = 4xy = 4x\sqrt{25 - x^2}$ where $x \in [0, 5]$.

6. Maximize $A(x) = 4x\sqrt{25 - x^2}$ on $[0, 5]$. Both 0 and 5 produce a minimum area of zero, so the maximum occurs when $A'(x) = 0$.

$$A'(x) = (4x)\tfrac{1}{2}(25 - x^2)^{-1/2}(-2x)$$
$$+ (25 - x^2)^{1/2}(4)$$
$$= 4\left(-\frac{x^2}{\sqrt{25 - x^2}} + \sqrt{25 - x^2}\right)$$
$$= 4\left(\frac{-x^2 + (25 - x^2)}{\sqrt{25 - x^2}}\right)$$
$$= \frac{4(25 - 2x^2)}{\sqrt{25 - x^2}}.$$

$A'(x) = 0$ when $25 - 2x^2 = 0$

$2x^2 = 25$

$x^2 = \dfrac{25}{2}$

$x = \dfrac{5}{\sqrt{2}} = \dfrac{5\sqrt{2}}{2}$ cm.

$y = \sqrt{25 - x^2} = \sqrt{25 - \dfrac{25}{2}} = \sqrt{\dfrac{25}{2}}$

$= \dfrac{5\sqrt{2}}{2}$ cm.

As you may well have guessed, a square peg fits in a round hole.

4) Suppose that a paint store has determined that to sell x gallon per day of its base coat primer it must set its price per gallon at $29.50 - 0.05x$ dollars. The store has determined that the cost to stock and sell those x gallons is $0.20x^2 + 5.50x + 200$. What should the store set for the price per gallon of the primer so that the store maximizes its profits for the primer?

The question asks for the maximum of the amount of the primer sales profit $P(x)$, where x is the number of gallons of primer sold. The domain is $[0, \infty)$.

The cost is $C(x) = 0.20x^2 + 5.50x + 200$.

The demand is $p(x) = 29.50 - 0.05x$, so the revenue is $R(x) = 29.50x - 0.05x^2$.

The profit is $P(x) = R(x) - C(x)$
$$= 29.50x - 0.05x^2 - (0.20x^2 + 5.50x + 200)$$
$$= -0.25x^2 + 24x - 200.$$

$P(x)$ is maximum when $P'(x) = 0$.

$P'(x) = -0.50x + 24 = 0$

$0.50x = 24$

$x = 48$.

$P''(x) = -0.50$ so $x = 48$ is a maximum. At $x = 48$ gallons, the price is $29.50 - 0.05(48) = \$27.10$.

The price should be set at \$27.10 per gallon to maximize the profit.

B. Technology Plus. Use a computer algebra system or a graphing calculator to solve.

T-1) Gordon is standing on one side of a canal that is 1 kilometer wide. A lighthouse is 1 kilometer away downstream from the opposite side of the canal (see the figure). Gordon can run 4 km/hr and swim 1 km/hr. Since Gordon needs to get to the lighthouse as fast as he can, he will run along the canal part way, then swim directly to the lighthouse.

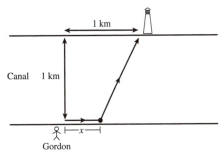

a) Write a function $T(x)$ for the time it takes Gordon to make the entire trip, where x is the distance that he runs. (Ignore any current in the water.)

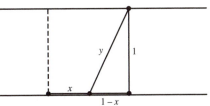

The time it takes to run x km is $\frac{x}{4}$.
The distance he swims (y) satisfies
$$y^2 = (1 - x)^2 + 1^2.$$
Thus, $y = \sqrt{(1 - x)^2 + 1}$.
Therefore, the total trip time is

$$T(x) = \frac{x}{4} + \frac{\sqrt{(1 - x)^2 + 1}}{1}$$

$$= \frac{x}{4} + \sqrt{(1 - x)^2 + 1}.$$

b) Use a graphing calculator to estimate the distance that he should run to minimize his trip time.

With a window of $[0, 1]$ by $[1, 1.5]$, the graph of $T(x)$ is

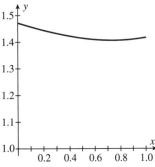

The minimum value of $T(x)$ occurs at approximately $x = 0.7418$. Gordon should run about three-quarters of the way and then swim the rest of the way.

Section 4.8   Newton's Method

Suppose you make a guess at the solution to $f(x) = 0$. Newton's method is a means to obtain a more accurate estimate based on your initial guess. Use a calculator to make sure your initial guess is somewhat near the exact solution.

Concepts to Master

Approximate solutions to $f(x) = 0$ using Newton's Method

Summary and Focus Questions

Page 338

Let f be a continuously differentiable function on an open interval with a real root. **Newton's Method** says that if x_1 is an estimate of the root, then

$$x_2 = x_1 - \frac{f(x_1)}{f'(x_1)}$$

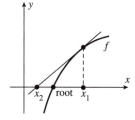

is often (but not always) a closer approximation to the root. The point x_2 is where the tangent line to f at x_1 crosses the x-axis.

The process of obtaining x_2 may be repeated using x_2 in the role of x_1 to obtain another approximation x_3. In this way we can generate a sequence of approximations $x_1, x_2, x_3, \ldots$ which can approach the value of the root. In general, to obtain a desired degree of accuracy, repeat Newton's Method until the difference between successive x_i is within that accuracy.

1) Indicate the point x_2 determined by Newton's Method in each:

a)

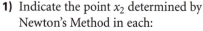

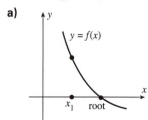

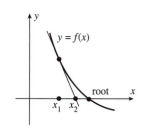

b)

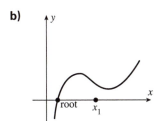

2) Use Newton's Method three times to approximate a root of $x^3 - 2x - 2 = 0$. Use an initial estimate of 1.0.

3) Estimate the x coordinates of the points where the functions $y = (x - 3)^2$ and $y = \sqrt{x}$ intersect.

   a) Estimate the intersection points using graphs.

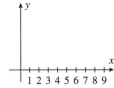

   b) Use Newton's Method once for each x estimate to improve your answers.

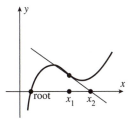

In this example x_2 is not a better estimate than x_1.

$f(x) = x^3 - 2x - 2, f'(x) = 3x^2 - 2.$
$x_1 = 1.0$, so

$$x_2 = 1.0 - \frac{f(1.0)}{f'(1.0)} = 1.0 - \frac{-3.0}{1.0} = 4.0.$$

$$x_3 = 4.0 - \frac{f(4.0)}{f'(4.0)} = 4.0 - \frac{54}{46} \approx 2.826.$$

$$x_4 = 2.826 - \frac{f(2.826)}{f'(2.826)}$$
$$= 2.826 - \frac{14.917}{21.959}$$
$$\approx 2.147.$$

First graph the functions:

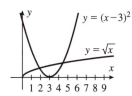

We estimate the intersections to be at $x = 2$ and $x = 5$.

Let $f(x) = (x - 3)^2 - \sqrt{x}$. The points of intersection are solutions to $f(x) = 0$. They are approximately $x = 2$ and $x = 5$.
$$f'(x) = 2(x - 3) - \frac{1}{2\sqrt{x}}.$$
For $x_1 = 2$, $f(2) = (2 - 3)^2 - \sqrt{2} \approx -0.414$ and
$$f'(2) = 2(2 - 3) - \frac{1}{2\sqrt{2}} \approx -2.354. \text{ Thus,}$$

$$x_2 = 2 - \frac{-0.414}{-2.353} \approx 1.824.$$

For $x_1 = 5, f(5) = (5 - 3)^2 - \sqrt{5} \approx 1.764$ and $f'(5) = 2(5 - 3) - \dfrac{1}{2\sqrt{5}} \approx 3.776$. Thus,

$$x_2 = 5 - \frac{1.764}{3.776} \approx 4.533.$$

4) Approximate a solution to $\cos x = x$ to six decimals. Use Newton's Method with an initial estimate of 0.

Let $f(x) = \cos x - x$ and $x_1 = 0$. Then $f'(x) = -\sin x - 1$. This table shows successive values for x using Newton's Method.

	x	$x - \dfrac{f(x)}{f'(x)}$
$x_1 =$	0	1
$x_2 =$	1	0.7503639
$x_3 =$	0.7503639	0.7391129
$x_4 =$	0.7391129	0.7390851
$x_5 =$	0.7390851	0.7390851
$x_6 =$	0.7390851	0.7390851

To six decimals, the solution to $\cos x = x$ is 0.739085.

Section 4.9 Antiderivatives

In this section we begin to learn how to "undo" derivatives; that is, given a derivative, recover the function from where it came. We will see that there is not just one antiderivative function, but rather a whole collection of functions, each of which differ by a constant from one another.

Concepts to Master

A. General and particular antiderivatives of a function
B. Differential equations

Summary and Focus Questions

Page 344

A. $F(x)$ is an **antiderivative** of the function $f(x)$ means $F'(x) = f(x)$.

For example, $F(x) = 3x^4$ is an antiderivative of $f(x) = 12x^3$ because $(3x^4)' = 12x^3$.

If F is an antiderivative of f then *all* other antiderivatives of f have the form $F(x) + C$, where C is a constant.

Here are some antidifferentiation formulas that you need to know well:

Function	Form of All Antiderivatives
x^n (except $n = -1$)	$\frac{x^{n+1}}{n+1} + C$
x^{-1}	$\ln|x| + C$
$\sin x$	$-\cos x + C$
$\cos x$	$\sin x + C$
$\sec^2 x$	$\tan x + C$
$\sec x \tan x$	$\sec x + C$
e^x	$e^x + C$
$\frac{1}{\sqrt{1-x^2}}$	$\sin^{-1}x + C$
$\frac{1}{1+x^2}$	$\tan^{-1}x + C$

The antiderivative of $f \pm g$ is the antiderivative of f plus or minus the antiderivative of g, and the antiderivative of $cf(x)$ is c times the antiderivative of $f(x)$ (c, any constant).

Examples:
The antiderivative of x^3 is $\frac{x^4}{4} + C$.

The antiderivative of $x + \sin x$ is $\frac{x^2}{2} - \cos x + C$.

The antiderivative of $10x^4$ is $10\left(\frac{x^5}{5}\right) + C = 2x^5 + C$.

The antiderivative of $6e^x - 2\sec^2 x$ is $6e^x - 2\tan x + C$.

In every case it is easy to check your answer—just take the derivative of your answer and see whether it is the function that you started with.

1) If $g(x)$ is an antiderivative of $h(x)$, then

　　_____$'(x) =$ _____(x).

$g'(x) = h(x)$.

2) Find all antiderivatives of:

　a) $f(x) = x^7$

$\frac{x^8}{8} + C$.

　b) $f(x) = \cos x - \sec^2 x$

$\sin x - \tan x + C$.

　c) $f(x) = x + x^{-2}$

$\frac{x^2}{2} - \frac{1}{x} + C$.

　d) $f(x) = x - e^x$

$\frac{x^2}{2} - e^x + C$.

　e) $f(x) = \dfrac{4}{\sqrt{1 - x^2}}$

$4 \sin^{-1} x + C$.

　f) $f(x) = \dfrac{2}{x}$

$2 \ln|x| + C$.

　g) $f(x) = 12x^2 + 8x + 4$

$4x^3 + 4x^2 + 4x + C$.

3) True or False:

If $f'(x) = g'(x)$, then $f(x) = g(x)$.

False. All we can say is that f and g differ by a constant.

B. The general solution to the **differential equation** $\dfrac{dy}{dx} = f(x)$ is all antiderivatives of f. To find a particular solution, first find the general solution, then substitute given values to determine the constant C.

Example: The general solution to $\dfrac{dy}{dx} = 10x$ is $y = 5x^2 + C$. If we are also given that $y = 4$ when $x = 2$, then substituting these values we have

　　$4 = 5(2)^2 + C$

　　$C = -16$.

The particular solution is $y = 5x^2 - 16$.

4) Find $f(x)$ where $f(2) = 3$ and $f'(x) = 4x + 5$.

$f(x) = 2x^2 + 5x + C$.
$f(2) = 2(2)^2 + 5(2) + C = 18 + C$.
$18 + C = 3$ so $C = -15$.
$f(x) = 2x^2 + 5x - 15$.

5) A particle moves along a scale with velocity $v = 3t + 7$. If the particle is at 4 on the scale at time $t = 1$, find the position function $s(t)$.

$s(t)$ is an antiderivative of $v(t) = 3t + 7$.

$s(t) = \frac{3}{2}t^2 + 7t + C$.

$4 = \frac{3}{2}(1)^2 + 7(1) + C$

$C = -\frac{9}{2}$.

$s(t) = \frac{3}{2}t^2 + 7t - \frac{9}{2}$.

6) Solve $\frac{dy}{dx} = 2 + x$, given that $y = 2$ when $x = 0$.

From $y' = 2 + x$, the general solution is

$y = 2x + \frac{x^2}{2} + C$.

At $x = 0, y = 2$. Therefore,

$2 = 0 + 0 + C$

$2 = C$.

The particular solution is

$y = \frac{x^2}{2} + 2x + 2$.

7) Solve $\frac{dy}{dx} = x - \sec^2 x$ given that $y = 0$ when $x = 0$.

From $y' = x - \sec^2 x$, the general solution is

$y = \frac{x^2}{2} - \tan x + C$.

At $x = 0, y = 0$. Therefore

$0 = \frac{0^2}{2} - \tan 0 + C$

$0 = 0 + C$

$C = 0$.

The particular solution is

$y = \frac{x^2}{2} - \tan x$.

Chapter 5 — Integrals

Section 5.1   Areas and Distances

Chapters 5, 6, 7, and 8 cover the concepts and techniques of integration, that branch of calculus used to find areas, volumes, total change, etc. (We will see soon how this relates to derivatives.) This chapter introduces integration with two types of problems—finding areas under curves and finding distances traveled given velocities. Just as limits were used in Chapter 2 to find tangents and velocities, limits will be used here to determine areas and distances.

Concepts to Master

A. Approximation of the area under the graph of a function

B. Area as a limit of sums

C. Determining distances from velocities

Summary and Focus Questions

Page 360

A. This section develops a method to find the area that lies under the curve $y = f(x)$, above the x-axis, and between the vertical lines $x = a$ and $x = b$, as shaded at the right. The way to approximate this area is to cover it with adjacent vertical rectangles (one rectangle is in the figure), find the area of each rectangle, and then add the areas. If the rectangles do a good job of covering the shaded area, the sum of their areas will be a good approximation for the actual area.

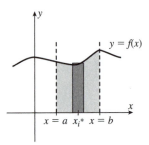

Thus, to approximate the area described above:

i) Partition $[a, b]$ in n subintervals, each of width
$$\Delta x = \frac{b - a}{n}.$$

ii) Select a **sample point** x_i^* from each subinterval $[x_{i-1}, x_i]$. Usually the choice for x_i^* is made in a systematic way, such as "chose the left endpoint of each subinterval."

iii) Calculate the sum

$$\sum_{i=1}^{n} f(x_i^*)\Delta x = f(x_1^*)\Delta x + f(x_2^*)\Delta x + f(x_3^*)\Delta x + \ldots + f(x_n^*)\Delta x.$$

Sums using the right endpoint of each subinterval are designated R_n. M_n and L_n refer to sums using the midpoint and left endpoint, respectively.

Example: Use 3 subintervals to estimate the area under the curve $y = x^3 + x$ between the lines $x = 2$ and $x = 8$. Use the midpoint of each subinterval for sample points.

For $n = 3$, $\Delta x = \dfrac{8 - 2}{3} = 2$. We calculate R_3 with a table.

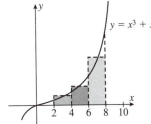

i	subint.	x_i^*	$f(x_i^*)$	$f(x_i^*)\Delta x$
1	[2, 4]	3	30	60
2	[4, 6]	5	130	260
3	[6, 8]	7	350	700

The total of the last column, 1020, is an approximation of the area.

1) Estimate the area under $f(x) = 20x - x^2$ that lies above the x-axis and between the lines $x = 4$ and $x = 16$. Use three subintervals of equal width and midpoints of the subintervals as sample points. Sketch the area and draw the approximating rectangles.

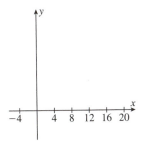

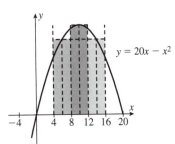

$n = 3$ and $\Delta x = \dfrac{16 - 4}{3} = 4$.

i	subint.	x_i^*	$f(x_i^*)$	$f(x_i^*)\Delta x$
1	[4, 8]	6	84	336
2	[8, 12]	10	100	400
3	[12, 16]	14	84	336

The area is approximately
$M_3 = 336 + 400 + 336 = 1072$.

$n = 4$ and $\Delta x = \dfrac{8 - 2}{4} = 1.5$.

2) Use the left endpoints as sample points to estimate the area under the function $f(x) = 1 + 4x^2$ that lies above the x-axis between $x = 2$ and $x = 8$. Use 4 subintervals of equal width.

i	subint.	x_i^*	$f(x_i^*)$	$f(x_i^*)\Delta x$
1	[2, 3.5]	2	17	25.5
2	[3.5, 5]	3.5	50	75
3	[5, 6.5]	5	101	151.5
4	[6.5, 8]	6.5	170	255

The area is approximately
$L_4 = 25.5 + 75 + 151.5 + 255 = 507$.

3) Estimate the area under the curve given below between $x = 1$ and $x = 6$. Use right endpoints for sample points and intervals of width 1.

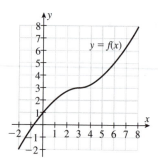

<Use the graph to estimate function values $f(x)$.>

$n = 5$ and $\Delta x = 1$.

i	subint.	x_i^*	$f(x_i^*)$	$f(x_i^*)\Delta x$
1	[1, 2]	2	2.7	2.7
2	[2, 3]	3	3	3
3	[3, 4]	4	3.3	3.3
4	[4, 5]	5	4	4
5	[5, 6]	6	5	5

The area is approximately
$R_5 = 2.7 + 3 + 3.3 + 4 + 5 = 18$.

4) Repeat question 3, this time using left endpoints of each subinterval.

$n = 5$ and $\Delta x = 1$.

i	subint.	x_i^*	$f(x_i^*)$	$f(x_i^*)\Delta x$
1	[1, 2]	1	2	2
2	[2, 3]	2	2.7	2.7
3	[3, 4]	3	3	3
4	[4, 5]	4	3.3	3.3
5	[5, 6]	5	4	4

The area is approximately
$L_5 = 2 + 2.7 + 3 + 3.3 + 4 = 15$.

5) Why is the estimation of the area by using left endpoints of each subinterval in question 4 smaller than your answer to question 3?

$R_5 \geq L_5$ because the function is increasing, the function values using the left endpoint are smaller than the function values using the right endpoints.

Page 365

B. Let f be a continuous nonnegative function on the interval $[a, b]$. Partition $[a, b]$ into n subintervals of equal length $\Delta x = \dfrac{b - a}{n}$ and from the ith subinterval choose x_i^* to be the right endpoint. We have seen that

$$R_n = \sum_{i=1}^{n} f(x_i^*)\Delta x = f(x_1^*)\Delta x + f(x_2^*)\Delta x + \ldots + f(x_n^*)\Delta x$$

is approximately the area under f, above the x-axis, and between the lines $x = a$ and $x = b$. Also, as n gets larger and larger, these sums get closer and closer to the actual area.

Therefore, we define the **area** under f, above the x-axis, and between the lines $x = a$ and $x = b$, to be

$$\text{Area} = \lim_{n \to \infty} R_n = \lim_{n \to \infty} \sum_{i=1}^{n} f(x_i^*)\Delta x.$$

The definition could have been stated in terms of sums with midpoints for x_i^*, or left endpoints for x_i^*, or any other choices for x_i^*.

The method to find the area is:

i) Partition $[a, b]$ in n subintervals, each of width $\Delta x = \dfrac{b - a}{n}$.

ii) Select a sample point x_i^* from each subinterval $[x_{i-1}, x_i]$ and write each x_i^* in terms of n and i.

iii) Find the approximating sum $\displaystyle\sum_{i=1}^{n} f(x_i^*)\Delta x$ and rewrite in terms of n.

iv) Determine the limit as $n \to \infty$ of the expression found in step iii).

To make step **iv)** easier, you should be consistent in your choice of x_i^* in step **ii)**. If you are calculating L_n, always using x_i^* as the left endpoint of each subinterval means that

$$x_i^* = a + (i - 1)\Delta x.$$

If you are calculating R_n, always using x_i^* as the right endpoint of each subinterval means that

$$x_i^* = a + i\Delta x.$$

Example: Using the left endpoint of each subinterval for x_i^*, find the area under $y = 2x$, above the x-axis and between lines $x = 0$ and $x = 3$.

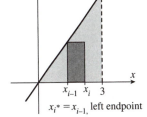

i) $\Delta x = \dfrac{3 - 0}{n} = \dfrac{3}{n}.$

ii) Let $x_i^* = 0 + (i - 1)\Delta x = (i - 1)\dfrac{3}{n}.$

Then $f(x_i^*)\Delta x = 2\left((i - 1)\dfrac{3}{n}\right)\Delta x = 2\left((i - 1)\dfrac{3}{n}\right)\dfrac{3}{n} = \dfrac{18}{n^2}(i - 1).$

iii) $L_n = \displaystyle\sum_{i=1}^{n} f(x_i^*)\Delta x = \sum_{i=1}^{n} \dfrac{18}{n^2}(i - 1)$

$$= \dfrac{18}{n^2}\sum_{i=1}^{n}(i - 1)$$

$$= \dfrac{18}{n^2}\left(\sum_{i=1}^{n} i - \sum_{i=1}^{n} 1\right)$$

$$= \dfrac{18}{n^2}\left(\dfrac{n(n + 1)}{2} - n\right)$$

$$= \dfrac{9(n + 1)}{n} - \dfrac{18}{n}.$$

iv) $\displaystyle\lim_{n \to \infty} L_n = \lim_{n \to \infty}\left(\dfrac{9(n + 1)}{n} - \dfrac{18}{n}\right) = 9(1) - 0 = 9.$

6) True or False:
The area of a region bounded by a non-negative continuous function $y = f(x)$, $y = 0$, $x = a$, $x = b$ is the limit of the sums of the areas of approximating rectangles.

True.

7) Find the area under $y = 10 - 2x$ between $x = 1$ and $x = 4$. Take x_i^* to be the right endpoint. Use subintervals of equal width.

With n subintervals we have
$$\Delta x = \frac{b - a}{n} = \frac{4 - 1}{n} = \frac{3}{n}.$$

$$\underset{\substack{1 \\ = x_0}}{\vphantom{|}} \quad \Delta x \ \ \Delta x \ \ \Delta x \ \ \Delta x \quad \underset{\substack{4 \\ = x_n}}{\vphantom{|}}$$
with points $x_1 \ x_2 \ x_3$

$x_1 = 1 + \Delta x$, $x_2 = 1 + 2\Delta x$, and in general $x_i = 1 + i\Delta x$. Selecting x_i^* to be the right endpoint means $x_i^* = x_i$. Thus
$$x_i^* = 1 + i\Delta x = 1 + i\left(\frac{3}{n}\right) = 1 + \frac{3i}{n}.$$
Hence $f(x_i^*) = 10 - 2\left(1 + \frac{3i}{n}\right) = 8 - \frac{6i}{n}.$
The approximating sum is
$$R_n = \sum_{i=1}^{n} f(x_i^*)\Delta x = \sum_{i=1}^{n}\left(8 - \frac{6i}{n}\right)\left(\frac{3}{n}\right)$$
$$= \sum_{i=1}^{n}\left(\frac{24}{n} - \frac{18i}{n^2}\right)$$
$$= \sum_{i=1}^{n}\frac{24}{n} - \sum_{i=1}^{n}\frac{18i}{n^2}$$
$$= \frac{24}{n}\sum_{i=1}^{n}1 - \frac{18}{n^2}\sum_{i=1}^{n}i$$
$$= \frac{24}{n}(n) - \frac{18}{n^2}\left(\frac{n(n + 1)}{2}\right).$$
$$= 24 - 9\frac{(n + 1)}{n}.$$
Finally,
$$\lim_{n \to \infty}\left(24 - 9\frac{(n + 1)}{n}\right) = 24 - 9(1) = 15.$$

C. If an object is moving along an axis in a positive direction and has a velocity of $v = f(t)$ at time t for $a \le t \le b$, then $\lim\limits_{n \to \infty}\sum\limits_{i=1}^{n} f(t_i^*)\Delta t$ is the **distance** the object travels during the time interval $[a, b]$.

8) Shu Ping rode her bike from home to the park between 10:00 and 10:25 A.M. She recorded the velocity reading on her speedometer every 5 minutes. See the table below. Estimate how far she traveled.

time	velocity (km/hr)
10:00	0 (start)
10:05	12
10:10	13
10:15	16
10:20	8
10:25	0 (reaches park)

$n = 5$ and $\Delta t = .0833$ (5 minutes).
We use the left endpoints for sample points.

i	subint.	t_i^*	$f(t_i^*)$	$f(t_i^*)\Delta t$
1	[10:00, 10:05]	10:00	0	0.000
2	[10:05, 10:10]	10:05	12	1.000
3	[10:10, 10:15]	10:10	13	1.083
4	[10:15, 10:20]	10:15	16	1.333
5	[10:20, 10:25]	10:20	8	0.667

The distance traveled is approximately
$$L_5 = 0 + 1 + 1.083 + 1.333 + 0.667$$
$$= 4.083 \text{ km.}$$

Section 5.2   The Definite Integral

This section defines the definite integral as a limit of sums like those discussed in the previous section. Some properties of definite integrals (such as the integral of the sum of two functions is the sum of the integrals) are discussed. If you think of definite integrals as areas under a graph and above the *x*-axis, then many of the properties represent well-known facts about areas.

Concepts to Master

A. Definition of $\int_a^b f(x)\, dx$; Integrability; Riemann sum

B. Evaluating definite integrals

C. Properties of definite integrals

D. Technology Plus

Summary and Focus Questions

Page
372

A. Let *f* be a continuous nonnegative function on the interval $[a, b]$. If we choose a partition of $[a, b]$ into *n* subintervals of equal length $\Delta x = \dfrac{b - a}{n}$ and choose $x_i{}^*$ to be any point from the *i*th subinterval, we define the **definite integral of *f* from *a* to *b*** as

$$\int_a^b f(x)\, dx = \lim_{n \to \infty} \sum_{i=1}^n f(x_i{}^*)\Delta x.$$

We say *f* is **integrable** on $[a, b]$ when this limit exits. The function *f* is the **integrand**. The **limits of integration** are *a* (the **lower limit**) and *b* (the **upper limit**). The sum $\sum_{i=1}^n f(x_i{}^*)\Delta x$ is called a **Riemann sum**. Finding the value of a definite integral is called **integration**.

The definite integral $\int_a^b f(x)\, dx$ exists in cases where *f* is continuous on $[a, b]$ or where *f* has a finite number of jump discontinuities on $[a, b]$.

If $f(x) \geq 0$ for $x \in [a, b]$, then $\int_a^b f(x)\, dx$ is the **area** under the curve $y = f(x)$, above the *x*-axis, and between the vertical lines $x = a$ and $x = b$.

1) If $\Delta x = \frac{3-1}{n}$ on the interval $[1, 3]$ and

$x_i^* \in [x_{i-1}, x_i]$ for each i then

$$\lim_{n \to \infty} \sum_{i=1}^{n} (x_i^*)^4 \Delta x = \underline{\quad}.$$

2) Does $\int_{1}^{6} [\![x]\!]\, dx$ exist?

($[\![x]\!]$ is the greatest integer function.)

3) Construct a Riemann sum for $f(x) = x^2$ on the interval $[2, 10]$ using the partition $[2, 4], [4, 6], [6, 8], [8, 10]$.

$$\int_{1}^{3} x^4\, dx.$$

Yes, $[\![x]\!]$ has only a finite number (5) of jump discontinuities on $[1, 6]$.

<*We must select a sample point from each subinterval.*>

$n = 4$ and $\Delta x = 2$. Select x_i^* to be the midpoint of each subinterval. (Your x_i^* choices may be different.)

The resulting Riemann sum for these choices of x_i^* is:

i	subint.	x_i^*	$f(x_i^*)$	$f(x_i^*)\Delta x_i$
1	[2, 4]	3	9	18
2	[4, 6]	5	25	50
3	[6, 8]	7	49	98
4	[8, 10]	9	81	162

$$M_4 = \sum_{i=1}^{4} f(x_i^*)\Delta x = 18 + 50 + 98 + 162$$

$$= 328.$$

$$\int_{-1}^{3} \left(\frac{x^3}{3} - x^2 + 2 \right) dx.$$

4) Write a definite integral for the shaded area.

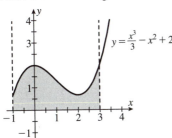

$$y = \frac{x^3}{3} - x^2 + 2$$

5) Sketch a graph that corresponds to the number $\int_{-\pi/4}^{\pi/2} \cos x\, dx$.

Since $\cos x \geq 0$ for $-\frac{\pi}{4} \leq x \leq \frac{\pi}{2}$, the integral is the area under $y = \cos x$, above the x-axis, and between $x = -\frac{\pi}{4}$ and $x = \frac{\pi}{2}$.

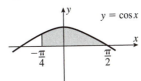

$y = \cos x$

B. Given a continuous function f on the interval $[a, b]$ that is partitioned into n subintervals with $\Delta x = \dfrac{b - a}{n}$, the use of consistent choices of the sample points x_i^* will simplify the expression for the Riemann sum.

Rule	Choice of x_i^*	Riemann sum expression
Right Endpoint	$x_i^* = f(x_i) = a + i\frac{b-a}{n}$	$R_n = \sum\limits_{i=1}^{n} f\left(a + i\frac{b-a}{n}\right)\Delta x$
Left Endpoint	$x_i^* = f(x_{i-1}) = a + (i - 1)\frac{b-a}{n}$	$L_n = \sum\limits_{i=1}^{n} f\left(a + (i - 1)\frac{b-a}{n}\right)\Delta x$
Midpoint	$x_i^* = \bar{x}_i = \dfrac{x_{i-1} + x_i}{2}$	$M_n = \sum\limits_{i=1}^{n} f(\bar{x}_i)\Delta x$

We have seen that to determine the exact value of $\lim\limits_{n\to\infty} \sum\limits_{i=1}^{n} f(x_i^*)\Delta x$ we need to rewrite sigma notation sums as expressions involving only the upper limit n. Some common such expressions for sums are:

$$\sum_{i=1}^{n} 1 = n.$$

$$\sum_{i=1}^{n} i = \frac{n(n + 1)}{2}.$$

$$\sum_{i=1}^{n} i^2 = \frac{n(n + 1)(2n + 1)}{6}.$$

$$\sum_{i=1}^{n} i^3 = \left(\frac{n(n + 1)}{2}\right)^2.$$

These may be used to simplify sums as in this example:

$$\sum_{i=1}^{n} (i^2 + 4i) = \sum_{i=1}^{n} i^2 + \sum_{i=1}^{n} 4i = \sum_{i=1}^{n} i^2 + 4\sum_{i=1}^{n} i$$

$$= \frac{n(n + 1)(2n + 1)}{6} + 4\frac{n(n + 1)}{2} = \frac{n(n + 1)(2n + 13)}{6}.$$

Appendix E in the text has a more complete discussion on sigma notation.

5) Write the following as an expression in n:

$$\sum_{i=1}^{n} (12i^2 - 4ni).$$

$$\sum_{i=1}^{n} (12i^2 - 4ni) = \sum_{i=1}^{n} 12i^2 - \sum_{i=1}^{n} 4ni$$

$$= 12\sum_{i=1}^{n} i^2 - 4n\sum_{i=1}^{n} i$$

$$= 12\frac{n(n + 1)(2n + 1)}{6} - (4n)\frac{n(n + 1)}{2}$$

$$= n(n + 1)(2n + 2) = 2n(n + 1)^2.$$

6) Approximate $\int_1^4 x^3 \, dx$ using the Midpoint Rule with $n = 3$.

Let $f(x) = x^3$, $n = 3$, and
$$\Delta x = \frac{b - a}{n} = \frac{4 - 1}{3} = 1.$$

i	subint.	$\overline{x}_i$	$f(\overline{x}_i)$	$f(\overline{x}_i) \, \Delta x$
1	[1, 2]	1.5	3.375	3.375
2	[2, 3]	2.5	15.625	15.625
3	[3, 4]	3.5	42.875	42.875

$$M_3 = \sum_{i=1}^{3} f(\overline{x}_i) \Delta x$$
$$= 3.375 + 15.625 + 42.875 = 61.875.$$

7) Use a partition of 4 equal-sized subintervals and right endpoints for sample points to approximate

$$\int_{-2}^{4} 3x^2 \, dx.$$

Let $f(x) = 3x^2$, $n = 4$, and
$$\Delta x = \frac{4 - (-2)}{4} = 1.5.$$

i	subint.	x_i^*	$f(x_i^*)$	$f(x_i^*)\Delta x$
1	[−2, −0.5]	−0.5	0.750	1.125
2	[−0.5, 1]	1	3.000	4.500
3	[1, 2.5]	2.5	18.750	28.125
4	[2.5, 4]	4	48.000	72.000

$$R_4 = \sum_{i=1}^{4} f(x_i^*)\Delta x =$$
$$1.125 + 4.500 + 28.125 + 72.000 = 105.75.$$

8) Find $\int_0^5 (x^3 + 2x^2) \, dx$ using the definition. Use right endpoints for x_i^*.

Let $f(x) = x^3 + 2x^2$. With n subintervals, $\Delta x = \frac{5 - 0}{n} = \frac{5}{n}$. Right endpoints for x_i^* means $x_i^* = a + i\Delta x = 0 + i\Delta x =$
$$i\Delta x = i\left(\frac{5}{n}\right) = \frac{5i}{n}.$$

$$R_n = \sum_{i=1}^{n} f(x_i^*)\Delta x = \sum_{i=1}^{n} \left(\left(\frac{5i}{n}\right)^3 + 2\left(\frac{5i}{n}\right)^2\right)\left(\frac{5}{n}\right)$$
$$= \sum_{i=1}^{n} \frac{625i^3}{n^4} + \frac{250i^2}{n^3}$$
$$= \frac{625}{n^4} \sum_{i=1}^{n} i^3 + \frac{250}{n^3} \sum_{i=1}^{n} i^2$$
$$= \frac{625}{n^4} \left(\frac{n(n + 1)}{2}\right)^2 + \frac{250}{n^3} \left(\frac{n(n + 1)(2n + 1)}{6}\right)$$
$$= \frac{625}{4}\left(\frac{n + 1}{n}\right)^2 + \frac{250}{6}\left(\frac{(n + 1)(2n + 1)}{n^2}\right).$$

Finally,
$$\lim_{n \to \infty} \left[\frac{625}{4}\left(\frac{n + 1}{n}\right)^2 + \frac{250}{6}\left(\frac{(n + 1)(2n + 1)}{n^2}\right)\right]$$
$$= \frac{625}{4}(1) + \frac{250}{6}(2) = \frac{2875}{12} \approx 239.5833.$$

9) Suppose $f(x)$ is continuous and increasing on $[a, b]$. Let L be the value of a Riemann sum for $\int_a^b f(x)\, dx$ using the left endpoint of each subinterval. Which is true?

a) $L \le \int_a^b f(x)\, dx.$

b) $L = \int_a^b f(x)\, dx.$

c) $L \ge \int_a^b f(x)\, dx.$

10) Evaluate $\int_3^6 (12 - x)\, dx.$
(Hint: What area is it?)

a), since the left endpoint will have the smallest value of $f(x)$ in each subinterval. Therefore, each subrectangle will underestimate the corresponding area under the graph.

The integral represents the shaded trapezoid area below.

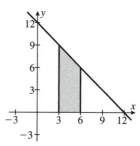

This area is $\frac{1}{2}(9 + 6)3 = 22.5.$

Page
379

C. Several properties of definite integrals help in evaluating or approximating their values. If f and g are integrable functions on an interval containing a, b, and c, then

$$\int_a^b f(x)\, dx = -\int_b^a f(x)\, dx.$$

$$\int_a^b f(x)\, dx = \int_a^c f(x)\, dx + \int_c^b f(x)\, dx.$$

$$\int_a^b (f(x) + g(x))\, dx = \int_a^b f(x)\, dx + \int_a^b g(x)\, dx.$$

$$\int_a^b (f(x) - g(x))\, dx = \int_a^b f(x)\, dx - \int_a^b g(x)\, dx.$$

$$\int_a^b k f(x)\, dx = k \int_a^b f(x)\, dx \text{ (where } k \text{ is a constant)}.$$

If $f(x) \ge g(x)$ for all $x \in [a, b]$, then $\int_a^b f(x)\, dx \ge \int_a^b g(x)\, dx.$

If $m \le f(x) \le M$ for $x \in [a, b]$, then $m(b - a) \le \int_a^b f(x)\, dx \le M(b - a).$

Some of these properties may be interpreted in terms of areas. For example, if $f(x) \geq 0$ and $a < b < c$, then

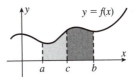

$$\int_a^b f(x)dx = \int_a^c f(x)dx + \int_c^b f(x)dx$$

says the area under f from a to b is the sum of the area under f from a to c and the area under f from c to b.

11) What are the missing limits of integration?

$$\int_4^7 x^3 \, dx = \int x^3 \, dx + \int_5^7 x^3 \, dx.$$

$$\int_4^7 x^3 \, dx = \int_4^5 x^3 \, dx + \int_5^7 x^3 \, dx.$$

12) True or False:

a) $\displaystyle\int_0^1 (\sin x + \cos x) \, dx$
$$= \int_0^1 \sin x \, dx + \int_0^1 \cos x \, dx.$$

True.

b) $\displaystyle\int_0^1 x(x^2 + 1) \, dx$
$$= \left(\int_0^1 x \, dx \right)\left(\int_0^1 (x^2 + 1) \, dx \right).$$

False.

c) $\displaystyle\int_0^1 x(x^2 + 1) \, dx = x \int_0^1 (x^2 + 1) \, dx.$

False.

d) $\displaystyle\int_0^1 x(t^2 + 1) \, dt = x \int_0^1 (t^2 + 1) \, dt.$

True. (x is a constant here because t is the variable of integration.)

e) $\displaystyle\int_0^{\pi/4} \tan x \, dx \geq 0.$

True. For $0 \leq x \leq \frac{\pi}{4}$, $\tan x \geq 0$.

f) $\displaystyle\int_0^1 2^x \, dx \leq \int_0^1 x \, dx.$

False. For $0 \leq x \leq 1$, $x < 2^x$. Therefore
$$\int_0^1 x \, dx \leq \int_0^1 2^x \, dx.$$

g) $\displaystyle\int_1^3 \frac{1}{x} \, dx = \int_1^5 \frac{1}{x} \, dx + \int_5^3 \frac{1}{x} \, dx.$

True.

13) Using $7 \leq x^2 + 1 \leq 10$ for $2.5 \leq x \leq 3$, find upper and lower estimates for
$$\int_{2.5}^3 (x^2 + 1) \, dx.$$

Since $7 \leq x^2 + 1 \leq 10$ on $[2.5, 3]$,

$$7(3 - 2.5) \leq \int_2^3 (x^2 + 1) \, dx \leq 10(3 - 2.5).$$

$$3.5 \leq \int_{2.5}^3 (x^2 + 1) \, dx \leq 5. \text{ (Later we will}$$

see the integral is $\dfrac{103}{24} \approx 4.29$.)

D. Technology Plus. Use a computer algebra system or a graphing calculator to solve.

T-1) Use a calculator or spreadsheet to estimate $\int_0^5 f(x)\, dx$, where $f(x) = 2^x + x^2$.

a) Use the Midpoint Rule with 10 intervals.

$\Delta x = \dfrac{5-0}{10} = 0.5.$

i	$\overline{x}_i$	$f(\overline{x}_i)$
1	0.25	1.252
2	0.75	2.244
3	1.25	3.941
4	1.75	6.426
5	2.25	9.819
6	2.75	14.290
7	3.25	20.076
8	3.75	27.517
9	4.25	37.090
10	4.75	49.471
		172.126

Thus, $\displaystyle\int_0^5 f(x)\, dx \approx (0.5)(172.126) = 86.063.$

b) Use the left endpoint for a sample point with 20 equal intervals.

$\Delta x = \dfrac{5-0}{20} = 0.25.$

i	x_i^*	$f(x_i^*)$
1	0.00	1.000
2	0.25	1.252
3	0.50	1.664
4	0.75	2.244
5	1.00	3.000
6	1.25	3.941
7	1.50	5.078
8	1.75	6.426
9	2.00	8.000
10	2.25	9.819
11	2.50	11.907
12	2.75	14.290
13	3.00	17.000
14	3.25	20.076
15	3.50	23.564
16	3.75	27.517
17	4.00	32.000
18	4.25	37.090
19	4.50	42.877
20	4.75	49.471
		318.216

Thus, $\displaystyle\int_0^5 f(x)\, dx \approx (0.25)(318.216)$
$= 79.554.$

c) $\displaystyle\int_0^5 f(x)\,dx = \frac{31}{\ln 2} + \frac{125}{3} \approx 86.390.$

Why is the answer for part **a)** closer to the actual value than the answer to **b)** even though **b)** uses twice as many rectangles?

The function $f(x) = 2^x + x^2$ is increasing, so the left endpoint of each subinterval will give the smallest possible value for $f(x)$ in each interval. Thus, the area of each subrectangle will underestimate the area it is intended to approximate. The Midpoint Rule, however, uses rectangles that give reasonable approximations to the areas they estimate.

Section 5.3   The Fundamental Theorem of Calculus

So far, it may seem that derivatives and definite integrals have nothing to do with each other. The Fundamental Theorem of Calculus gives a direct connection between the two concepts. As its name suggests, the importance of the theorem cannot be overemphasized. The Fundamental Theorem has two parts; each part is a different way to express the relationship between derivatives and integrals.

Concepts to Master

A. The Fundamental Theorem of Calculus, Parts 1 and 2

B. Technology Plus

Summary and Focus Questions

Page 386

A. The **Fundamental Theorem of Calculus** has two different, yet equivalent forms. The first part says that if f is a continuous function on $[a, b]$, we can create another function g so that $g' = f$.

Part 1 (FTC1): If f is continuous on $[a, b]$ and if g is a function defined by

$$g(x) = \int_a^x f(t)\, dt,$$

then g is continuous on $[a, b]$ and $g'(x) = f(x)$ for all $x \in (a, b)$.

Example: If $g(x) = \int_2^x (3t^2 + 2t)\, dt$, the Fundamental Theorem says $g'(x) = 3x^2 + 2x$.

Part 1 will be useful in later chapters when we define functions using definite integrals.

For the second part of the Fundamental Theorem, recall from Chapter 4 that F is an *antiderivitive* of f means that $F' = f$. The second part relates definite integrals and antiderivatives.

Part 2 (FTC2): If f is continuous on $[a, b]$ and F is an antiderivative of f, then

$$\int_a^b f(x)\, dx = F(b) - F(a).$$

Part 2 provides a handy way to evaluate $\int_a^b f(x)\, dx$ without resorting to limits of Riemann sums:

i) Find an (in fact, any!) antiderivative F of f.

ii) Calculate $F(b) - F(a)$.

For shorthand, we write $F(x) \Big]_a^b = F(b) - F(a)$.

Example: To find $\int_1^3 (3x^2 + 2x)\, dx$, we see that $F(x) = x^3 + x^2$ is an antiderviative of $f(x) = 3x^2 + 2x$, $F(3) = 3^3 + 3^2 = 36$ and $F(1) = 1^3 + 1^2 = 2$.

Thus, $\int_1^3 (3x^2 + 2x)\, dx = F(3) - F(1) = 36 - 2 = 34$.

1) If $h(x) = \int_4^x \sqrt{t}\, dt$, then $h'(x) =$ _____.

$h'(x) = \sqrt{x}$. (Remember that h is a function of x, not t.)

2) If $f(x)$ is continuous on $[a, b]$, then for all $x \in (a, b)$, $\dfrac{d}{dx} \int_a^x f(t)\, dt =$ _____.

$f(x)$. This is Part 1 of the Fundamental Theorem of Calculus.

3) If $f'(x)$ is continuous on $[a, b]$, then for all $x \in [a, b]$, $f(a) + \int_a^x f'(t)\, dt =$ _____.

$f(x)$. This is the second part of the Fundamental Theorem of Calculus.

4) What do questions 2) and 3) say about differentiation and integration?

Question 2) says differentiation will "undo" integration while 3) says integration will "undo" taking a derivative. Thus, integration and differentiation are inverse processes.

5) Evaluate

a) $\int_0^1 (1 - x)\, dx$

An antiderivative of $1 - x$ is $F(x) = x - \dfrac{x^2}{2}$.

Thus $\int_0^1 (1 - x)\, dx = F(x) \Big]_0^1$

$$= F(1) - F(0)$$
$$= \left(1 - \tfrac{1}{2}\right) - \left(0 - \tfrac{0}{2}\right)$$
$$= \tfrac{1}{2}.$$

b) $\int_{-1}^1 e^x\, dx$

An antiderivative of e^x is $F(x) = e^x$. Thus

$\int_{-1}^1 e^x\, dx = F(x) \Big]_{-1}^1 = F(1) - F(-1) = e^1 - e^{-1}$

$$= \frac{e^2 - 1}{e}.$$

c) $\int_{-1}^1 \dfrac{1}{x^2}\, dx$

We cannot use FTC2 for this integral because $f(x) = \dfrac{1}{x^2}$ is not continuous on $[-1, 1]$. We must wait until a later chapter.

6) For each $x \in [1, 3]$, let $H(x)$ be the value of the area shaded below:

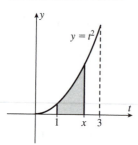

Find $H'(x)$.

$H(x) = \int_1^x t^2 \, dt$. Therefore, by FTC1

$H'(x) = x^2$.

B. Technology Plus. Use a computer algebra system or a graphing calculator to solve.

T-1) Graph $f(x) = x + 2x^2$ and $y = 7x$ for $0 \le x \le 3$ on the same screen and use it to estimate $\int_0^3 f(x) \, dx$.

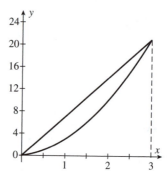

The area under $f(x) = x + 2x^2$ appears to be about $\frac{3}{4}$ of the area under $y = 7x$. That triangular area is $\frac{1}{2}(3)(21) = \frac{63}{2}$.

Our estimate of $\int_0^3 (x + 2x^2) \, dx$ is

$\frac{3}{4}\left(\frac{63}{2}\right) = 23.625.$

We note that the area is exactly

$$\int_0^3 (x + 2x^2) \, dx = \left(\frac{x^2}{2} + \frac{2x^3}{3}\right)\Big]_0^3$$
$$= \frac{45}{2} = 22.5.$$

Section 5.4   Indefinite Integrals and the Net Change Theorem

This section shows you how to evaluate definite integrals by applying Part 2 of the Fundamental Theorem of Calculus. This is a basic technique that must be learned well. The section concludes by observing that the Fundamental Theorem of Calculus says that the total change of a function from a to b is the integral of the instantaneous rate of change of the function.

Concepts to Master

A. Indefinite integrals

B. Evaluation of definite integrals using the Fundamental Theorem of Calculus

C. The Net Change Theorem

Summary and Focus Questions

Page 397

A. The **indefinite integral of f with respect to x** is written $\int f(x)\,dx$ and is the antiderivative of f. In other words, $\int f(x)\,dx = F(x)$ means $F'(x) = f(x)$.

Example: Some antiderivatives of $6x$ are: $3x^2$, $3x^2 + 7$, $3x^2 - 2.4$, etc. All have the form $3x^2 + C$, where C is a constant called the **constant of integration.** Thus, $\int 6x\,dx = 3x^2 + C$.

Example: $\int \sec^2 x\,dx = \tan x + C$ because $\dfrac{d}{dx}(\tan x + C) = \sec^2 x$.

Here are some properties of indefinite integrals. Each corresponds to a property of differentiation.

The derivative of $\int f(x)\,dx$ is $f(x)$, by definition.

$\int c f(x)\,dx = c \int f(x)\,dx$ (c, a constant)

$\int (f(x) + g(x))\,dx = \int f(x)\,dx + \int g(x)\,dx$

$\int (f(x) - g(x))\,dx = \int f(x)\,dx - \int g(x)\,dx$

Here are some fundamental indefinite integral formulas:

$\int x^n\,dx = \dfrac{x^{n+1}}{n+1} + C$ for any number n except -1

$\int \dfrac{1}{x}\,dx = \ln|x| + C$

$\int \sin x\,dx = -\cos x + C$ $\qquad$ $\int \sec x \tan x\,dx = \sec x + C$

$$\int \cos x \, dx = \sin x + C \qquad\qquad \int \csc^2 x \, dx = -\cot x + C$$

$$\int \sec^2 x \, dx = \tan x + C \qquad\qquad \int \csc x \cot x \, dx = -\csc x + C$$

$$\int e^x \, dx = e^x + C \qquad\qquad \int a^x \, dx = \frac{a^x}{\ln a} + C$$

$$\int \frac{1}{\sqrt{1 - x^2}} \, dx = \sin^{-1}x + C \qquad\qquad \int \frac{1}{1+x^2} \, dx = \tan^{-1}x + C$$

$$\int \sinh x \, dx = \cosh x + C \qquad\qquad \int \cosh x \, dx = \sinh x + C$$

Each integration formula may be checked by taking the derivative of the right hand side and comparing the result to the integrand.

1) Evaluate

a) $\int x^4 \, dx$

$\frac{x^5}{5} + C.$

b) $\int 12x^2 \, dx$

$\int 12x^2 \, dx = 12 \int x^2 \, dx = 12\frac{x^3}{3} + C = 4x^3 + C.$

c) $\int (6x + 12x^3) \, dx$

$3x^2 + 3x^4 + C.$

d) $\int (2 \cos x - \sec^2 x) \, dx$

$2 \sin x - \tan x + C.$

e) $\int (x - \sin x) \, dx$

$\frac{x^2}{2} + \cos x + C.$

f) $\int 2e^x \, dx$

$2e^x + C.$

2) If $\int h(x) \, dx = \int g(x) \, dx$, does $h(x) = g(x)$?

Yes. If two functions are equal then their derivatives are equal.

3) True or False:

a) $\int 5 f(x) \, dx = 5 \int f(x) \, dx.$

True.

b) $\int [f(x)]^2 \, dx = \left(\int f(x) \, dx \right)^2.$

False.

c) $\int f(x)g(x) \, dx = \int f(x) \, dx \int g(x) \, dx.$

False. For example, the statement is false when $f(x) = x^3$ and $g(x) = x^4$.
$\int x^7 \, dx \neq \int x^3 \, dx \cdot \int x^4 \, dx.$

d) $\frac{d}{dx}\left(\int f(x) \, dx \right) = f(x).$

True.

e) $\int 5dx = 5 + C.$

False. $\int 5 \, dx = 5x + C.$

B. Part 2 of the Fundamental Theorem of Calculus provides a handy procedure for evaluating some definite integrals without calculating limits of Riemann

Page 397

sums. To find $\int_a^b f(x)\,dx$:

i) Find $\int f(x)\,dx = F(x)$ (any antiderivative of f).

ii) Compute $F(x)\Big]_a^b = F(b) - F(a)$.

Remember that this method may be applied when f is continuous on $[a, b]$.

4) Determine

a) $\displaystyle\int_1^4 (2x + 3)\,dx$

$$\int_1^4 (2x + 3)\,dx = (x^2 + 3x)\Big]_1^4$$
$$= (4^2 + 3(4)) - (1^2 + 3(1))$$
$$= 24.$$

b) $\displaystyle\int_0^1 \sqrt{x}\,dx$

$$\int_0^1 x^{1/2}\,dx = \frac{x^{3/2}}{3/2} = \frac{2}{3}x^{3/2}\Big]_0^1$$
$$= \frac{2}{3}(1^{3/2} - 0^{3/2}) = \frac{2}{3}.$$

c) $\displaystyle\int_{-1}^3 x^{-3}\,dx$

The Fundamental Theorem does not apply because x^{-3} is not continuous at $x = 0$.

d) $\displaystyle\int_{\pi/4}^{\pi/3} \sec x \tan x\,dx$

$$\int_{\pi/4}^{\pi/3} \sec x \tan x = \sec x \Big]_{\pi/4}^{\pi/3}$$
$$= \sec \frac{\pi}{3} - \sec \frac{\pi}{4} = 2 - \sqrt{2}.$$

e) $\displaystyle\int_1^3 \frac{1}{x}\,dx$

$$\int_1^3 \frac{1}{x}\,dx = \ln |x| \Big]_1^3 = \ln 3 - \ln 1$$
$$= \ln 3 - 0 = \ln 3.$$

f) $\displaystyle\int_0^\pi (\cos x - \sin x)\,dx$

$$\int_0^\pi (\cos x - \sin x)\,dx = (\sin x + \cos x)\Big]_0^\pi$$
$$= (\sin \pi + \cos \pi)$$
$$- (\sin 0 + \cos 0)$$
$$= (0 - 1)$$
$$- (0 + 1) = -2.$$

C. For $y = F(x)$, $F'(x)$ represents the change in F at x and $\int_a^b F'(x)\,dx$ is the net

Page 401

change of F over the interval $[a, b]$. Therefore, we can restate the Fundamental Theorem, Part 2.

The Net Change Theorem: The net change of $F(x)$ over $[a, b]$ is

$$\int_a^b F'(x)\,dx = F(b) - F(a).$$

The Net Change Theorem says that for functions that are not nonnegative, the definite integral represents a kind of "net area."

For example, for the function graphed here,

$$\int_a^b f(x)\, dx = -A_1 + A_2 - A_3 + A_4,$$ where

A_1, A_2, A_3, A_4 are the indicated areas.

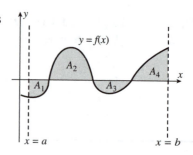

5) Evaluate $\int_{-2}^2 x^3\, dx$. What does this number represent?

$$\int_{-2}^2 x^3\, dx = \left.\frac{x^4}{4}\right|_{-2}^2 = \frac{2^4}{4} - \frac{(-2)^4}{4} = 0.$$

The net area is 0 which means that the area above the x-axis and the area below the x-axis are the same.

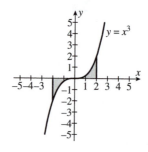

6) Find a definite integral expression for $A_1 - A_2$, where A_1 and A_2 are the areas below.

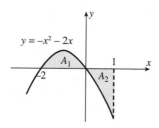

$$A_1 - A_2 = \int_{-2}^1 (-x^2 - 2x)\, dx.$$

7) If $R(x)$ is the revenue obtained by selling x units, then the marginal revenue is $R'(x)$.

What is $\int_{x_1}^{x_2} R'(x)\, dx$?

This is the total change in revenue when the number of units sold increases from x_1 to x_2.

Section 5.5   The Substitution Rule

This section gives us a technique for evaluating indefinite integrals where the integrand is a composition of functions. Keep in mind that the substitution process is just the Chain Rule stated in terms of indefinite integrals.

Concepts to Master

A. Evaluation of integrals using substitution

B. Integrals involving symmetric functions

Summary and Focus Questions

Page 407

A. The **Substitution Rule** may be used when the integrand involves composition: If $u = g(x)$ is a differentiable function whose range is an interval I and f is continuous on I, then

$$\int f(g(x))(g'(x))\, dx = \int f(u)\, du.$$

If we treat dx and du as differentials, then from $u = g(x)$ we can deduce that

$$du = g'(x)\, dx.$$

The Substitution Rule is then indeed a substitution of $f(u)$ for $f(g(x))$ and du for $g'(x)\, dx$ in order to write the given integral in the simplier form: $\int f(u)\, du$. The key step in using the Substitution Rule is to identify the substitution variable $u = g(x)$ and make certain that the integrand also includes the factor $g'(x)\, dx$.

Example. The integral $\int (x^2 + 1)^3 2x\, dx$ has as best choice for substitution variable $u = x^2 + 1$ because we can look further and see that $du = 2x\, dx$ occurs as a factor. Therefore, we can write

$$\int \underbrace{(x^2 + 1)^3}_{u}\, \underbrace{2x\, dx}_{du} = \int u^3\, du.$$

Since $\int u^3\, du = \dfrac{u^4}{4} + C$, we can make the substitution $u = x^2 + 1$, perform the integration, and rewrite the solution in terms of the original variable x.

$$\int (x^2 + 1)^3 2x\, dx = \int u^3\, du = \frac{u^4}{4} + C = \frac{(x^2 + 1)^4}{4} + C.$$

Example. The integral $\int (x^2 + 1)^3 x\, dx$ is almost like the one above but has no factor of 2. Again we will use $u = x^2 + 1$ and $du = 2x\, dx$, but we note that from $du = 2x\, dx$ we can write $x\, dx = \dfrac{du}{2}$.

We make the substitution $u = x^2 + 1$ and $\dfrac{du}{2} = x\,dx$, and then pull the constant $\dfrac{1}{2}$ out front of the integral:

$$\int (x^2 + 1)^3 x\,dx = \int u^3 \frac{du}{2} = \frac{1}{2}\int u^3\,du = \frac{1}{2}\frac{u^4}{4} + C = \frac{(x^2 + 1)^4}{8} + C.$$

In the case of a definite integral, such as $\displaystyle\int_0^2 (x^2 + 1)^3 x\,dx$, the preferred method changes the limits of integration to fit u:

Since $u(x) = x^2 + 1$, $u(2) = 5$ and $u(0) = 1$.

$$\int_0^2 (x^2 + 1)^3 x\,dx = \frac{1}{2}\int_1^5 u^3\,du = \frac{u^4}{8}\Big]_1^5 = \frac{625}{8} - \frac{1}{8} = 78.$$

Example: Evaluate $\displaystyle\int_0^1 x(3x^2 + 1)^4\,dx$.

Let $u = 3x^2 + 1$.
$du = 6x\,dx$.
$x\,dx = \dfrac{1}{6}\,du$.
At $x = 0$, $u = 3(0)^2 + 1 = 1$. At $x = 1$, $u = 3(1)^2 + 1 = 4$.
$$\int_0^1 x(3x^2 + 1)^4\,dx = \int_1^4 u^4 \frac{1}{6}\,du \quad \text{(Notice the change of limits.)}$$

$$= \frac{1}{6}\int_1^4 u^4\,du = \frac{1}{6}\frac{u^5}{5}\Big]_1^4 = \frac{u^5}{30}\Big]_1^4 = \frac{4^5}{30} - \frac{1^5}{30} = \frac{1023}{30} = 34.1.$$

Example: Evaluate $\displaystyle\int_0^2 4xe^{x^2}\,dx$.

Let $u = x^2$.
$du = 2x\,dx$.
$x\,dx = \dfrac{1}{2}\,du$.
At $x = 0$, $u = 0^0 = 0$. At $x = 2$, $u = 2^2 = 4$.
$$\int_0^1 4xe^{x^2}\,dx = 4\int_0^1 xe^{x^2}\,dx = 4\int_0^4 e^u \frac{1}{2}\,du = \frac{4}{2}\int_0^4 e^u\,du = 2e^u\Big]_0^4$$
$$= 2e^4 - 2e^0 = 2e^4 - 2.$$

1) Use the substitution $u = x^2 + 3$ to evaluate $\displaystyle\int x\sqrt{x^2 + 3}\,dx$.

First note $\displaystyle\int x\sqrt{x^2 + 3}\,dx = \int (x^2 + 3)^{1/2} x\,dx$.
Let $u = x^2 + 3$.
$du = 2x\,dx$
$x\,dx = \dfrac{du}{2}$.
$\displaystyle\int x(x^2 + 3)^{1/2}\,dx = \int u^{1/2} \frac{du}{2} = \frac{1}{2}\int u^{1/2}\,du$
$= \dfrac{1}{2} \cdot \dfrac{2}{3} u^{3/2} + C = \dfrac{1}{3}(x^2 + 3)^{3/2} + C.$

2) Find a substitution $u = g(x)$ that may be used for each of the following. Write the simplified integral and evaluate.

a) $\int (x^3 + 3x)^4 (x^2 + 1)\, dx$

$u = $ _____

$du = $ _____

Let $u = x^3 + 3x$.

$du = (3x^2 + 3)dx = 3(x^2 + 1)dx$.

$(x^2 + 1)dx = \frac{du}{3}$.

$\int (x^3 + 3x)^4 (x^2 + 1)\, dx = \int u^4 \frac{du}{3}$

$= \frac{1}{3}\int u^4\, du = \frac{1}{3}\frac{u^5}{5} + C = \frac{1}{15}u^5 + C$

$= \frac{1}{15}(x^3 + 3x)^5 + C.$

b) $\int x^2 \sin x^3\, dx$

$u = $ _____

$du = $ _____

Let $u = x^3$.

$du = 3x^2\, dx$.

$x^2\, dx = \frac{du}{3}$.

$\int x^2 \sin x^3\, dx$

$= \int \sin u \frac{du}{3} = \frac{1}{3}\int \sin u\, du$

$= -\frac{1}{3}\cos u + C = -\frac{1}{3}\cos x^3 + C.$

c) $\int \tan^2 x \sec^2 x\, dx$

$u = $ _____

$du = $ _____

Let $u = \tan x$.

$du = \sec^2 x\, dx$.

$\int \tan^2 x \sec^2 x\, dx$

$= \int (\tan x)^2 \sec^2 x\, dx = \int u^2\, du$

$= \frac{1}{3}u^3 + C = \frac{1}{3}\tan^3 x + C.$

3) Evaluate each:

a) $\int_0^1 \sqrt{2x + 1}\, dx.$

<This looks like $\int \sqrt{u}\, du$ so $u = 2x + 1$ is a good first choice.>

Let $u = 2x + 1$.

$du = 2\, dx$

$dx = \frac{du}{2}$.

At $x = 0$, $u = 1$. At $x = 1$, $u = 3$.

$\int_0^1 \sqrt{2x + 1}\, dx = \int_0^1 (2x + 1)^{1/2}\, dx$

$= \int_1^3 u^{1/2} \frac{du}{2}$

$= \frac{1}{2}\int_1^3 u^{1/2}\, du = \frac{1}{2} \cdot \frac{2}{3}u^{3/2}\Big]_1^3$

$= \frac{1}{3}(3^{3/2} - 1) = \sqrt{3} - \frac{1}{3}.$

b) $\int_0^{\pi/3} (\cos^3 x + 1)\sin x \, dx$

<*"sin x dx" looks like a good candidate for du, so let u = cos x.*>

Let $u = \cos x$.

$du = -\sin x \, dx$.

$\sin x \, dx = -du$.

At $x = 0$, $u = \cos 0 = 1$.

At $x = \frac{\pi}{3}$, $u = \cos \frac{\pi}{3} = \frac{1}{2}$.

$\int_0^{\pi/3} (\cos^3 x + 1)\sin x \, dx$

$= \int_1^{1/2} (u^3 + 1)(-du)$

$= -\int_1^{1/2} (u^3 + 1) \, du$

$= \int_{1/2}^1 (u^3 + 1) \, du = \left(\frac{u^4}{4} + u\right)\Big]_{1/2}^1$

$= \left(\frac{1}{4} + 1\right) - \left(\frac{1}{64} + \frac{1}{2}\right) = \frac{47}{64}$.

c) $\int x^5(x^3 + 2)^2 \, dx$.

The first substitution to try is

$u = x^3 + 2$.

$du = 3x^2 \, dx$.

$x^2 \, dx = \frac{du}{3}$.

Since we have an x^5 term it appears we may need a different substitution. But

$x^5 = x^3 \cdot x^2 = (u - 2)x^2$ since

$u = x^3 + 2$.

$\int x^5(x^3 + 2)^2 \, dx = \int (u - 2)u^2 \, \frac{du}{3}$

$= \frac{1}{3}\int (u^3 - 2u^2)du = \frac{1}{3}\left(\frac{u^4}{4} - \frac{2}{3}u^3\right) + C$

$= \frac{u^4}{12} - \frac{2u^3}{9} + C$

$= \frac{(x^3 + 2)^4}{12} - \frac{2(x^3 + 2)^3}{9} + C$.

d) $\int \dfrac{x^3}{2 + x^4}\, dx$

Let $u = 2 + x^4$.

$du = 4x^3\, dx$.

$x^3\, dx = \dfrac{du}{4}$.

$$\int \frac{x^3}{2 + x^4}\, dx = \int \frac{1}{2 + x^4} x^3\, dx$$

$$= \int \frac{1}{u} \frac{du}{4} = \frac{1}{4} \int \frac{1}{u}\, du$$

$$= \frac{1}{4} \ln |u| + C$$

$$= \frac{1}{4} \ln(2 + x^4) + C$$

$$= \ln \sqrt[4]{2 + x^4} + C.$$

Note that absolute value signs are not needed because $2 + x^4 > 0$ for all x.

e) $\int 7^{-x}\, dx$

Let $u = -x$.

$du = -dx$.

$dx = -du$.

$$\int 7^{-x}\, dx = \int 7^u(-du) = -\int 7^u\, du$$

$$= -\frac{7^u}{\ln 7} + C = -\frac{7^{-x}}{\ln 7} + C.$$

4) Find the area of the shaded region:

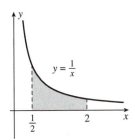

The area is

$$\int_{1/2}^{2} \frac{1}{x}\, dx = \ln |x| \,\Big]_{\frac{1}{2}}^{2} = \ln 2 - \ln \frac{1}{2}$$

$$= \ln 2 - \ln 2^{-1} = \ln 2 + \ln 2$$

$$= 2 \ln 2 \approx 1.39.$$

5) Evaluate

a) $\displaystyle\int_{1}^{\sqrt{3}} \frac{1}{1 + x^2}\, dx$

<Recognize this as one of the basic integrals.>

$$\int_{1}^{\sqrt{3}} \frac{1}{1 + x^2}\, dx = \tan^{-1} x \,\Big]_{1}^{\sqrt{3}}$$

$$= \tan^{-1}(\sqrt{3}) - \tan^{-1}(1)$$

$$= \frac{\pi}{3} - \frac{\pi}{4} = \frac{\pi}{12}.$$

b) $\int_0^1 \dfrac{1}{\sqrt{4 - x^2}} \, dx$

This is almost the form needed for $\sin^{-1} x$ except we have $\sqrt{4 - x^2}$ instead of $\sqrt{1 - x^2}$. Make the constant 1:

$$\sqrt{4 - x^2} = \sqrt{4\left(1 - \frac{x^2}{4}\right)} = 2\sqrt{1 - \left(\frac{x}{2}\right)^2}.$$

Let $u = \frac{x}{2}$.

$du = \dfrac{dx}{2}$.

$u(0) = 0,\ u(1) = \frac{1}{2}$.

$$\int_0^1 \dfrac{1}{2\sqrt{1 - \left(\frac{x}{2}\right)^2}} \, dx$$

$$= \int_0^{\frac{1}{2}} \dfrac{1}{\sqrt{1 - u^2}} \, du = \sin^{-1} u \Big]_0^{\frac{1}{2}}$$

$$= \sin^{-1} \tfrac{1}{2} - \sin^{-1} 0$$

$$= \tfrac{\pi}{6} - 0 = \tfrac{\pi}{6}.$$

c) $\int \dfrac{2x}{\sqrt{1 - x^2}} \, dx$

Although this may appear to involve $\sin^{-1}$, this is a simple substitution.

Let $u = 1 - x^2$.

$du = -2x\, dx$.

$2x\, dx = -du$.

$$\int \dfrac{2x}{\sqrt{1 - x^2}} \, dx = \int 2x(1 - x^2)^{-1/2} \, dx$$

$$= \int (u)^{-1/2} (-du) = -\int (u)^{-1/2} (du)$$

$$= \dfrac{-u^{1/2}}{1/2} + C = -2u^{1/2} + C$$

$$= -2\sqrt{1 - x^2} + C.$$

d) $\int \csc ax \cot ax \, dx$, where $a \neq 0$.

<Remove the factor a by substitution.>

Let $u = ax$.

$du = a\, dx$.

$dx = \dfrac{du}{a}$.

$$\int \csc ax \cot ax \, dx = \int \csc u \cot u \, \dfrac{du}{a}$$

$$= \frac{1}{a} \int \csc u \cot u \, du = \frac{1}{a} (-\csc u) + C$$

$$= -\dfrac{\csc ax}{a} + C.$$

e) $\int_{\frac{\pi}{8}}^{\frac{\pi}{6}} \tan^4 2x \sec^2 2x\, dx.$

$<$*Recognize* $\sec^2 2x$ *as part of du.*$>$

Let $u = \tan 2x.$
$du = (\sec^2 2x)(2)dx.$
$\sec^2 2x\, dx = \dfrac{du}{2}.$

$u\left(\dfrac{\pi}{8}\right) = \tan\left(2\dfrac{\pi}{8}\right) = \tan\dfrac{\pi}{4} = 1.$

$u\left(\dfrac{\pi}{6}\right) = \tan\left(2\dfrac{\pi}{6}\right) = \tan\dfrac{\pi}{3} = \sqrt{3}.$

$\int_{\frac{\pi}{8}}^{\frac{\pi}{6}} \tan^4 2x \sec^2 2x\, dx = \int_{1}^{\sqrt{3}} u^4 \dfrac{du}{2} = \dfrac{1}{2}\int_{1}^{\sqrt{3}} u^4\, du$

$= \dfrac{u^5}{10}\Big]_{1}^{\sqrt{3}} = \dfrac{\left(\sqrt{3}\right)^5 - 1}{10} = \dfrac{9\sqrt{3}-1}{10}.$

Page 412

B. Recall that a function f is *even* if $f(-x) = f(x)$ for all x in its domain and *odd* if $f(-x) = -f(x)$ for all x in its domain. Suppose f is continuous on $[-a, a]$.

If f is even, $\displaystyle\int_{-a}^{a} f(x)\, dx = 2\int_{0}^{a} f(x) = dx.$

If f is odd, $\displaystyle\int_{-a}^{a} f(x)\, dx = 0.$

6) Evaluate

a) $\displaystyle\int_{-\pi/2}^{\pi/2} \sin x\, dx.$

Since $\sin x$ is odd, this integral is 0.

b) $\displaystyle\int_{-3}^{3} (x^4 + x^2)\, dx.$

Since $x^4 + x^2$ is even,

$\displaystyle\int_{-3}^{3} (x^4 + x^2)\, dx = 2\int_{0}^{3} (x^4 + x^2)\, dx$

$= 2\left(\dfrac{x^5}{5} + \dfrac{x^3}{3}\right)\Big]_{0}^{3}$

$= 2\left[\left(\dfrac{3^5}{5} + \dfrac{3^3}{3}\right) - 0\right]$

$= \dfrac{576}{5}.$

Chapter 6 — Applications of Integration

Section 6.1   Areas between Curves

This chapter contains several applications of definite integrals. It will help to remember that each application arises naturally as a limit of Riemann sums—the limit might be the sum of areas of several adjacent rectangles, the sum of the volumes of several parallel slices of a solid, the sum of the volumes of several concentric cylinders, and so on. In this section we determine the areas between two curves. There will be cases where it is better to treat y as a function of x and others where we should treat x as a function of y.

Concepts to Master

A. Area between $y = f(x)$, $y = g(x)$, $x = a$, $x = b$

B. Area between $x = f(y)$, $x = g(y)$, $y = c$, $y = d$

C. Area enclosed by two curves

D. Technology Plus

Summary and Focus Questions

Page 422

A. Suppose a region is bounded by continuous functions $y = f(x)$ and $y = g(x)$ with $f(x) \geq g(x)$ on $[a, b]$ and vertical lines $x = a$ and $x = b$. The area of the region may be approximated by a Riemann sum with terms $[f(x_i{}^*) - g(x_i{}^*)]\Delta x$.

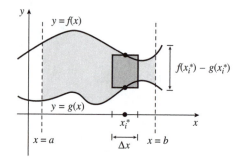

The exact area is

$$\lim_{n \to \infty} \sum_{i=1}^{n} [f(x_i{}^*) - g(x_i{}^*)]\Delta x$$

$$= \int_a^b [f(x) - g(x)] \, dx.$$

Example: Find a definite integral for the area of the shaded region in the graph at the right.

The "top" function is $y = 6 - \frac{x}{3}$ and the "bottom" function

is $y = x^2$. The area is $\int_1^2 \left((6 - \frac{x}{3}) - x^2 \right) dx \left(= \frac{19}{6} \right).$

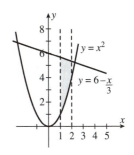

1) Set up a definite integral for the area shaded below.

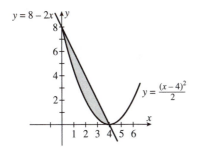

2) Set up a definite integral for the area of the region bounded by $y = x^3, y = 3 - x,$ $x = 0, x = 1.$

The "bottom" function is $f(x) = \frac{(x - 4)^2}{2}$ and the "top" function is $g(x) = 8 - 2x$. The area is

$$\int_0^4 \left((8 - 2x) - \frac{(x - 4)^2}{2} \right) dx \left(= \frac{16}{3} \right).$$

First sketch a graph.

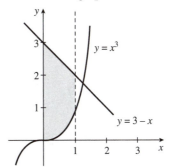

For $0 \le x \le 1, 3 - x \ge x^3$. The area is
$$\int_0^1 [3 - x - x^3] \, dx \left(= \frac{9}{4} \right).$$

Page 426

B. Suppose a region R is bounded by continuous functions $x = f(y)$, $x = g(y)$, with $f(y) \geq g(y)$ on $[c, d]$ and horizontal lines $y = c$ and $y = d$. The area of the region is found in a manner similar to the method in part **A**. Here, however, x is a function of y, f is the "right" function and g is the "left" function, and the area is

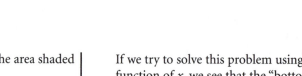
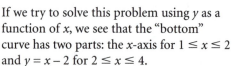

$$\lim_{n\to\infty}\sum_{i=1}^{n} [f(y_i^*) - g(y_i^*)]\Delta y$$

$$= \int_c^d [f(y) - g(y)]\, dy.$$

3) Set up a definite integral for the area shaded below.

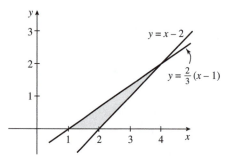

If we try to solve this problem using y as a function of x, we see that the "bottom" curve has two parts: the x-axis for $1 \leq x \leq 2$ and $y = x - 2$ for $2 \leq x \leq 4$.

To avoid this complication, we treat x as a function of y.

From $y = x - 2$, $x = y + 2$. The "right" function is $x = y + 2$.

From $y = \frac{2}{3}(x - 1)$, $x = \frac{3}{2}y + 1$. The "left" function is $x = \frac{3}{2}y + 1$.

The area is

$$\int_0^2 \left((y + 2) - \left(\frac{3}{2}y + 1\right)\right) dy \;(= 1).$$

4) Set up a definite integral for the area of the region bounded by $x = 4 - y^2$, $x = y - 2$, $y = 2$, $y = -1$.

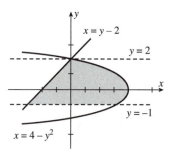

For $-1 \leq y \leq 2$, $4 - y^2 \geq y - 2$.

The area is

$$\int_{-1}^{2} [4 - y^2 - (y - 2)]\, dy$$

$$= \int_{-1}^{2} [6 - y^2 - y]\, dy \left(= \frac{27}{2}\right).$$

Page 426

C. Some regions may be described so that either the $\int \dots dx$ form or the $\int \dots dy$ form may be used. You should select the form with the integrand for which finding an antiderivative is easier.

Example: Find a definite integral for the area of the shaded region in the graph at the right.

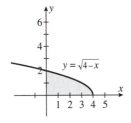

With x as the independent variable the "top" function is $y = \sqrt{4-x}$ and the "bottom" function is $y = 0$. The area is

$$\int_0^4 \left(\sqrt{4-x}\right)dx \left(= \tfrac{16}{3}\right).$$

With y as the independent variable we solve for x:

$$y = \sqrt{4-x}$$
$$y^2 = 4-x$$
$$x = 4-y^2.$$

The "right" function is $x = 4 - y^2$ and the "left" function is $x = 0$. The area is

$$\int_0^2 (4-y^2)dy \left(= \tfrac{16}{3}\right).$$

The second integral is a little easier to calculate because the integration does not involve the Substitution Rule.

To find the area of regions such as the types sketched below, the x coordinates of the points of intersection a, b, c, $\dots$ must be found. This is done by setting $f(x) = g(x)$ and solving for x. Often a sketch of the functions will help or a graphing calculator can be used to find approximate values for the points of intersection.

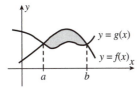

Area =

$$\int_a^b [f(x) - g(x)]\, dx$$

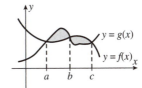

Area =

$$\int_a^b [g(x) - f(x)]dx + \int_b^c [f(x) - g(x)]\, dx$$

Example: Find a definite integral for the area of the shaded region in the graph at the right.

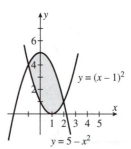

$y = 5 - x^2$ and $y = (x - 1)^2$ intersect at

$$5 - x^2 = (x - 1)^2$$
$$5 - x^2 = x^2 - 2x + 1$$
$$2x^2 - 2x - 4 = 0$$
$$2(x - 2)(x + 1) = 0, \text{ so } x = 2 \text{ and } x = -1.$$

With x as the independent variable, the "top" function is $y = 5 - x^2$ and the "bottom" function is $y = (x - 1)^2$. The area is

$$\int_{-1}^{2} \left((5 - x^2) - (x - 1)^2 \right) dx \ (= 9).$$

5) Write an expression involving definite integrals for the area of each:

   a) the region bounded by $y = (x + 1)^2$ and $y = x + 3$.

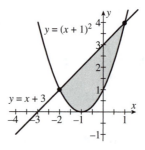

The "top" curve is $y = x + 3$ and "bottom" curve is $y = (x + 1)^2$. First find where they intersect:

$$(x + 1)^2 = x + 3$$
$$x^2 + 2x + 1 = x + 3$$
$$x^2 + x - 2 = 0$$
$$(x - 1)(x + 2) = 0 \text{ at } x = -2, 1.$$

The area is $\displaystyle\int_{-2}^{1} [(x + 3) - (x + 1)^2] dx \left(= \frac{9}{2} \right).$

b) the shaded region.

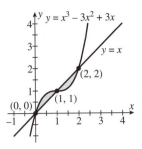

<After drawing the graph, it helps to notice the region consists of two areas and, therefore, we need two integrals.>

For $0 \leq x \leq 1$, $x^3 - 3x^2 + 3x \geq x$ while for $1 \leq x \leq 2$, $x \geq x^3 - 3x^2 + 3x$.
The area is the sum of two integrals:

$$\int_0^1 ((x^3 - 3x^2 + 3x) - x)) \, dx$$

$$+ \int_1^2 (x - (x^3 - 3x^2 + 3x)) \, dx$$

$$= \int_0^1 (x^3 - 3x^2 + 2x) \, dx$$

$$+ \int_1^2 (-x^3 + 3x^2 - 2x) \, dx$$

$$\left(= \frac{1}{4} + \frac{1}{4} = \frac{1}{2} \right).$$

c) the shaded region.

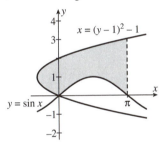

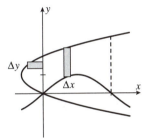

Using the y-axis to divide the region into two parts, the total area is the sum of two areas.

The area to the left of the y-axis is found using x as a function of y. The area to the right of the y-axis is found using y as a function of x: solving $x = (y - 1)^2 - 1$ for y gives $y = 1 + \sqrt{x + 1}$.
The total area is

$$\int_0^2 0 - ((y - 1)^2 - 1) \, dy$$

$$+ \int_0^\pi ((1 + \sqrt{x + 1}) - \sin x) \, dx$$

$$= \int_0^2 (1 - (y - 1)^2) \, dy$$

$$+ \int_0^\pi (1 + \sqrt{x + 1} - \sin x) \, dx$$

$$\left(= \frac{2(\pi + 1)^{3/2}}{3} + \pi - \frac{4}{3} \right).$$

6) Write an expression involving definite integrals for the area of the region bounded by:

a) $y = 2x$ and $y = 8 - x^2$.

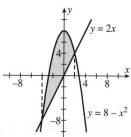

First find the points of intersection
$$2x = 8 - x^2$$
$$x^2 + 2x - 8 = 0$$
$$(x - 2)(x + 4) = 0, \text{ at } x = 2, -4.$$

The area is $\displaystyle\int_{-4}^{2} (8 - x^2 - 2x)dx \ (=36)$.

b) the shaded area between $y = \sin x$ and $y = \cos x$.

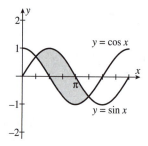

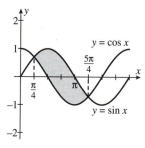

First find the points of intersection.

$$\sin x = \cos x; \ \frac{\sin x}{\cos x} = 1; \ \tan x = 1.$$

So $x = \frac{\pi}{4}$ and $x = \frac{5\pi}{4}$.

$$\text{Area} = \int_{\pi/4}^{5\pi/4} (\sin x - \cos x)dx \ (= 2\sqrt{2}).$$

c) $y = x, y = -x, y = 2x - 3.$

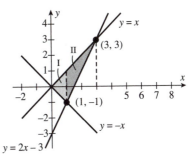

Divide the area into two regions:
Region I is bounded by $x = 0$, $x = 1$, $y = x$,
and $y = -x$.
Region II is bounded by $x = 1$, $x = 3$,
$y = x$, and $y = 2x - 3$.
The area is

$$\int_0^1 [(x - (-x)]dx + \int_1^3 [x - (2x - 3)] \, dx$$

$$= \int_0^1 2x \, dx + \int_1^3 (3 - x)dx \ (=3).$$

D. Technology Plus. Use a computer algebra system or a graphing calculator to solve.

T-1) Use a graphing calculator to graph
$y = 8 - x^2$ and $y = x^2 - 2x + 2$. Then
zoom in to find the points of intersection.
Then find the area between the two curves.

Using a $[-3, 3]$ by $[-1, 9]$ window we see

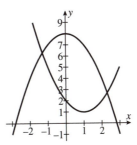

The graphs intersect at approximately
$x = -1.30$ and $x = 2.30$.

The area is

$$\int_{-1.30}^{2.30} (8 - x^2) - (x^2 - 2x + 2) \, dx$$

$$= \int_{-1.30}^{2.30} (-2x^2 + 2x + 6) \, dx$$

$$= \left(\frac{-2}{3}x^3 + x^2 + 6x \right)\Big|_{-1.30}^{2.30} \approx 15.624.$$

T-2) Use a graphing calculator to graph $y = 4 - x^2$ and $x = (y - 2)^2$ then decide whether the area between the two curves should be found using $\int_a^b \ldots dx$ or $\int_c^d \ldots dy$.

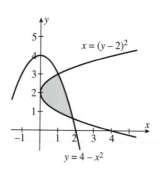

$\int_c^d \ldots dy$ is preferred because the "right" curve would be $x = \sqrt{4 - y}$ and the "left" curve would be $x = (y - 2)^2$.

Using $\int_a^b \ldots dx$ would require splitting the area into two regions.

Section 6.2 Volumes

This section and the next show you ways to determine the volume of a solid. The solids include irregular ones (such as a potato) and solids of revolution obtained by rotating an area about an axis (such as rotating a right triangle about one of its two shorter sides to obtain a cone). In this section we visualize the solid as being composed of many thin slices (cross sections). The sum of the volumes of these cross sections will lead to an integral for the volume of the solid.

Concepts to Master

A. Determine the volume of a solid by the slicing method

B. Volume of a solid of revolution

C. Technology Plus

Summary and Focus Questions

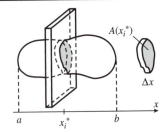

Page 430

A. Suppose the cross-sectional area of a solid cut by planes perpendicular to an x-axis is known to be $A(x)$ for each $x \in [a, b]$. The volume of a typical slice is approximately $A(x_i^*)\Delta x$ and thus the total volume of all slices is $\sum_{i=1}^{n} A(x_i^*)\Delta x$.

The limit of these sums by this **method of slicing** is the volume of the solid

$$V = \int_a^b A(x)\, dx.$$

The x-axis must be positioned carefully so that the function $A(x)$ can be determined and is integrable.

Example: Set up a definite integral for the volume of a 1-meter-high cone with base diameter of 2 meters. See the figure at the right.

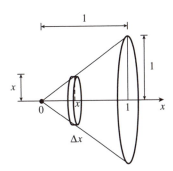

Place an x-axis through the middle of the cone. For $0 \leq x \leq 1$, the slice through the cone is a circular slab of radius x (by similar triangles) whose area is $A(x) = \pi x^2$ and volume is $\pi x^2\, \Delta x$. The volume is

$$\int_0^1 \pi x^2\, dx \left(= \frac{\pi}{3}\right).$$

1) Set up a definite integral for the volume of a right triangular solid that is 1 m, 1 m, and 2 m on its edges (see the figure).

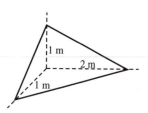

1 m
2 m
1 m

Draw an x-axis along the 2-meter side with the origin in the corner.

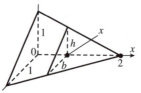

A typical slice is a triangle perpendicular to the x-axis. At point x, the slice area is $A(x) = \frac{1}{2}bh$. By similar triangles, $\frac{2-x}{2} = \frac{h}{1}$, so $h = 1 - \frac{x}{2}$. Likewise, $b = 1 - \frac{x}{2}$, so $A(x) = \frac{1}{2}\left(1 - \frac{x}{2}\right)^2$.

The volume is $\int_0^2 \frac{1}{2}\left(1 - \frac{x}{2}\right)^2 dx \left(= \frac{1}{3}\right)$.

2) The Great Pyramid of Giza is approximately 146.5 m high and has a square base of about 230.4 m on each side. Set up a definite integral for the volume of the Great Pyramid.

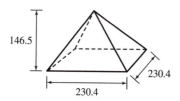

146.5
230.4
230.4

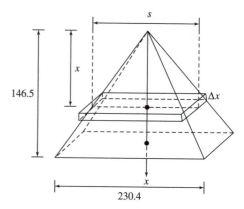

Place an x-axis down through the middle of the pyramid, with 0 at the apex. For $0 \le x \le 146.5$, the slice through the pyramid is a square slab of side s. By similar right triangles,

$$\frac{x}{146.5} = \frac{\frac{s}{2}}{\frac{230.4}{2}}, \text{ so } s = \frac{230.4}{146.5}x.$$

The area of the square with side s is

$$A(x) = \left(\frac{230.4}{146.5}x\right)^2 \text{ and volume of the slab is}$$

$\left(\frac{230.4}{146.5}x\right)^2 \Delta x$. The volume of the pyramid is

$$\int_0^{146.5} \left(\frac{230.4}{146.5}x\right)^2 dx \ (\approx 2{,}592{,}276.5 \text{ m}^3).$$

B. A solid of revolution is an object obtained by rotating a planar region about a line. The line may be an x-axis, y-axis, or some other line. Sometimes the resulting solid will have a cross section in the shape of a disk; other times the cross section will be washer shaped. The volume will be approximated by a sum of volumes of disks or washers which will suggest the definite integral for the exact volume.

Page 432

If the region under $y = f(x)$ is rotated about the x-axis, the volume may be approximated by a sum of volumes of disks. The volume of a typical disk is $\pi(\text{radius})^2(\text{thickness}) = \pi[f(x_i^*)]^2 \Delta x$.

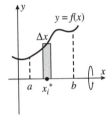
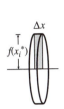

The sum of volume of disks is

$$\sum_{i=1}^{n} \pi[f(x_i^*)]^2 \Delta x. \text{ The exact volume is}$$

$$V = \int_a^b \pi[f(x)]^2 \, dx.$$

If the region between $y = f(x)$ and $y = g(x)$ is rotated about the x-axis, the volume may be approximated by a sum of volumes of washers. The volume of a typical washer is
$\pi(\text{outer radius})^2(\text{thickness})$
$\quad - \pi(\text{inner radius})^2(\text{thickness})$
$\quad = \pi[(f(x_i^*))^2 - (g(x_i^*))^2]\Delta x.$

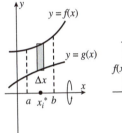
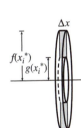

The sum of the volume of washers is

$$\sum_{i=1}^{n} \pi\big[(f(x_i^*))^2 - (g(x_i^*))^2\big]\Delta x. \text{ The exact volume is}$$

$$V = \int_a^b \pi[(f(x))^2 - (g(x))^2] \, dx.$$

We repeat the first example of this section, this time treating the problem as a solid of revolution.

Example: The region bounded by $y = x$, $y = 0$, $x = 0$, and $x = 1$ is rotated about the x-axis to form a cone 1 meter high with base diameter of 2 meters. Set up a definite integral for the volume of the cone.

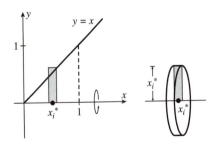

For $0 \le x \le 1$, the rectangle rotated around the x axis is a disk of radius x (since $f(x) = x$) whose area is $\pi(f(x))^2 = \pi x^2$ and volume is $\pi x^2 \Delta x$.

The volume of the cone is $\int_0^1 \pi x^2 \, dx \left(= \frac{\pi}{3}\right)$.

3) Set up a definite integral for the volume of each solid of revolution.

a) Rotate the region $y = 4 - x^2$, $y = 0$, $x = 0$, $x = 2$ about the x-axis.

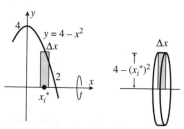

Volume of disk $= \pi(4 - (x_i^*)^2)^2 \Delta x$.

Volume of solid $= \displaystyle\int_0^2 \pi(4 - x^2)^2 \, dx$
$$\left(= \frac{256}{15}\pi\right).$$

b) The solid obtained by rotating about the y-axis the region bounded by $y = \sqrt{x}$, $x = 0$, $y = 4$.

<Write x as a function of y.>
From $y = \sqrt{x}$, $x = y^2$.

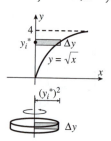

Volume of disk $= \pi((y_i^*)^2)^2 \Delta y$
$$= \pi(y_i^*)^4 \Delta y.$$

Volume of solid $= \displaystyle\int_0^4 \pi y^4 \, dy \left(= \frac{1024\pi}{5}\right).$

c) The solid obtained by rotating about the line $y = 3$ the region bounded by $y = 10 - x^2$, $x = 0$, $x = 2$, $y = 3$.

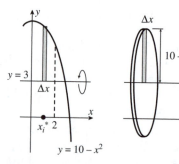

Volume of disk $= \pi(10 - (x_i^*)^2 - 3)^2 \Delta x$

Volume of solid $= \displaystyle\int_0^2 \pi(10 - x^2 - 3)^2 \, dx$

$$= \int_0^2 \pi(7 - x^2)^2 \, dx \left(= \frac{1006}{15}\pi\right).$$

4) Set up a definite integral for the volume of each solid of revolution.

a) The solid obtained by rotation about the x-axis of the region bounded by the y-axis, $y = x^3$, and $y = 8$.

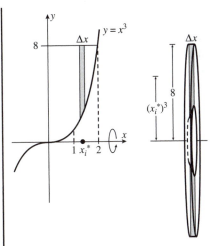

Volume of washer
$$= \pi[8^2 - ((x_i{}^*)^3)^2]\Delta x.$$

Volume of solid
$$= \int_0^2 \pi[8^2 - (x^3)^2]\ dx \left(= \frac{768\pi}{7}\right).$$

b) Set up a definite integral for the volume of the solid obtained by rotation about the x-axis of the region between $x = 1$, $x = 2$, $y = x^2$, and $y = x^3$.

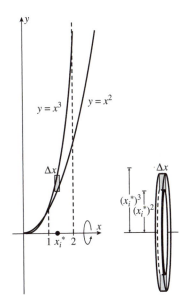

Volume of washer
$$= \pi[((x_i{}^*)^3)^2 - ((x_i{}^*)^2)^2]\Delta x.$$

Volume of solid
$$= \int_1^2 \pi[(x^3)^2 - (x^2)^2]\ dx$$
$$= \pi\int_1^2 (x^6 - x^4)dx \left(= \frac{418}{35}\pi\right).$$

C. Technology Plus. Use a computer algebra system or a graphing calculator to solve.

T-1) Use a spreadsheet or a calculator to find the Riemann sum for estimating the volume of the solid obtained by rotating $y = e^x$, $0 \leq x \leq 2$, about the x-axis. Use $n = 20$ and the midpoint of each subinterval.

$$\Delta x = \frac{2 - 0}{20} = 0.10.$$

n	x_i^*	$f(x_i^*)$	$\pi(f(x_i^*))^2\Delta x$
1	0.05	1.051	0.347
2	0.15	1.162	0.424
3	0.25	1.284	0.518
4	0.35	1.419	0.633
5	0.45	1.568	0.773
6	0.55	1.733	0.944
7	0.65	1.916	1.153
8	0.75	2.117	1.408
9	0.85	2.340	1.720
10	0.95	2.586	2.100
11	1.05	2.858	2.565
12	1.15	3.158	3.133
13	1.25	3.490	3.827
14	1.35	3.857	4.675
15	1.45	4.263	5.710
16	1.55	4.711	6.974
17	1.65	5.207	8.518
18	1.75	5.755	10.404
19	1.85	6.360	12.707
20	1.95	7.029	15.520
			84.053

$$\int_0^2 \pi(e^x)^2 \, dx \approx 84.053.$$

Section 6.3 Volumes by Cylindrical Shells

This section presents another way to determine a definite integral for a solid of revolution. This time the volume is approximated by a sum of volumes of concentric cylindrical shells. As in the previous section, the key will be to visualize the solid as composed of many shells whose volumes can be determined.

Concepts to Master

Volumes of solids of revolution by the shell method

Summary and Focus Questions

Page 441

The volume of the solid obtained by rotating the region bounded by $y = f(x)$, $y = 0$, $x = a$, $x = b$ about the y-axis may be approximated by a sum of volumes of shells. The volume of a typical shell is

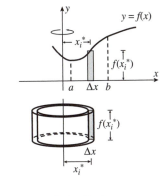

$$2\pi(\text{radius})(\text{height})(\text{thickness})$$
$$= 2\pi x_i{}^* f(x_i{}^*)\Delta x.$$

The total volume of the solid by this **method of (cylindrical) shells** is

$$V = \int_a^b 2\pi x f(x) \, dx.$$

Volumes of solids of revolution about the x-axis are computed similarly.

Example: Use the shell method to set up a definite integral for the volume of solid obtained by rotating about the y-axis the region bounded by $x = 1$, $x = 2$, $y = 0$, and $y = \dfrac{1}{x}$.

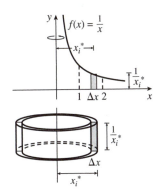

For $1 \le x_i{}^* \le 2$, the cylindrical shell of a rectangular region has volume

$$2\pi\left(x_i{}^*\right)\left(\frac{1}{x_i{}^*}\right)\Delta x = 2\pi\Delta x.$$

The volume of the solid is $\displaystyle\int_1^2 2\pi \, dx \ (= 2\pi)$.

1) Use the shell method to set up a definite integral for the volume of the solid obtained by rotating about the y-axis the region bounded by $y = \sqrt{x}, y = 0, x = 1, x = 4$.

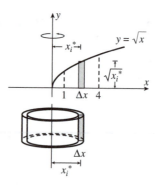

Volume of shell $= 2\pi x_i{}^*(\sqrt{x_i{}^*})\Delta x$
$$= 2\pi(x_i{}^*)^{3/2}\Delta x.$$

Volume of solid $= \displaystyle\int_1^4 2\pi x^{3/2}\,dx \left(= \dfrac{124\pi}{5}\right)$.

2) Use the shell method to set up a definite integral for the volume of the solid obtained by rotating about the x-axis the region bounded by $x = y^2, y = 0, x = 4$.

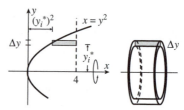

Volume of shell $= 2\pi y_i{}^*(4 - (y_i{}^*)^2)\Delta y$
$$= 2\pi(4y_i{}^* - (y_i{}^*)^3)\Delta y.$$

Volume of solid $= \displaystyle\int_0^2 2\pi(4y - y^3)\,dy$
$$(= 8\pi).$$

3) A solid "dog dish" is obtained by rotating the region bounded by $y = 0, y = x^3, x = 0, x = 1$ about the y-axis. Set up two definite integrals for the volume—one by the slicing method and the other using shells.

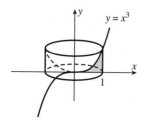

i) Using the slicing method, the slices along the y-axis are washers:

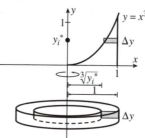

Volume of washer $= \pi[1^2 - (\sqrt[3]{y})^2]\Delta y$.

Volume of solid $= \displaystyle\int_0^1 \pi(1 - y^{2/3})\,dy \left(= \dfrac{2\pi}{5}\right)$.

ii) Using shells:

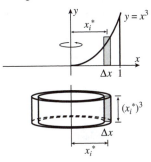

Volume of shell $= 2\pi x_i{}^* (x_i{}^*)^3 \Delta x$
$= 2\pi (x_i{}^*)^4 \Delta x.$

Volume of solid $= \displaystyle\int_0^1 2\pi x^4 \, dx \left(= \dfrac{2\pi}{5} \right).$

Section 6.4   Work

Work is a measure of the effort expended by a force moving an object from one point to another. Think of the total work done as a sum of the amounts of work done moving the object over several small consecutive intervals. Thus, the total amount of work done is determined by a definite integral.

Concepts to Master

Work done by a varying amount of force along an axis

Summary and Focus Questions

Page
446

The work done to move an object a distance d with a constant force F is

$$W = Fd.$$

If the force is measured in pounds and the distance in feet, the units of work are foot-pounds. In the metric system, the unit of measure for work is a Newton-meter (kg m²/s²), which is a called Joule (J).

Example: The amount of work done lifting a box that weighs 3 kg off the ground 2 meters is $W = (3 \text{ kg})(9.8 \text{ m/s}^2)(2 \text{ m}) = 58.8 \text{ kg m}^2/\text{s}^2 = 58.8 \text{ J}$, where 9.8 m/s^2 is the acceleration constant due to gravity.

If an object moves along an x-axis from a to b by a varying force $F(x)$, the work done may be approximated by the sum of work done over subintervals between a and b. The work over a subinterval is approximately $F(x_i^*)\Delta x$—force times distance. The total work is approximately $\sum_{i=1}^{n} F(x_i^*)\Delta x$.

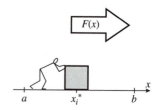

Thus the total work done moving from a to b is

$$W = \int_a^b F(x)\, dx.$$

Example: A definite integral for the amount of work done by pushing an object along a track from meter marker 1 to meter marker 10 if the force applied at meter marker x is $x + x^2$ kg is

$$\int_1^{10} (x + x^2)dx = 382.5 \text{ J}.$$

1) How much work is done lifting an 800 kilogram piano upward 6 meters?

The force is a constant 800 kg and the distance is 6 m. The work is
$W = (800)(6) = 4800 \text{ J}.$

2) How much work is done moving a particle along an *x*-axis from 0 to $\frac{\pi}{2}$ feet if at each point *x* the amount of force is cos *x* pounds?

$$W = \int_0^{\pi/2} \cos x \, dx = \sin x \Big]_0^{\pi/2} = 1 \text{ ft-lb.}$$

3) Find an integral for how much work is done in pumping over the edge all the liquid of density ρ out of a 2-foot-long trough in the shape of an equilateral triangle with sides of 1 foot.

<A typical water slice needs to be lifted out over the edge of the trough.>

The work done to pump out a slice of the liquid is (density · gravitational constant · volume · height raised).

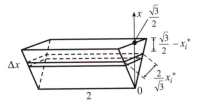

The volume of a slice is (length · width · height). The end of the trough is given in this figure.

From the end, the slice looks like this:

By similar triangles the length is $\frac{2}{\sqrt{3}}x_i^*$.

Thus the volume of a slice is

$$\left(\frac{2}{\sqrt{3}}x_i^*\right)(2)(\Delta x) = \frac{4}{\sqrt{3}}x_i^*\Delta x.$$

The density is ρ; *g* is the gravitational constant.

The height raised is $\frac{\sqrt{3}}{2} - x_i^*$.

The work done lifting the slice is

$$\rho g\left(\frac{4}{\sqrt{3}}x_i^*\Delta x\right)\left(\frac{\sqrt{3}}{2} - x_i^*\right).$$

The total work is

$$= \int_0^{\sqrt{3}/2} \rho g\frac{4}{\sqrt{3}}x\left(\frac{\sqrt{3}}{2} - x\right) dx \left(= \frac{\rho g}{4}\right).$$

Section 6.5   Average Value of a Function

As the graph of a function f varies from $x = a$ to $x = b$, $f(x)$ may assume many different values. This section shows how the average of these values may be determined using $\int_a^b f(x)\, dx$. This section also contains a restatement of the Mean Value Theorem—this time in terms of definite integrals.

Concepts to Master

A. Average value of a function

B. Mean Value Theorem for Integrals

Summary and Focus Questions

Page 451

A. If we divide the area of a rectangle by its width, we obtain its height. In a similar way, if $f(x) \geq 0$ for $a \leq x \leq b$ and we think of $\int_a^b f(x)\, dx$ as the area under $y = f(x)$ from a to b, then when we divide $\int_a^b f(x)\, dx$ by its width $(b - a)$, we will obtain the typical value (average height) of $f(x)$.

The **average value** of a continuous function $y = f(x)$ over a closed interval $[a, b]$ is

$$f_{\text{ave}} = \frac{1}{b - a} \int_a^b f(x)\, dx.$$

Example: The average value for the function $f(x) = x^3$ over the interval $[0, 2]$ is

$$f_{\text{ave}} = \frac{1}{2 - 0} \int_0^2 x^3 dx = \frac{1}{2} \cdot \frac{x^4}{4} \Big]_0^2 = \frac{1}{2} \cdot 4 = 2.$$

Therefore, as x ranges from 0 to 2, the $f(x)$ values range from 0 to 8 with an "average" $f(x)$ value of 2.

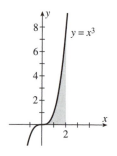

1) Find the average value of $f(x) = 2x + 6x^2$ over $[1, 4]$.

$$f_{\text{ave}} = \frac{1}{4 - 1} \int_1^4 (2x + 6x^2)\, dx$$

$$= \frac{1}{3}(x^2 + 2x^3) \Big]_1^4 = 47.$$

2) Find the average value of cos x over $[0, \pi]$.

$$f_{ave} = \frac{1}{\pi - 0} \int_0^\pi \cos x \, dx = \frac{1}{\pi} \sin x \Big]_0^\pi$$

$$= \frac{1}{\pi}(0 - 0) = 0.$$

Intuitively this makes sense because the "right half" of the cosine curve (the part between $\frac{\pi}{2}$ and π) is below the x-axis and is symmetric with the other half above the x-axis. The cosine values average zero over the interval $[0, \pi]$.

**Page
452**

B. The Mean Value Theorem for Integrals:

If f is continuous on a closed interval $[a, b]$ there exists $c \in [a, b]$ such that

$$\int_a^b f(x) \, dx = f(c)(b - a).$$

When $f(x) \geq 0$ for $x \in [a, b]$, we can interpret

i) $\int_a^b f(x) \, dx$ as the area under $y = f(x)$, and

ii) $f(c)(b - a)$ as the area of the rectangle with width $(b - a)$ and height $f(c)$.

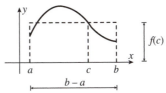

Therefore, the Mean Value Theorem for Integrals says there is some point $c \in [a, b]$ where the area of the rectangle equals the area under the curve.

Example: Find the number c guaranteed by the Mean Value Theorem for Integrals for $f(x) = 10 - x$ on the interval $[2, 6]$.

$$\int_2^6 (10 - x) dx = \left(10x - \frac{x^2}{2}\right)\Big]_2^6 = 24. \text{ Now solve } 24 = f(c)(6 - 2).$$

$24 = 4(10 - c)$, so $10 - c = 6$. Therefore $c = 4$.

3) Mark on the graph of $y = f(x)$ a number c guaranteed by the Mean Value Theorem for Integrals:

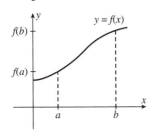

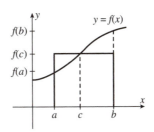

The number c is chosen so that the area under the graph of $y = f(x)$ from a to b equals the area of the rectangle with width $b - a$ and height $f(c)$.

4) For $f(x) = x^2$ on $[1, 4]$, find a value for c that satisfies the Mean Value Theorem for Integrals.

$\int_1^4 x^2 \, dx = \frac{x^3}{3} \Big|_1^4 = 21$. So $21 = f(c)(4 - 1)$.

$f(c) = 7, c^2 = 7, c = \sqrt{7}$.

c is not $-\sqrt{7}$ since it is not in the interval $[1, 4]$.

5) Let $F'(x) = f(x)$ be continuous on $[a, b]$. State the conclusion of the Mean Value Theorem for Integrals in terms of the function $F(x)$. Use the Fundamental Theorem of Calculus.

The conclusion is

"there is a number c such that

$\int_a^b f(x) \, dx = f(c)(b - a)$."

By the Fundamental Theorem of Calculus

$\int_a^b f(x) \, dx = F(b) - F(a)$.

Since $f(c) = F'(c)$,

$f(c)(b - a) = F'(c)(b - a)$.

Therefore, the conclusion, in terms of $F(x)$, becomes

"there is a number c such that

$F(b) - F(a) = F'(c)(b - a)$."

This is precisely the conclusion of the Mean Value Theorem of Chapter 4!

6) True, False. If $y = f(x)$ is continuous and increasing on $[a, b]$, then there is exactly one value c in $[a, b]$ that satisfies the conclusion of the Mean Value Theorem for Integrals.

True.

Chapter 7 — Techniques of Integration

Section 7.1   Integration by Parts

Recall that the Product Rule for derivatives says that if u and v are functions of x, then $(uv)' = uv' + v'u$. The integration by parts formula in this section is simply a restatement of the Product Rule in terms of antiderivatives.

Concepts to Master

Evaluate integrals using integration by parts

Summary and Focus Questions

Page 464

For differentiable functions u and v, the **integration by parts** formula is

$$\int u \, dv = uv - \int v \, du.$$

Integration by parts is a rule for replacing the task of evaluating $\int u \, dv$ with hopefully an easier task of evaluating $\int v \, du$. The key to successful use is to identify the two terms u and v' in the integrand (remember $u \, dv = uv' \, dx$), and to check that $\int v \, du$ is indeed simpler to evaluate.

Example: Find $\int x \ln x \, dx$.

Initially, we might decide $u = x$ because $du = dx$ is simple. However, the companion choice $dv = \ln x \, dx$ would mean $v = \int dv = \int \ln x \, dx$ would be rather difficult to obtain and $\int v \, du$ would be worse.

So we choose $u = \ln x$, $dv = x \, dx$. Then $du = \frac{1}{x} \, dx$ and $v = \int dv = \int x \, dx = \frac{x^2}{2}$.

Then

$$\int x \ln x \, dx = \underbrace{(\ln x)}_{u} \underbrace{\left(\frac{x^2}{2}\right)}_{v} - \int \underbrace{\frac{x^2}{2}}_{v} \underbrace{\left(\frac{1}{x}\right) dx}_{du} = \frac{1}{2}x^2 \ln x - \frac{1}{2}\int x \, dx$$

$$= \frac{1}{2}x^2 \ln x - \frac{1}{4}x^2 + C.$$

235

In some problems you may need to use integration by parts more than once in order to obtain an integral $\int v\,du$ that can be evaluated by other means.

1) Differentiate both sides of the integration by parts formula with respect to x. What result do you get? (Remember that $udv = uv'\,dx$ and $vdu = vu'\,dx$.)

From $\int udv = uv - \int vdu$, we have $\int uv'\,dx = uv - \int vu'\,dx$. Taking the derivative with respect to x gives $uv' = (uv)' - vu'$. Solving for $(uv)'$ we have $(uv)' = uv' + vu'$, the Product Rule for derivatives!

2) Give an appropriate choice for u and dv in each by parts integral and evaluate:

a) $\int x^2 \ln x\, dx$

$u = $ _____

$dv = $ _____

$u = \ln x,\ du = \frac{1}{x}\,dx.$

$dv = x^2\,dx,\ v = \frac{x^3}{3}.$

$\int x^2 \ln x\,dx = \ln x\left(\frac{x^3}{3}\right) - \int \frac{x^3}{3}\left(\frac{1}{x}\right)dx$

$= (\ln x)\frac{x^3}{3} - \int \frac{x^2}{3}\,dx$

$= \frac{x^3}{3}\ln x - \frac{1}{3}\int x^2\,dx = \frac{x^3}{3}\ln x - \frac{1}{9}x^3 + C.$

b) $\int x \cos 2x\, dx$

$u = $ _____

$dv = $ _____

$u = x,\ du = dx.$

$dv = \cos 2x\,dx,\ v = \frac{1}{2}\sin 2x.$

$\int x \cos 2x\,dx = x\left(\frac{1}{2}\sin 2x\right) - \int \frac{1}{2}\sin 2x\,dx$

$= \frac{1}{2}x \sin 2x - \frac{1}{2}\int \sin 2x\,dx$

$= \frac{1}{2}x \sin 2x - \frac{1}{2}\left(-\frac{1}{2}\cos 2x\right)$

$= \frac{1}{2}x \sin 2x + \frac{1}{4}\cos 2x + C.$

3) Evaluate $\int x^2 e^{5x}\, dx.$

Let $u = x^2$ and $dv = e^{5x}\,dx.$

Then $du = 2x\,dx,\ v = \frac{1}{5}e^{5x}$, and

$v\,du = \frac{1}{5}e^{5x}\,2x\,dx = \frac{2}{5}xe^{5x}\,dx.$

$\int x^2 e^{5x}\,dx = x^2\left(\frac{1}{5}e^{5x}\right) - \int \frac{1}{5}e^{5x}(2x)\,dx$

$= x^2\left(\frac{1}{5}e^{5x}\right) - \int \frac{2}{5}xe^{5x}\,dx$

$= \frac{1}{5}x^2\,e^{5x} - \frac{2}{5}\int x\,e^{5x}\,dx.$

The last integral is easier to solve than the original so we are on the right track. Apply by parts to it:

$u = x,\ du = dx,$

$dv = e^{5x}\,dx,\ v = \frac{1}{5}e^{5x}.$

$$\int x\, e^{5x}\, dx = x\left(\tfrac{1}{5}e^{5x}\right) - \int \tfrac{1}{5}e^{5x}\, dx$$

$$= \tfrac{1}{5}xe^{5x} - \tfrac{1}{25}e^{5x} + C.$$

Thus, $\int x^2 e^{5x}\, dx$

$$= \tfrac{1}{5}x^2 e^{5x} - \tfrac{2}{5}\left(\tfrac{1}{5}xe^{5x} - \tfrac{1}{25}e^{5x} + C\right)$$

$$= \tfrac{1}{5}x^2 e^{5x} - \tfrac{2}{25}xe^{5x} + \tfrac{2}{125}e^{5x} + C.$$

4) Evaluate $\displaystyle\int_1^e \ln x\, dx$.

Let $u = \ln x$ and $dv = 1\, dx$.
$du = \dfrac{1}{x}\, dx, v = x.$
$$\int \ln x\, dx = x \ln x - \int x\left(\tfrac{1}{x}\right) dx$$

$$= x \ln x - \int dx$$

$$= x \ln x - x + C.$$

Therefore, $\displaystyle\int_1^e \ln x\, dx = (x \ln x - x)\Big]_1^e$

$$= (e \ln e - e) - (1 \ln 1 - 1) = 1.$$

5) Evaluate $\displaystyle\int \sin x \cos x\, dx$ by parts.

$u = \sin x, du = \cos x\, dx,$
$dv = \cos x\, dx, v = \sin x.$

$$\int \sin x \cos x\, dx$$
$$= (\sin x)(\sin x) - \int (\sin x)\cos x\, dx.$$

Thus

$$\int \sin x \cos x\, dx = \sin^2 x - \int \sin x \cos x\, dx.$$

By adding $\int \sin x \cos x\, dx$ to both sides, we

have $2\displaystyle\int \sin x \cos x\, dx = \sin^2 x,$

so $\displaystyle\int \sin x \cos x\, dx = \tfrac{1}{2} \sin^2 x + C.$

Note: This integral could have been
evaluated with the simple substitution.

$u = \sin x, du = \cos x\, dx$ to yield $\int u\, du$. We
could also have solved this by using the
identity $\sin x \cos x = \tfrac{1}{2} \sin 2x$. The resulting
integral $\int \tfrac{1}{2} \sin 2x\, dx$ is solved using the
substitution $u = 2x$. We chose the above
method to illustrate integration by parts.

Section 7.2   Trigonometric Integrals

This section will show you how to evaluate integrals involving products of trigonometric functions. We already know how to solve some, such as $\int (\sin^4 x \cos x)dx$. It may be evaluated using the substitution $u = \sin x$, $du = \cos x \, dx$, which results in $\int u^4 \, du$. This section uses trigonometric identities to rewrite integrands into forms for which substitution may be used.

Concepts to Master

A. Evaluate integrals whose integrands are powers of sine and cosine, tangent and secant, or cotangent and cosecant

B. Evaluate integrals whose integrands are products of sines and/or cosines with multiples of the independent variable

Summary and Focus Questions

Page
471

A. The table on the next page shows how to use trigonometric identities to rewrite integrals of any of these forms:

$$\int (\sin^m x \cos^n x)dx$$

$$\int (\tan^m x \sec^n x)dx$$

$$\int (\cot^m x \csc^n x)dx.$$

Note that m and n must be integers. If the conditions(s) are satisfied, then the original integral may be rewritten and simplified as indicated.

These techniques may be used to find two more integrals to add to our list of basic integrals:

$$\int \tan x \, dx = \ln|\sec x| + C$$

$$\int \sec x \, dx = \ln|\sec x + \tan x| + C.$$

Integrals of the form $\int \sin^m x \cos^n x \, dx$

If m is odd, $m = 2k + 1$:

   i) Factor out one $\sin x$ and use the identity $\sin^2 x = 1 - \cos^2 x$.

   ii) Write the remaining $\sin^{m-1} x = \sin^{2k} x = (\sin^2 x)^k$ as $(1 - \cos^2 x)^k$:

$$\int \sin^m x \cos^n x \, dx = \int (1 - \cos^2 x)^k \cos^n x \sin x \, dx$$

   iii) Substitute $u = \cos x$, $du = -\sin x \, dx$.

If n is odd, $n = 2k + 1$:

   i) Factor out one $\cos x$ and use the identity $\cos^2 x = 1 - \sin^2 x$.

   ii) Write the remaining $\cos^{n-1} x = \cos^{2k} x = (\cos^2 x)^k$ as $(1 - \sin^2 x)^k$:

$$\int \sin^m x \cos^n x \, dx = \int \sin^m x (1 - \sin^2 x)^k \cos x \, dx$$

   iii) Substitute $u = \sin x$, $du = \cos x \, dx$.

If m and n are both even:

   i) Use the half angle identities: $\sin^2 x = \dfrac{1 - \cos 2x}{2}$ and/or $\cos^2 x = \dfrac{1 + \cos 2x}{2}$.

   ii) Substitute the identities to get started:

$$\int \sin^m x \cos^n x \, dx = \int \left(\frac{1 - \cos 2x}{2} \right)^{m/2} \left(\frac{1 + \cos 2x}{2} \right)^{n/2} dx$$

   iii) Multiply out, collect terms, and evaluate. Sometimes the identity $\sin x \cos x = \frac{1}{2} \sin 2x$ may be used to simplify the integrand.

Integrals of the form $\int \tan^m x \sec^n x \, dx$

If m is odd, $m = 2k + 1$:

   i) Factor out one $\sec x \tan x$ and use the identity $\tan^2 x = \sec^2 x - 1$.

   ii) Write the remaining $\tan^{m-1} x = \tan^{2k} x = (\tan^2 x)^k$ as $(\sec^2 x - 1)^k$:

$$\int \tan^m x \sec^n x \, dx = \int (\sec^2 x - 1)^k \sec^{n-1} x \sec x \tan x \, dx$$

   iii) Substitute $u = \sec x$, $du = \sec x \tan x \, dx$.

If n is even, $n = 2k$:

   i) Factor out one $\sec^2 x$ and use the identity $\sec^2 x = \tan^2 x + 1$.

   ii) Write the remaining $\sec^{n-2} x = \sec^{2k-2} x = (\sec^2 x)^{k-1}$ as $(\tan^2 x + 1)^{k-1}$:

$$\int \tan^m x \sec^n x \, dx = \int \tan^m x (\tan^2 x + 1)^{k-1} \sec^2 x \, dx$$

   iii) Substitute $u = \tan x$, $du = \sec^2 x \, dx$.

If m is even and n is odd:

   This particular combination of m and n may use the identity $\tan^2 x = \sec^2 x - 1$, use integration by parts, and/or algebraic techniques.

Integrals of the form $\int \cot^m x \csc^n x \, dx$

These integrals are handled in the same manner as integrals of the form $\int \tan^m x \sec^n x \, dx$ using the identity $\cot^2 x = \csc^2 x - 1$.

Example: Evaluate $\int \cos^5 x\, dx$.

Since cosine has an odd power, factor out one cosine and use the identity $\cos^2 x = 1 - \sin^2 x$.

$$\int \cos^5 x\, dx = \int (\cos^4 x)\cos x\, dx = \int (\cos^2 x)^2 \cos x\, dx = \int (1 - \sin^2 x)^2 \cos x\, dx.$$

$$= \int (1 - u^2)^2\, du \quad \text{(Substitute } u = \sin x,\ du = \cos x\, dx)$$

$$= \int (1 - 2u^2 + u^4)\, du = u - \tfrac{2}{3}u^3 + \tfrac{1}{5}u^5 + C = \sin x - \tfrac{2}{3}\sin^3 x + \tfrac{1}{5}\sin^5 x + C.$$

Example: Evaluate $\int \sin^2 x \cos^2 x\, dx$.

Since both sine and cosine have even powers, use the half angle identities.

$$\int \sin^2 x \cos^2 x\, dx = \int \left(\frac{1 - \cos 2x}{2} \right)\left(\frac{1 + \cos 2x}{2} \right) dx = \tfrac{1}{4}\int \left(1 - \cos^2 2x\right) dx$$

$$= \tfrac{1}{4}\int \left(1 - \frac{1 + \cos 4x}{2}\right) dx = \tfrac{1}{8}\int (1 - \cos 4x)\, dx$$

$$= \tfrac{1}{8}\left(x - \tfrac{1}{4}\sin 4x\right) + C = \tfrac{1}{32}(4x - \sin 4x) + C.$$

1) Evaluate $\int \sin^5 x\, dx$.

This fits the form $\int \sin^m x \cos^n x\, dx$ with $m = 5, n = 0$. Since m is odd,

$$\int \sin^5 x\, dx = \int \sin^4 x \sin x\, dx$$

$$= \int (1 - \cos^2 x)^2 \sin x\, dx$$

$$(u = \cos x,\ du = -\sin x\, dx)$$

$$= -\int (1 - u^2)^2\, du$$

$$= -\int (1 - 2u^2 + u^4)\, du$$

$$= -u + \tfrac{2}{3}u^3 - \tfrac{1}{5}u^5 + C$$

$$= -\cos x + \tfrac{2}{3}\cos^3 x - \tfrac{1}{5}\cos^5 x + C.$$

2) Evaluate $\int \tan x \sec^4 x\, dx$.

Here $m = 1$ and $n = 4$ so we have our choice of identities. We select the case where n is even.

$$\int \tan x \sec^4 x\, dx$$

$$= \int \tan x(\sec^2 x)\sec^2 x\, dx$$

$$= \int \tan x(\tan^2 x + 1)\sec^2 x\, dx$$

$$= \int (\tan^3 x + \tan x)\sec^2 x\, dx$$

$$(u = \tan x,\ du = \sec^2 x\, dx)$$

$$= \int (u^3 + u)\, du = \frac{u^4}{4} + \frac{u^2}{2}$$

$$= \tfrac{1}{4}\tan^4 x + \tfrac{1}{2}\tan^2 x + C.$$

3) Evaluate $\int \sec 6x\, dx$.

Let $u = 6x$, $du = 6\, dx$, $dx = \dfrac{du}{6}$.

$$\int \sec 6x\, dx = \int \sec u\, \frac{du}{6}$$

$$= \frac{1}{6} \int \sec u\, du$$

$$= \frac{1}{6} \ln|\sec u + \tan u| + C$$

$$= \frac{1}{6} \ln|\sec 6x + \tan 6x| + C.$$

4) Evaluate $\int \cos^4 x\, dx$.

Here $m = 0$ and $n = 4$ are both even.

$$\int \cos^4 x\, dx = \int (\cos^2 x)^2\, dx$$

$$= \int \left(\frac{1 + \cos 2x}{2}\right)^2 dx$$

$$= \frac{1}{4} \int (1 + 2\cos 2x + \cos^2 2x)\, dx$$

$$= \frac{1}{4} \int \left(1 + 2\cos 2x + \frac{1 + \cos 4x}{2}\right) dx$$

$$= \frac{1}{4} \int \left(\frac{3}{2} + 2\cos 2x + \frac{1}{2}\cos 4x\right) dx$$

$$= \frac{1}{4}\left(\frac{3}{2}x + \sin 2x + \frac{1}{2}\frac{\sin 4x}{4}\right) + C$$

$$= \frac{3}{8}x + \frac{1}{4}\sin 2x + \frac{1}{32}\sin 4x + C.$$

5) Evaluate $\int \tan^2 x \sec x\, dx$.

Here $m = 2$ is even and $n = 1$ is odd.

Use the identity $\tan^2 x = \sec^2 x - 1$.

$$\int \tan^2 x \sec x\, dx = \int (\sec^2 x - 1)\sec x\, dx$$

$$= \int (\sec^3 x - \sec x)\, dx$$

$$= \int \sec^3 x\, dx - \int \sec x\, dx$$

$$= \frac{1}{2}(\sec x \tan x + \ln|\sec x + \tan x|)$$

$$\qquad - \ln|\sec x + \tan x| + C$$

$$= \frac{1}{2}(\sec x \tan x - \ln|\sec x + \tan x|) + C.$$

$(\int \sec^3 x\, dx$ is evaluated by parts, where $u = \sec x$, $dv = \sec^2 x\, dx$.)

6) Evaluate $\displaystyle\int \frac{\sin^3 x}{\sqrt{\cos x}}\, dx.$

Here $m = 3$ and $n = -\frac{1}{2}$. Although n is not an integer, the same technique works.

$$\int \frac{\sin^3 x}{\sqrt{\cos x}}\, dx = \int (\cos x)^{-1/2} \sin^2 x \sin x\, dx$$
$$= \int (\cos x)^{-1/2}(1 - \cos^2 x) \sin x\, dx$$

$(u = \cos x,\ du = -\sin x\, dx)$

$$= -\int u^{-1/2}(1 - u^2)\, du$$
$$= -\int (u^{-1/2} - u^{3/2})\, du$$
$$= -(2u^{1/2} - \tfrac{2}{5}u^{5/2}) + C$$
$$= \tfrac{2}{5}\cos^{5/2} x - 2\sqrt{\cos x} + C.$$

B. The table below shows how to use trigonometric identities to rewrite integrals of any of these forms:

Page
476

$$\int (\sin mx \sin nx)\, dx$$
$$\int (\cos mx \cos nx)\, dx$$
$$\int (\sin mx \cos nx)\, dx.$$

Note that m and n may be any real numbers. If the conditions(s) are satisfied, then the original integral may be rewritten and simplified as indicated.

Integrals of the form $\int \sin mx \sin nx\, dx$

i) Use the identity $\sin A \sin B = \dfrac{\cos(A - B) - \cos(A + B)}{2}$

ii) Rewrite: $\int \sin mx \sin nx\, dx = \int \frac{1}{2}[\cos(m - n)x - \cos(m + n)x]\, dx$

iii) Simplify the integrand and evaluate.

Integrals of the form $\int \cos mx \cos nx\, dx$

i) Use the identity $\cos A \cos B = \dfrac{\cos(A - B) + \cos(A + B)}{2}$

ii) Rewrite: $\int \cos mx \cos nx\, dx = \int \frac{1}{2}[\cos(m - n)x + \cos(m + n)x]\, dx$

iii) Simplify the integrand and evaluate.

Integrals of the form $\int \sin mx \cos nx\, dx$

i) Use the identity $\sin A \cos B = \dfrac{\sin(A - B) + \sin(A + B)}{2}$

ii) Rewrite: $\int \sin mx \cos nx\, dx = \int \frac{1}{2}[\sin(m - n)x + \sin(m + n)x]\, dx$

iii) Simplify the integrand and evaluate.

Example: Evaluate $\int \cos 5x \cos 2x \, dx$.

$$\int \cos 5x \cos 2x \, dx = \int \tfrac{1}{2}(\cos(5x - 2x) + \cos(5x + 2x)) \, dx = \tfrac{1}{2}\int(\cos 3x + \cos 7x) \, dx$$

$$\text{(Use two substitutions: } u = 3x \text{ and } u = 7x.)$$

$$= \tfrac{1}{2}\left(\tfrac{1}{3}\sin 3x + \tfrac{1}{7}\sin 7x\right) + C = \tfrac{1}{6}\sin 3x + \tfrac{1}{14}\sin 7x + C.$$

7) Evaluate $\int \sin 6x \cos 2x \, dx$.

The form of the integrand is $\sin A \cos B$ where $A = 6x$ and $B = 2x$.

$$\int \sin 6x \cos 2x \, dx$$

$$= \int \tfrac{1}{2}(\sin 4x + \sin 8x) \, dx$$

$$= \tfrac{1}{2}\int \sin 4x \, dx + \tfrac{1}{2}\int \sin 8x \, dx$$

(Two substitutions: $u = 4x$ and $u = 8x$.)

$$= \tfrac{1}{2}\left(\tfrac{-\cos 4x}{4}\right) + \tfrac{1}{2}\left(\tfrac{-\cos 8x}{8}\right) + C$$

$$= -\tfrac{1}{8}\cos 4x - \tfrac{1}{16}\cos 8x + C.$$

8) Evaluate $\int \sin x \sin(-x) \, dx$.

The form of the integrand is $\sin A \sin B$ where $A = x$ and $B = -x$.

$$\int \sin x \sin(-x) \, dx$$

$$= \int \tfrac{1}{2}\left[\cos(x - (-x)) - \cos(x + (-x))\right] dx$$

$$= \int \tfrac{1}{2}(\cos 2x - \cos 0) \, dx$$

$$= \tfrac{1}{2}\int \cos 2x \, dx - \int \tfrac{1}{2} \, dx$$

$$\text{(Let } u = 2x, \, du = 2 \, dx, \, dx = \tfrac{du}{2}.)$$

$$= \tfrac{1}{2}\int \cos u \, \tfrac{du}{2} - \int \tfrac{1}{2} \, dx$$

$$= \tfrac{1}{4}\int \cos u \, du - \int \tfrac{1}{2} \, dx$$

$$= \tfrac{1}{4}\sin u - \tfrac{1}{2}x + C$$

$$= \tfrac{1}{4}\sin 2x - \tfrac{1}{2}x + C.$$

Section 7.3 Trigonometric Substitution

Until now we have been using direct substitutions ($u = \ldots$, $du = \ldots dx$) to simplify an integrand. This section describes a form of "inverse" substitution to evaluate $\int f(x)\, dx$ by replacing x (and dx) by a function of another variable.

Concepts to Master

Evaluate integrals containing $\sqrt{a^2 - x^2}$, $\sqrt{a^2 + x^2}$, and $\sqrt{x^2 - a^2}$

Summary and Focus Questions

Page 478

"Trigonometric substitution" is a technique that may be used for integrals involving one of the radicals $\sqrt{a^2 - x^2}$, $\sqrt{a^2 + x^2}$, or $\sqrt{x^2 - a^2}$. We write the variable x as a trigonometric function of another variable θ, determine dx and the radical in terms of θ, and substitute the results into the integral using one of these:

$$x = a \sin\theta \text{ and } dx = a \cos\theta\, d\theta$$
$$x = a \tan\theta \text{ and } dx = a \sec^2\theta\, d\theta$$
$$x = a \sec\theta \text{ and } dx = a \sec\theta \tan\theta\, d\theta.$$

The resulting integrals are then evaluated by the techniques of Section 7.2.

Here are the choices for the trigonometric functions and substitutions.

Integrand contains	Substitution	Mnemonic triangle
$\sqrt{a^2 - x^2}$	For $-\frac{\pi}{2} \le \theta \le \frac{\pi}{2}$, let $x = a \sin\theta$. $dx = a \cos\theta\, d\theta$ $\sqrt{a^2 - x^2} = a \cos\theta$.	
$\sqrt{a^2 + x^2}$	For $-\frac{\pi}{2} < \theta < \frac{\pi}{2}$, let $x = a \tan\theta$. $dx = a \sec^2\theta\, d\theta$ $\sqrt{a^2 + x^2} = a \sec\theta$.	
$\sqrt{x^2 - a^2}$	For $0 \le \theta < \frac{\pi}{2}$ or $\pi \le \theta < \frac{3\pi}{2}$, let $x = a \sec\theta$. $dx = a \sec\theta \tan\theta\, d\theta$ $\sqrt{a^2 + x^2} = a \sec\theta$.	

The figures describe the substitutions and relationships only when $0 < \theta < \frac{\pi}{2}$ but they will help you remember them. The expressions for $\sqrt{a^2 - x^2}$, $\sqrt{a^2 + x^2}$, $\sqrt{x^2 - a^2}$ are based on the identities $\sin^2\theta + \cos^2\theta = 1$ and $\tan^2\theta = \sec^2\theta - 1$, and are valid only for the values of θ indicated.

Example: Evaluate $\int \sqrt{10 - x^2}\, dx$.

The integrand has the form $\sqrt{a^2 - x^2}$ with $a = \sqrt{10}$.

For $-\dfrac{\pi}{2} \le \theta \le \dfrac{\pi}{2}$,

let $x = \sqrt{10} \sin \theta$. Then

$$dx = \sqrt{10} \cos \theta\, d\theta$$

$$\sqrt{10 - x^2} = \sqrt{10 - (\sqrt{10} \sin \theta)^2} = \sqrt{10}\sqrt{1 - \sin^2 \theta}$$

$$= \sqrt{10}\sqrt{\cos^2 \theta} = \sqrt{10}\,|\cos \theta| = \sqrt{10} \cos \theta \ \left(\text{because } -\tfrac{\pi}{2} \le \theta \le \tfrac{\pi}{2}\right).$$

$$\int \sqrt{10 - x^2}\, dx = \int (\sqrt{10} \cos \theta)\sqrt{10} \cos \theta\, d\theta = 10 \int \cos^2 \theta\, d\theta$$

$$= 10 \int \left(\frac{1 + \cos 2\theta}{2}\right) d\theta = 5 \int (1 + \cos 2\theta)\, d\theta = 5\left(\theta + \frac{1}{2}\sin 2\theta\right) + C$$

$$= 5\theta + \frac{5}{2}\sin 2\theta + C = 5\theta + \frac{5}{2}(2 \sin \theta \cos \theta) + C$$

$$= 5\theta + 5 \sin \theta \cos \theta + C$$

$$= 5 \sin^{-1} \frac{x}{\sqrt{10}} + 5\frac{x}{\sqrt{10}}\frac{\sqrt{10 - x^2}}{\sqrt{10}} + C$$

$$= 5 \sin^{-1} \frac{x}{\sqrt{10}} + \frac{1}{2}x\sqrt{10 - x^2} + C.$$

It may be necessary to first use algebra to transform an integrand before substitution. One technique is completing the square:

Example: Evaluate $\int \dfrac{dx}{\sqrt{x^2 + 4x + 5}}$.

The integrand $\dfrac{1}{\sqrt{x^2 + 4x + 5}}$ does not have the form $\dfrac{1}{\sqrt{1 + u^2}}$, but will after completing the square:

$$x^2 + 4x + 5 = x^2 + 4x + 4 + 5 - 4 = (x + 2)^2 + 1.$$

Thus, $\displaystyle \int \frac{dx}{\sqrt{x^2 + 4x + 5}} = \int \frac{dx}{\sqrt{1 + (x + 2)^2}}$.

Let $u = x + 2$. Then $du = dx$ and $\displaystyle \int \frac{dx}{\sqrt{1 + (x + 2)^2}} = \int \frac{1}{\sqrt{1 + u^2}}\, du$.

We now use a trigonometric substitution:

For $-\dfrac{\pi}{2} < \theta < \dfrac{\pi}{2}$,

let $x = \tan \theta$. Then

$$dx = \sec^2 \theta\, d\theta$$

$$\sqrt{1 + x^2} = \sec \theta \ \left(\text{because } -\tfrac{\pi}{2} < \theta < \tfrac{\pi}{2}\right).$$

$$\int \frac{1}{\sqrt{1 + u^2}}\, du = \int \frac{1}{\sec \theta} \sec^2 \theta\, d\theta = \int \sec \theta\, d\theta = \ln|\sec \theta + \tan \theta| + C$$

$$= \ln\left|\sqrt{1 + u^2} + u\right| + C = \ln\left|\sqrt{x^2 + 4x + 5} + x + 2\right| + C.$$

1) Rewrite each of the following using a trigonometric substitution:

a) $\int x^2 \sqrt{4 - x^2}\, dx$

$\frac{x}{2} = \sin \theta$

$\frac{\sqrt{4 - x^2}}{2} = \cos \theta$

Let $x = 2 \sin \theta$ for $-\frac{\pi}{2} \le \theta \le \frac{\pi}{2}$. Then

$dx = 2 \cos \theta\, d\theta$ and

$\sqrt{4 - x^2} = 2 \cos \theta$.

$\int x^2 \sqrt{4 - x^2}\, dx$

$= \int (2 \sin \theta)^2 (2 \cos \theta)(2 \cos \theta\, d\theta)$

$= 16 \int \sin^2 \theta \cos^2 \theta\, d\theta.$

b) $\int x^2 \sqrt{x^2 - 4}\, dx$

$\frac{x}{2} = \sec \theta$

$\frac{\sqrt{x^2 - 4}}{2} = \tan \theta$

Let $x = 2 \sec \theta$ for $0 \le \theta < \frac{\pi}{2}$ or

$\pi \le \theta < \frac{3\pi}{2}$. Then $dx = 2 \sec \theta \tan \theta\, d\theta$ and

$\sqrt{x^2 - 4} = 2 \tan \theta$.

$\int x^2 \sqrt{x^2 - 4}\, dx$

$= \int (2 \sec \theta)^2 (2 \tan \theta)(2 \sec \theta \tan \theta\, d\theta)$

$= 16 \int \sec^3 \theta \tan^2 \theta\, d\theta.$

c) $\int \frac{x + 1}{\sqrt{4 + x^2}}\, dx$

$\frac{x}{2} = \tan \theta$

$\frac{\sqrt{4 + x^2}}{2} = \sec \theta$

Let $x = 2 \tan \theta$ for $-\frac{\pi}{2} < \theta < \frac{\pi}{2}$. Then

$dx = 2 \sec^2 \theta\, d\theta$ and

$\sqrt{4 + x^2} = 2 \sec \theta$.

$\int \frac{x + 1}{\sqrt{4 + x^2}}\, dx$

$= \int \frac{2 \tan \theta + 1}{2 \sec \theta} \cdot 2 \sec^2 \theta\, d\theta$

$= \int (2 \tan \theta + 1) \sec \theta\, d\theta$

$= \int 2 \tan \theta \sec \theta\, d\theta + \int \sec \theta\, d\theta.$

2) $\int$ Evaluate $\dfrac{x}{1 + x^2}\, dx$.

While a trigonometric substitution will lead you to a solution, a direct substitution is all that is needed:
$$u = 1 + x^2,\ du = 2x\, dx$$

$$\int \frac{x}{1 + x^2}\, dx = \frac{1}{2}\int u^{-1}\, du = \frac{1}{2}\ln|u| + C$$

$$= \frac{1}{2}\ln(1 + x^2) + C = \ln\sqrt{1 + x^2} + C.$$

3) Rewrite $\int \dfrac{x^2}{\sqrt{13 - 6x + x^2}}\, dx$

as a trigonometric integral.

$\sqrt{13 - 6x + x^2}$ does not look friendly until we complete the square:
$$13 - 6x + x^2 = 13 + (x^2 - 6x + 9 - 9)$$
$$= x^2 - 6x + 9 + 4$$
$$= (x - 3)^2 + 4.$$
Let $u = x - 3$. Then $du = dx$, $x = u + 3$, and $13 - 6x + x^2 = u^2 + 4$.

Then $\displaystyle\int \frac{x^2}{\sqrt{13 - 6x + x^2}}\, dx = \int \frac{(u + 3)^2}{\sqrt{u^2 + 4}}\, du$.

The integral is now ready for the trigonometric substitution
$u = 2\tan\theta$ for $-\dfrac{\pi}{2} < \theta < \dfrac{\pi}{2}$. Then
$du = 2\sec^2\theta\, d\theta$.

$\dfrac{u}{2} = \tan\theta$

$\dfrac{\sqrt{u^2 + 4}}{2} = \sec\theta$

$$\int \frac{(u + 3)^2}{\sqrt{u^2 + 4}}\, du = \int \frac{(2\tan\theta + 3)^2}{2\sec\theta}\, (2\sec^2\theta)\, d\theta$$

$$= \int (4\tan^2\theta + 12\tan\theta + 9)\sec\theta\, d\theta$$

$$= \int (4(\sec^2\theta - 1) + 12\tan\theta + 9)\sec\theta\, d\theta$$

$$= \int (4\sec^3\theta + 12\sec\theta\tan\theta + 5\sec\theta)\, d\theta.$$

Section 7.4   Integration of Rational Functions by Partial Fractions

This section shows you how to rewrite a rational function so that it may be integrated. There are no new calculus concepts in this section—only algebraic techniques and substitutions for rewriting the integrands.

Concepts to Master

A. Evaluate integrals containing rational functions by partial fractions; Method of determining coefficients

B. Evaluate integrals using rationalizing substitutions

C. Technology Plus

Summary and Focus Questions

Page 484

A. A **rational function** has the form $\frac{P(x)}{Q(x)}$ where $P(x)$ and $Q(x)$ are polynomials. The **method of partial fractions** for evaluating $\int \frac{P(x)}{Q(x)} \, dx$ involves writing the integrand $\frac{P(x)}{Q(x)}$ as a sum of rational functions each of which you are able to integrate. For example, later we will see that the rational function

$$f(x) = \frac{x^4 + 1}{x^2 - 1} \text{ may be written}$$

$$\frac{x^4 + 1}{x^2 - 1} = x^2 + 1 + \frac{1}{x - 1} + \frac{-1}{x + 1}.$$

We know how to integrate each of the four terms on the right.

Here are the steps in the method of partial fractions.

Step 1: If the degree of $P(x)$ is greater than or equal to the degree of $Q(x)$, use polynomial long division to write $\frac{P(x)}{Q(x)} = H(x) + \frac{R(x)}{Q(x)}$, where the degree of $R(x)$ is less than the degree of $Q(x)$.

Step 2: Write $Q(x)$ in factored form as a product of powers of distinct linear $(x - a)$ and irreducible (not factorable) quadratics $(ax^2 + bx + c)$.

Step 3: Write $\frac{R(x)}{Q(x)} = T_1(x) + T_2(x) + \dots + T_n(x)$

where the $T_i(x)$ are terms corresponding to the distinct factors of $Q(x)$ found in Step 2. The forms of $T_i(x)$ are given in the table on the next page.

Type of $Q(x)$ factor	Corresponding $T_i(x)$ term
$x - a$	$\dfrac{A}{x-a}$
$(x-a)^k$	$\dfrac{A_1}{x-a} + \dfrac{A_2}{(x-a)^2} + \cdots + \dfrac{A_k}{(x-a)^k}$
$ax^2 + bx + c$	$\dfrac{Ax+B}{ax^2+bx+c}$
$(ax^2+bx+c)^k$	$\dfrac{A_1 x + B_1}{ax^2+bx+c} + \dfrac{A_2 x + B_2}{(ax^2+bx+c)^2} + \cdots + \dfrac{A_k x + B_k}{(ax^2+bx+c)^k}$

Step 4: Write $R(x) = Q(x)[T_1(x) + T_2(x) + \ldots + T_n(x)]$. Then multiply out the right side, combining like terms.

Step 5: Equate coefficients of like terms in the equation in Step 4. This gives a system of linear equations.

Step 6: Solve the linear system from Step 5. This is frequently done by solving for one variable in one equation and using the result to eliminate that variable in the other equations.

Finally, write the integrand as a sum of terms that may be evaluated:

$$\frac{P(x)}{Q(x)} = H(x) + T_1(x) + T_2(x) + \ldots + T_n(x).$$

Shortcut: Sometimes some of the unknown coefficients can be found after Step 3 by judicious substitution of values for x. The rest of the coefficients may be found by continuing with Steps 4, 5, and 6.

Example: Evaluate: $\displaystyle\int \frac{x^4 + 1}{x^2 - 1}\, dx$.

Step 1: Since the degree of the numerator (4) is greater than the degree of the denominator (2), we use long division of polynomials:

$$
\begin{array}{r}
x^2 + 1 \\
x^2 - 1 \overline{\smash{\big)}\, x^4 + 1} \\
\underline{x^4 - x^2} \\
x^2 + 1 \\
\underline{x^2 - 1} \\
2
\end{array}
$$

Thus, $\dfrac{x^4 + 1}{x^2 - 1} = x^2 + 1 + \dfrac{2}{x^2 - 1}$.

$\displaystyle\int (x^2 + 1)\, dx = \frac{x^3}{3} + x + C.$ Use partial fractions to find $\displaystyle\int \frac{2}{x^2 - 1}\, dx$.

Step 2: $x^2 - 1 = (x - 1)(x + 1)$.

Step 3: $\dfrac{2}{(x-1)(x+1)} = \dfrac{A}{x-1} + \dfrac{B}{x+1}$.

Step 4: Multiply both sides in Step 3 by $(x - 1)(x + 1)$ to get
$$2 = A(x + 1) + B(x - 1).$$

Steps 5 and 6: At this point we could multiply out the terms and obtain two equations in A and B, but it is simpler to substitute judicious choices for x:

For $x = 1$: $A(2) + 0 = 2$. Therefore, $A = 1$.

For $x = -1$: $0 + B(-2) = 2$. Therefore, $B = -1$.

Thus, $\dfrac{2}{(x-1)(x+1)} = \dfrac{1}{x-1} + \dfrac{-1}{x+1}$.

Finally, $\displaystyle\int \frac{2}{x^2-1}\,dx = \int\left(\frac{1}{x-1} + \frac{-1}{x+1}\right)dx = \int \frac{1}{x-1}\,dx - \int \frac{1}{x+1}\,dx$

$$\text{(Two substitutions: } u = x - 1 \text{ and } u = x + 1.)$$

$$= \ln|x-1| - \ln|x+1| + C = \ln\left|\frac{x-1}{x+1}\right| + C.$$

Therefore, $\displaystyle\int \frac{x^4+1}{x^2-1}\,dx = \frac{x^3}{3} + x + \ln\left|\frac{x-1}{x+1}\right| + C.$

Example: Evaluate: $\displaystyle\int \frac{x^3 + 5x^2 - 4x + 4}{(x^2 - 2x + 1)(x^2 + x + 1)}\,dx.$

Step 1: Because the degree of the numerator is less than the degree of the denominator, skip this step.

Step 2: $Q(x) = (x^2 - 2x + 1)(x^2 + x + 1) = (x - 1)^2(x^2 + x + 1)$. Note that $x^2 + x + 1$ is cannot be factored using real numbers.

Step 3: $\dfrac{x^3 + 5x^2 - 4x + 4}{(x-1)^2(x^2+x+1)} = \dfrac{A}{(x-1)} + \dfrac{B}{(x-1)^2} + \dfrac{Cx + D}{x^2+x+1}.$

Step 4: Multiply by the denominator $Q(x)$:

$x^3 + 5x^2 - 4x + 4$

$\qquad = A(x-1)(x^2 + x + 1) + B(x^2 + x + 1) + (Cx + D)(x - 1)^2.$

Multiply out the right side:

$x^3 + 5x^2 - 4x + 4$

$\qquad = Ax^3 - A + Bx^2 + Bx + B + Cx^3 - 2Cx^2 + Cx + Dx^2 - 2Dx + D$

and collect like terms:

$x^3 + 5x^2 - 4x + 4$

$\qquad = (A + C)x^3 + (B - 2C + D)x^2 + (B + C - 2D)x + (-A + B + D).$

Step 5: Equate coefficients from the equality in Step 4:

$\quad 1 = A + C$

$\quad 5 = B - 2C + D$

$-4 = B + C - 2D$

$\quad 4 = -A + B + D$

Step 6: Solve the system.

From the first equation, $A = 1 - C$ and thus the others become:

$5 = B - 2C + D$	$5 = B - 2C + D$
$-4 = B + C - 2D$	or $\qquad -4 = B + C - 2D$
$4 = -(1 - C) + B + D$	$5 = B + C + D$

Subtracting the second equation from the third yields $9 = 3D$, or $D = 3$. From the first equation, using $C = 0$ and $D = 3$, $5 = B - 2(0) + 3$, or $B = 2$. Finally, from $A = 1 - C$, $A = 1$.

Therefore, $\dfrac{x^3 + 5x^2 - 4x + 4}{(x^2 - 2x + 1)(x^2 + x + 1)} = \dfrac{1}{x - 1} + \dfrac{2}{(x - 1)^2} + \dfrac{3}{x^2 + x + 1}.$

The original integral may be written as the sum of three integrals, each of which we know how to evaluate:

$$\int \dfrac{x^3 + 5x^2 - 4x + 4}{(x^2 - 2x + 1)(x^2 + x + 1)}\,dx = \int \dfrac{1}{x - 1}\,dx + \int \dfrac{2}{(x - 1)^2}\,dx + \int \dfrac{3}{x^2 + x + 1}\,dx.$$

$\displaystyle\int \dfrac{1}{x - 1}\,dx = \ln|x - 1| + C$ (substitute $u = x - 1$).

$\displaystyle\int \dfrac{2}{(x - 1)^2}\,dx = \dfrac{-2}{x - 1} + C$ (substitute $u = x - 1$).

$\displaystyle\int \dfrac{3}{x^2 + x + 1}\,dx = \int \dfrac{3}{(x + 1/2)^2 + 3/4}\,dx = 4\int \dfrac{1}{\left[\frac{2}{\sqrt{3}}\left(x + \frac{1}{2}\right)\right]^2 + 1}\,dx$

$$= 2\sqrt{3}\,\tan^{-1}\!\left[\dfrac{2x + 1}{\sqrt{3}}\right] + C.$$

Thus $\displaystyle\int \dfrac{x^3 + 5x^2 - 4x + 4}{(x^2 - 2x + 1)(x^2 + x + 1)}\,dx$

$$= \ln|x - 1| - \dfrac{2}{x - 1} + 2\sqrt{3}\,\tan^{-1}\!\left[\dfrac{2x + 1}{\sqrt{3}}\right] + C.$$

In this example we could have found $B = 2$ after step 4 by setting $x = 1$. The rest of the procedure would be followed to find A, C, and D.

1) Write the partial fraction form for each.

a) $\dfrac{2x + 1}{x(x + 8)}$

$\dfrac{A}{x} + \dfrac{B}{x + 8}.$

b) $\dfrac{4x^2 + 11}{(x + 1)^2(x^2 - 3x + 5)}$

$\dfrac{A}{x + 1} + \dfrac{B}{(x + 1)^2} + \dfrac{Cx + D}{x^2 - 3x + 5}.$

c) $\dfrac{3}{x(x^2 + 2x + 2)^2}$

$\dfrac{A}{x} + \dfrac{Bx + C}{x^2 + 2x + 2} + \dfrac{Dx + E}{(x^2 + 2x + 2)^2}.$

d) $\dfrac{x^3}{x^2 + 2x + 1}$

First use long division:

$$\begin{array}{r}
x - 2 \\
x^2 + 2x + 1\,\overline{\smash{\big)}\,x^3 } \\
\underline{x^3 + 2x^2 + x} \\
-2x^2 - x \\
\underline{-2x^2 - 4x - 2} \\
3x + 2
\end{array}$$

Therefore, $\dfrac{x^3}{x^2 + 2x + 1}$

$= x - 2 + \dfrac{3x + 2}{x^2 + 2x + 1} = x - 2 + \dfrac{3x + 2}{(x + 1)^2}.$

$= x - 2 + \dfrac{A}{x + 1} + \dfrac{B}{(x + 1)^2}.$

2) Evaluate $\int \frac{6 - x}{x(x + 3)} \, dx$.

Step 1: Not needed.

Step 2: $Q(x) = x(x + 3)$.

Step 3: $\frac{6 - x}{x(x + 3)} = \frac{A}{x} + \frac{B}{x + 3}$.

Step 4: Clear fractions:

$6 - x = A(x + 3) + Bx$.

Combine terms:

$6 - x = (A + B)x + 3A$.

Step 5: Equate coefficients:

$A + B = -1$

$3A = 6$

Step 6: Solving, we see $A = 2$ from the second equation. Thus $2 + B = -1, B = -3$.

Therefore, $\int \frac{6 - x}{x(x + 3)} \, dx = \int \left(\frac{2}{x} + \frac{-3}{x + 3} \right) dx$

$= 2 \ln |x| - 3 \ln |x + 3| + C$.

Note: At Step 4, after clearing fractions to get $6 - x = A(x + 3) + Bx$ we could substitute $x = 0$ to get

$6 = A(0 + 3) + 0$

$A = 2$,

and we could substitute $x = -3$ to get

$6 - (-3) = A(0) + B(-3)$

$9 = -3B$.

$B = -3$.

B. Sometimes the substitution $u = \sqrt[n]{g(x)}$ in the form $u^n = g(x)$ will transform an integrand involving radicals into a rational function.

Page 492

Example: Evaluate $\int \frac{x}{\sqrt{x - 2}} \, dx$.

Let $u = \sqrt{x - 2}$. Then $u^2 = x - 2, x = u^2 + 2$, and $dx = 2u \, du$.

$\int \frac{x}{\sqrt{x - 2}} \, dx = \int \frac{u^2 + 2}{u} \, 2u \, du = 2 \int (u^2 + 2) \, du = \frac{2u^3}{3} + 4u + C$

$= \frac{2}{3}(x - 2)^{3/2} + 4\sqrt{x - 2} + C$.

3) Evaluate $\int \dfrac{\sqrt[3]{x}}{\sqrt[3]{x}+1}\,dx$.

Let $u = \sqrt[3]{x}$. Then $x = u^3$ and $dx = 3u^2\,du$.

The integral becomes

$$\int \frac{u}{u+1}\,3u^2\,du = 3\int \frac{u^3}{u+1}\,du.$$

By long division

$$\frac{u^3}{u+1} = u^2 - u + 1 + \frac{-1}{u+1}.$$

$$3\int \frac{u^3}{u+1}\,du = 3\int \left(u^2 - u + 1 - \frac{1}{u+1}\right)du$$

$$= 3\left(\frac{u^3}{3} - \frac{u^2}{2} + u - \ln|u+1|\right) + C.$$

Since $u = \sqrt[3]{x}$, $\displaystyle\int \frac{\sqrt[3]{x}}{\sqrt[3]{x}+1}\,dx$

$$= 3\left(\frac{x}{3} - \frac{x^{2/3}}{2} + \sqrt[3]{x} - \ln\left|\sqrt[3]{x}+1\right|\right) + C.$$

4) Evaluate $\int \dfrac{\sqrt{x}}{1+\sqrt[3]{x}}\,dx$.

Choose $u = \sqrt[6]{x}$ because then $\sqrt[3]{x} = u^2$
and $\sqrt{x} = u^3$ are both integer powers of u.
(Note: 6 is the least common multiple of
2 and 3.)
Then $x = u^6$ and $dx = 6u^5\,du$. Using this
substitution and partial fractions,

$$\int \frac{\sqrt{x}}{1+\sqrt[3]{x}}\,dx = \int \frac{u^3}{1+u^2}(6u^5)\,du$$

$$= 6\int \frac{u^8}{1+u^2}\,du = 6\int \left(u^6 - u^4 + u^2 - 1 + \frac{1}{1+u^2}\right)du$$

$$= \frac{6}{7}x^{7/6} - \frac{6}{5}x^{5/6} + \frac{6}{3}x^{3/6} - 6x^{1/6} + 6\tan^{-1}x^{1/6} + C$$

$$= \frac{6}{7}\sqrt[6]{x^7} - \frac{6}{5}\sqrt[6]{x^5} + 2\sqrt{x} - 6\sqrt[6]{x} + 6\tan^{-1}\sqrt[6]{x} + C.$$

C. Technology Plus. Use a computer algebra system or a graphing calculator to
solve.

T-1) Use a CAS to find the partial fraction
decomposition of

$$\frac{12x^3}{x^6 + x^5 + x^4 - x^2 - x - 1}.$$

Using either a Texas Instruments calculator or
Maple, the expansion is

$$\frac{-4x - 8}{x^2 + x + 1} + \frac{6}{x^2 + 1} + \frac{3}{x+1} + \frac{1}{x-1}.$$

Section 7.5 Strategy for Integration

There are no new integration techniques in this section. Its purpose is to draw together all the techniques covered so far. In this section you will also encounter some rather simple looking functions which cannot be evaluated by any of the techniques that you have seen so far.

Concepts to Master

A. Review all the integration techniques

B. Functions that are not readily integrable

Summary and Focus Questions

Page 494

A. You should know well all twenty integration formulas listed at the beginning of this section of your textbook. Every integral we have seen so far can be solved using one or more of the twenty basic ones. To approach an integration problem you should

   i) Look for ways to simplify or rewrite the integrand.
   ii) Look for a substitution of the variable of integration.
   iii) Use integration by parts.

Here is a table of the basic strategies, with examples:

Method	Example	First step
i) Rewrite the integrand		
Algebra	$\int x^2 \left(1 + \dfrac{2}{\sqrt{x}}\right) dx$	$= \int \left(x^2 + 2x^{\frac{3}{2}}\right) dx$
Trigonometric Identity	$\int \sin^3 x \cos^2 x \, dx$	$= \int \sin^2 x \cos^2 x \sin x \, dx$
		$= \int (1 - \cos^2 x)\cos^2 x \sin x \, dx.\; u = \cos x \ldots$
Partial Fractions	$\int \dfrac{5}{(x+2)(x-3)} \, dx$	$= \int \left(\dfrac{-1}{x+2} + \dfrac{1}{x-3}\right) dx = \ldots$
ii) Substitution		
Direct	$\int \dfrac{x^2}{x^3+1} \, dx$	$u = x^3 + 1,\; du = 3x^2 \, dx \ldots$
Trigonometric	$\int \dfrac{x^2}{\sqrt{4-x^2}} dx$	$x = 2 \sin \theta,\; dx = 2 \cos \theta \, d\theta \ldots$
Rationalizing	$\int \dfrac{x}{\sqrt{x}+1} dx$	$u = \sqrt{x},\; u^2 = x,\; dx = 2u \, du, \ldots$
iii) Integration by Parts	$\int x^3 \ln x \, dx$	$u = \ln x,\; dv = x^3 \, dx, \ldots$

1) Indicate the technique to evaluate each:

a) $\int \dfrac{e^x}{e^x - 1}\, dx$

Direct substitution, $u = e^x - 1$.

b) $\int x^2 \sqrt{16 + x^2}\, dx$

Trigonometric substitution, $x = 4 \tan \theta$.

c) $\int \dfrac{e^x - 1}{e^x}\, dx$

Algebra, $\int \dfrac{e^x - 1}{e^x}\, dx = \int\left(1 - \dfrac{1}{e^x}\right)\, dx = \int\left(1 - e^{-x}\right)\, dx.$
Now use the substitution $u = -x$.

d) $\int \dfrac{2x + 1}{x^3(x + 1)}\, dx$

Partial fractions, $\dfrac{A}{x} + \dfrac{B}{x^2} + \dfrac{C}{x^3} + \dfrac{D}{x + 1}.$

e) $\int \tan^3 x \sec^4 x\, dx$

Trigonometric integral,
$$\int \tan^3 x \sec^2 x\, (\sec^2 x)\, dx$$
$$= \int \tan^3 x\, (\tan^2 x + 1)(\sec^2 x)\, dx.$$

f) $\int x \sec^2 x\, dx$

Integration by parts,
$u = x,\ dv = \sec^2 x\, dx.$

g) $\int \dfrac{\csc^2 x}{1 + \cot^2 x}\, dx$

Since $1 + \cot^2 x = \csc^2 x$, this one reduces to
$$\int 1\, dx = x + C.$$

**Page
498**

B. Every function you have encountered so far is **elementary,** not because it is necessarily easy, but because it is possible to express it as a combination (addition, subtraction, multiplication, division, composition) of polynomial, rational, exponential, logarithmic, trigonometric, inverse trigonometric, hyperbolic, and inverse hyperbolic functions.

The *derivative* of an elementary function is elementary, but the *antiderivative* of an elementary function need not be elementary.

Thus, there are some elementary functions for which there is no simple form for the antiderivative. $\int \dfrac{1}{\ln x}\, dx$ is one such integral. Chapter 11 discusses a method to evaluate these.

2) True or False:

a) $f(x) = e^{x^{x^x}}$ is elementary.

True.

b) If $f(x)$ is not elementary, then $\int f(x)\, dx$ is not elementary.

True.

c) If $f(x)$ is elementary, then $\int f(x)\, dx$ is elementary.

False.

3) Evaluate $\int \cos x^2\, dx.$

We have no techniques for this one. See Chapter 11.

Section 7.6 Using Tables of Integrals and Computer Algebra Systems

This section covers two common tools to assist in performing integration. To successfully use a table of integrals or a computer algebra system you need an understanding of the techniques of the previous sections.

Concepts to Master

A. Evaluate integrals using a table of integrals

B. Evaluate integrals using a computer algebra system

Summary and Focus Questions

Page
500

A. There is a table of 120 integration formulas on the inside back cover of your textbook. Most of these formulas are derived using the techniques you have already studied. The integrals are described with variable u and constants a, b, etc., and are grouped by form of the integrand.

Sometimes you must use algebra to adjust the integrand before determining the appropriate form. In a few cases, such as integral formula #58, the formula is a "reduction" formula that replaces the given integral by an expression containing another integral.

1) What integral formula from the Table of Integrals in the text may be used for each?

a) $\int \dfrac{dx}{x^2(4 + 5x)}$

#50, $a = 4$, $b = 5$.

b) $\int \dfrac{\sqrt{4x^2 - 9}}{x} \, dx$

#41, $u = 2x$, $a = 3$

c) $\int \dfrac{1}{5x^2 - 3} \, dx$

#20, $u = \sqrt{5}x$, $a = \sqrt{3}$

d) $\int \dfrac{\sqrt{25 + 4x^2}}{x} \, dx$

#23, $u = 2x$, $a = 5$

e) $\int 3x \cos^{-1} 2x \, dx$

#91, $u = 2x$

B. Computer algebra system (CAS) programs for computers and calculators are great tools for the symbolic manipulation involved in finding antiderivatives. Here are some cautions to observe when using a CAS:

i) There is usually no constant of integration displayed (answers to $\int 2x\,dx$ may appear as x^2 or $x^2 + 5$ rather than $x^2 + C$).

ii) An answer you obtain by hand may not look like the one produced by the CAS, yet they are the same. You may need to apply algebra and various identities to transform one to another. Here are a few common differences:

a) For an integral like $\int (2x + 5)^4\,dx$, some systems will expand an integrand into a polynomial rather than use the direct substitution $u = 2x + 5$, $du = 2\,dx$.

b) Fractional answers and constants are often grouped differently or factored out. For example, $\frac{1}{2}x + \frac{1}{4x}$ may be expressed as $\frac{2x^2 + 1}{4x}$.

c) Hyperbolic trigonometric functions are usually expressed with logarithms. You will often see $\ln(x + \sqrt{x^2 + 1})$ rather than $\sinh^{-1} x$.

2) Use a CAS to evaluate the integrals in question 1.

Here are answers using a Texas Instruments calculator:

a) $\dfrac{5x \ln |5x + 4| - 5x \ln |x| - 4}{16x}$.

b) $-\left(3 \tan^{-1} \dfrac{\sqrt{4x^2 - 9}}{3} - \sqrt{4x^2 - 9}\right)$.

c) $\dfrac{-\sqrt{15} \ln\left(\dfrac{|\sqrt{15x} + 3|}{|\sqrt{15x} - 3|}\right)}{30}$.

d) $\dfrac{10 \ln(\sqrt{4x^2 + 25} - 5) - 5 \ln x^2 + 2\sqrt{4x^2 + 25}}{2}$.

e) $\dfrac{-3\left((8x^2 - 1)\sin^{-1}(2x) + 2x(\sqrt{-(4x^2 - 1)} - 2\pi x)\right)}{16}$.

Here are answers using Maple:

a) $\dfrac{5}{16}\ln(4 + 5x) - \dfrac{5}{16}\ln(x) - \dfrac{1}{4}\dfrac{1}{x}$.

b) $\sqrt{4x^2 - 9} + 3 \arctan\left(\dfrac{3}{\sqrt{4x^2 - 9}}\right)$.

c) $-\dfrac{1}{15}\sqrt{15} \arctan h\left(\dfrac{1}{3}x\sqrt{15}\right)$.

d) $\sqrt{25 + 4x^2} - 5 \arctan h\left(\dfrac{5}{\sqrt{25 + 4x^2}}\right)$.

e) $\dfrac{3}{2}x^2 \arccos(2x) - \dfrac{3}{8}x\sqrt{1 - 4x^2} + \dfrac{3}{16}\arcsin(2x)$.

3) Use a CAS to find each.

a) $\int \dfrac{\sqrt{4x^2 - 1}}{x^2}\,dx$

Using a Texas Instruments calculator gives

$$\dfrac{2x\ln\left(\sqrt{4x^2 - 1} + 2x\right) - \sqrt{4x^2 - 1}}{x} + C.$$

Maple's answer is

$$\dfrac{(4x^2 - 1)^{\frac{3}{2}}}{x} - 4x\sqrt{4x^2 - 1} - \ln(x\sqrt{4} + \sqrt{4x^2 - 1})\sqrt{4},$$

which is the same.

b) $\int \dfrac{e^x}{x}\,dx$

The CAS will probably return something like "$\int \dfrac{e^x}{x}\,dx$," which means the integral is not elementary.

Section 7.7   Approximate Integration

We have seen that one way to approximate the value of a definite integral that represents an area is to cover the area with approximating rectangles and compute the corresponding Riemann sum. This section considers several ways to create sums that approximate the value of a given integral.

Concepts to Master

A. Approximate a definite integral using the Left Endpoint Rule, the Right Endpoint Rule, the Midpoint Rule, the Trapezoidal Rule, and Simpson's Rule

B. Estimate the maximum error

C. Technology Plus

Summary and Focus Questions

Page 506

A. Suppose f is an integrable function on a closed interval $[a, b]$ which has been partitioned into n equal subintervals of length $\Delta x = \dfrac{b-a}{n}$, with $x_i = a + i\Delta x$. In the figure at the right, $n = 4$. There are five rules in this section, three of which we have seen before, to approximate the value of $\displaystyle\int_a^b f(x)\, dx$.

Each rule replaces portions of the graph of $y = f(x)$ with lines segments or parabolas and computes the sum of the areas under the replacements. See the strategy and graph for $n = 4$ for each rule.

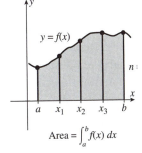

Area $= \displaystyle\int_a^b f(x)\, dx$

Left Endpoint Rule
Strategy: Replace each portion of the graph with a horizontal line segment through the point on the graph corresponding to the left endpoint of the subinterval. Sum the areas of the rectangles.

Each $x_i^* = x_{i-1}$, the left endpoint.

Approximating Sum:

$$L_n = \Delta x[f(a) + f(x_1) + f(x_2) + \dots + f(x_{n-1})].$$

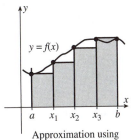

Approximation using
Left Endpoint Rule, $n = 4$

Right Endpoint Rule

Strategy: Replace each portion of the graph with a horizontal line segment through the point on the graph corresponding to the right endpoint of the subinterval. Sum the areas of the rectangles.

Each $x_i^* = x_i$, the right endpoint.

Approximating Sum:

$$R_n = \Delta x[f(x_i) + f(x_2) + \dots + f(x_{n-1}) + f(b)].$$

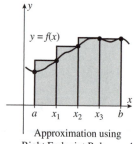

Approximation using
Right Endpoint Rule, $n = 4$

Midpoint Rule

Strategy: Replace each portion of the graph with a horizontal line segment through the point on the graph corresponding to the midpoint of the subinterval. Sum the areas of the rectangles.

Each $\overline{x}_i = \dfrac{x_{i-1} + x_i}{2}$, the midpoint.

Approximating Sum:

$$M_n = \Delta x[f(\overline{x_1}) + f(\overline{x_2}) + \dots + f(\overline{x_n})].$$

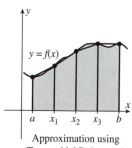

Approximation using
Midpoint Rule, $n = 4$

Trapezoidal Rule

Strategy: Replace each portion of the graph with a line segment through the points on the graph corresponding to both endpoints of the subinterval. Sum the areas of the trapezoids.
Each $x_i = a + i\Delta x, i = 0, 1, 2, \dots, n$.

Approximating Sum:

$$T_n = \frac{\Delta x}{2}[f(a) + 2f(x_1) + 2f(x_2) + \dots + 2f(x_{n-1}) + f(b)].$$

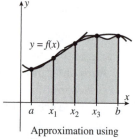

Approximation using
Trapezoidal Rule, $n = 4$

Simpson's Rule

Strategy: Replace pairs of portions of the graph with parabolas through the points on the graph corresponding to the endpoints of the pair of adjacent subintervals. Note: n must be even. Sum the areas of the regions under the parabolas.
Each $x_i = a + i\Delta x, i = 0, 1, 2, \dots, n$.

Approximating Sum:

Approximation using
Simpson's Rule, $n = 4$

$$S_n = \frac{\Delta x}{3}[f(a) + 4f(x_1) + 2f(x_2) + 4f(x_3) + \dots + 4f(x_{n-1}) + f(b)].$$

1) For each rule, how many terms are in the sum for n subintervals?

      Left Endpoint Rule _____

      Right Endpoint Rule _____

      Midpoint Rule _____

      Trapezoidal Rule _____

      Simpson's Rule _____

n.

n.

n.

$n + 1$.

$n + 1$.

2) Estimate $\int_{-1}^{5} 2^x \, dx$ to three decimal places with $n = 6$ subintervals using:

    a) the Left Endpoint Rule.

$$\Delta x = \frac{5 - (-1)}{6} = 1.$$

i	$x_i{}^*$	$f(x_i{}^*)$
1	−1	0.5
2	0	1
3	1	2
4	2	4
5	3	8
6	4	16

The approximation is
$1(0.5 + 1 + 2 + 4 + 8 + 16) = 31.5.$

    b) the Right Endpoint Rule.

i	$x_i{}^*$	$f(x_i{}^*)$
1	0	1
2	1	2
3	2	4
4	3	8
5	4	16
6	5	32

The approximation is
$1(1 + 2 + 4 + 8 + 16 + 32) = 63.$

    c) the Midpoint Rule.

i	$\overline{x}_i$	$f(\overline{x}_i)$
1	−0.5	.707
2	0.5	1.414
3	1.5	2.828
4	2.5	5.657
5	3.5	11.314
6	4.5	22.627

The approximation is
$1(.707 + 1.414 + 2.828 + 5.657$
$\qquad + 11.314 + 22.627) = 44.547.$

d) the Trapezoidal Rule.

i	x_i	$f(x_i)$	term
0	−1	0.5	0.5
1	0	1	2(1) = 2
2	1	2	2(2) = 4
3	2	4	2(4) = 8
4	3	8	2(8) = 16
5	4	16	2(16) = 32
6	5	32	32

The approximation is

$\frac{1}{2}(0.5 + 2 + 4 + 8 + 16 + 32 + 32)$
$= 47.250.$

e) Simpson's Rule.

i	x_i	$f(x_i)$	term
0	−1	0.5	0.5
1	0	1	4(1) = 4
2	1	2	2(2) = 4
3	2	4	4(4) = 16
4	3	8	2(8) = 16
5	4	16	4(16) = 64
6	5	32	32

The approximation is

$\frac{1}{3}(0.5 + 4 + 4 + 16 + 16 + 64 + 32)$
$= 45.500.$

$\left(\text{The actual value of } \displaystyle\int_{-1}^{5} 2^x \, dx \text{ is} \right.$
$\left. \dfrac{63}{\ln 4} \approx 45.445. \right)$

3) For $\displaystyle\int_{1}^{35} (40 - x) \, dx$ and any n, the Midpoint approximation will be:

a) too large

b) too small

c) exact

c) exact, because $y = 40 - x$ is linear. The area of the midpoint rectangle will equal the area under $y = 40 - x$.

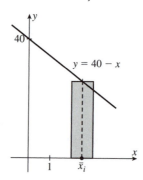

4) For $\int_1^6 (40 - x^2)\, dx$ and any n, the Trapezoidal Rule approximation will be:

   a) too large

   b) too small

   c) exact

b) too small. Since the graph of $y = 40 - x^2$ is concave downward for $1 \le x \le 6$, all inscribed trapezoids will have area less than the areas under $y = 40 - x^2$.

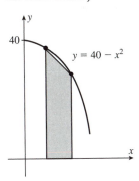

5) For $\int_1^6 (40 - x^2)\, dx$ and any even n, Simpson's Rule approximation will be:

   a) too large

   b) too small

   c) exact

c) exact, because Simpson's Rule replaces the function with portions of approximating parabolas and since the graph of $y = 40 - x^2$ is a parabola, the approximation is exact.

6) The distances across a small pond were measured at 10 m intervals and are given in the figure. Use the Trapezoidal Rule to estimate the area covered by the pond.

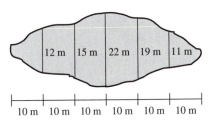

| 12 m | 15 m | 22 m | 19 m | 11 m |

| 10 m | 10 m | 10 m | 10 m | 10 m | 10 m |

$\Delta x = 10$ m.

i	dist	term
0	0	0
1	12	2(12) = 24
2	15	2(15) = 30
3	22	2(22) = 44
4	19	2(19) = 38
5	11	2(11) = 22
6	0	0

The pond area is approximately

$$\frac{10}{2}(0 + 24 + 30 + 44 + 38 + 22 + 0)$$
$$= 790 \text{ m}^2.$$

Page
508

B. The **error of estimation** is the difference between the actual value of $\int_a^b f(x)\,dx$ and the estimate. Here is a table that describes, under the conditions given, an upper bound on the error of estimation for the three methods in this section.

Rule	Condition for all $x \in [a, b]$	Maximum Error of Estimation
Midpoint	$\lvert f''(x) \rvert \le K$	$\dfrac{K(b-a)^3}{24n^2}$
Trapezoidal	$\lvert f''(x) \rvert \le K$	$\dfrac{K(b-a)^3}{12n^2}$
Simpson's	$\lvert f^{(4)}(x) \rvert \le K$	$\dfrac{K(b-a)^5}{180n^4}$

7) Find the maximum error in estimating $\int_{-1}^{3} x^3\,dx$ with $n = 8$ using the Trapezoidal Rule.

The maximum is $\dfrac{K(3-(-1))^3}{12(8)^2} = \dfrac{K}{12}$.

To find a value for K we note $f''(x) = 6x$.

For $-1 \le x \le 3$, $\lvert f''(x) \rvert = \lvert 6x \rvert \le 6 \cdot 3 = 18$.

We choose $K = 18$. Thus the maximum error of estimation is $\dfrac{K}{12} = \dfrac{18}{12} = 1.5$.

Note $\int_{-1}^{3} x^3\,dx = 20$ and the estimate using the Trapezoidal Rule is 20.5, which is within 1.5 of 20.

8) Find the maximum error in estimating $\int_{1}^{3} \sin 2x\,dx$ with $n = 4$ using Simpson's Rule.

$\dfrac{K(3-1)^5}{180(4)^4} = \dfrac{K}{1440}$.

For $f(x) = \sin 2x$, $f^{(4)}(x) = 16 \sin 2x$. For $1 \le x \le 3$, $\lvert \sin 2x \rvert \le 1$, so $\lvert f^{(4)}(x) \rvert \le 16$.

We choose $K = 16$. Thus, the maximum error is $\dfrac{K}{1440} = \dfrac{16}{1440} = \dfrac{1}{90} \approx 0.0111$.

To 4 decimals, $\int_{1}^{3} \sin 2x\,dx = -0.6882$. The Simpson Rule estimate is -0.6925, which is within 0.0111 of -0.6882.

C. Technology Plus. Use a computer algebra system or a graphing calculator to solve.

T-1) Use a spreadsheet or calculator to estimate $\int_0^5 (25x - x^3)dx$ by

    a) the Midpoint Rule with $n = 20$.

$\Delta x = \dfrac{5 - 0}{20} = 0.25.$

i	$\overline{x}_i$	$f(\overline{x}_i)$
1	0.125	3.123
2	0.375	9.322
3	0.625	15.381
4	0.875	21.205
5	1.125	26.701
6	1.375	31.775
7	1.625	36.334
8	1.875	40.283
9	2.125	43.529
10	2.375	45.979
11	2.625	47.537
12	2.875	48.111
13	3.125	47.607
14	3.375	45.932
15	3.625	42.990
16	3.875	38.689
17	4.125	32.936
18	4.375	25.635
19	4.625	16.693
20	4.875	6.018
		625.780

The integral is approximately
$(0.25)(625.780) \approx 156.445$.

b) the Trapezoidal Rule with $n = 20$.

$\Delta x = \dfrac{5 - 0}{20} = 0.25.$

i	x_i	$f(x_i)$	term
0	0.00	0.000	0.000
1	0.25	6.234	12.468
2	0.50	12.375	24.750
3	0.75	18.328	36.656
4	1.00	24.000	48.000
5	1.25	29.297	58.594
6	1.50	34.125	68.250
7	1.75	38.391	76.782
8	2.00	42.000	84.000
9	2.25	44.859	89.718
10	2.50	46.875	93.750
11	2.75	47.953	95.906
12	3.00	48.000	96.000
13	3.25	46.922	93.844
14	3.50	44.625	89.250
15	3.75	41.016	82.032
16	4.00	36.000	72.000
17	4.25	29.484	58.968
18	4.50	21.375	42.750
19	4.75	11.578	23.156
20	5.00	0.000	0.000
			1246.874

The integral is approximately
$\frac{1}{2}(0.25)(1246.874) \approx 155.86.$

c) Simpson's Rule with $n = 20$.

$$\Delta x = \frac{5 - 0}{20} = 0.25.$$

i	x_i	$f(x_i)$	term
0	0.00	0.000	0.000
1	0.25	6.234	24.936
2	0.50	12.375	24.750
3	0.75	18.328	73.312
4	1.00	24.000	48.000
5	1.25	29.297	117.188
6	1.50	34.125	68.250
7	1.75	38.391	153.564
8	2.00	42.000	84.000
9	2.25	44.859	179.436
10	2.50	46.875	93.750
11	2.75	47.953	191.812
12	3.00	48.000	96.000
13	3.25	46.922	187.688
14	3.50	44.625	89.250
15	3.75	41.016	164.064
16	4.00	36.000	72.000
17	4.25	29.484	117.936
18	4.50	21.375	42.750
19	4.75	11.578	46.312
20	5.00	0.000	0.000
			1874.998

The integral is approximately
$\frac{1}{3}(0.25)(1874.998) \approx 156.250$.

Section 7.8   Improper Integrals

This section covers improper integrals $\int_a^b f(x)\,dx$ where either a or b is infinity (or $-\infty$) or $y = f(x)$ grows infinitely large somewhere between a and b. Improper integrals are evaluated as limits of ordinary definite integrals.

Concepts to Master

A. Definitions of improper integrals; Type I and Type II integrals; Convergence; Divergence

B. Comparison Test for improper integrals

Summary and Focus Questions

Page 519

A. Improper integrals may take several forms. They may be of **Type I** where there is an infinite limit of intergration, or they may be of **Type II** and have a "trouble point" where the function is not defined and grows infinitely large or infinitely negative. Improper integrals are defined and evaluated as limits. If an improper integral exists, it is said to **converge;** otherwise it **diverges.** Improper integrals are evaluated as limits according to their definitions. The various Type I and Type II improper integrals are:

Definition	Necessary condition(s)	Example
Type 1		
$\int_a^\infty f(x)\,dx = \lim\limits_{t\to\infty}\int_a^t f(x)\,dx$	$\int_a^t f(x)\,dx$ exists for all $t \ge a$	$\int_1^\infty e^{-x}\,dx$
$\int_{-\infty}^b f(x)\,dx = \lim\limits_{t\to-\infty}\int_t^b f(x)\,dx$	$\int_t^b f(x)\,dx$ exists for all $t \le b$	$\int_{-\infty}^0 \sin x\,dx$
$\int_{-\infty}^\infty f(x)\,dx = \int_{-\infty}^a f(x)\,dx + \int_a^\infty f(x)\,dx$	Both integrals on the right exist	$\int_{-\infty}^\infty \dfrac{2x}{(1+x^2)^2}\,dx$
Type 2		
$\int_a^b f(x)\,dx = \lim\limits_{t\to a+}\int_t^b f(x)\,dx$	f is continuous on $(a, b]$ and not continuous at a; $\int_t^b f(x)\,dx$ exists for $t \in (a, b]$	$\int_0^1 x^{-2/3}\,dx$
$\int_a^b f(x)\,dx = \lim\limits_{t\to b-}\int_a^t f(x)\,dx$	f is continuous on $[a, b)$ and not continuous at b; $\int_a^t f(x)\,dx$ exists for all $t \in [a, b)$	$\int_0^1 \dfrac{1}{1-x}\,dx$
$\int_a^b f(x)\,dx = \int_a^c f(x)\,dx + \int_c^b f(x)\,dx$	f is continuous on $[a, c)$ and $(c, b]$ and not continuous at c; both improper integrals on the right exist	$\int_{-1}^1 x^{-2/3}\,dx$

One particular kind of improper integral of Type 1 will be used frequently:

$\int_1^\infty \dfrac{1}{x^p}\, dx$ is convergent for $p > 1$ and is divergent for $p \le 1$.

Examples: Evaluate each

a) $\int_1^\infty e^{-x} dx$ is a Type 1 improper integral with upper limit infinity.

$$\int_1^\infty e^{-x} dx = \lim_{t \to \infty} \int_1^t e^{-x}\, dx = \lim_{t \to \infty} (-1)e^{-x}\Big]_1^t = \lim_{t \to \infty} (-1)(e^{-t} - e^{-1})$$

$$= (-1)\left(0 - \tfrac{1}{e}\right) = \tfrac{1}{e} \approx 0.368.$$

b) $\int_{-\infty}^0 \sin x\, dx$ is a Type 1 improper integral with lower limit negative infinity.

$$\int_{-\infty}^0 \sin x\, dx = \lim_{t \to -\infty} \int_t^0 \sin x\, dx = \lim_{t \to -\infty} (-\cos x)]_t^0 = \lim_{t \to -\infty} (-\cos 0 + \cos t) = \lim_{t \to -\infty} (\cos t - 1),$$

which does not exist. Therefore, $\int_{-\infty}^0 \sin x\, dx$ diverges.

c) $\int_{-\infty}^\infty \dfrac{2x}{(1+x^2)^2}\, dx$ is a Type 1 improper integral with upper and lower infinite limits.

Choose intermediate point $a = 0$. Then $\int_{-\infty}^\infty \dfrac{2x}{(1+x^2)^2}\, dx = \int_{-\infty}^0 \dfrac{2x}{(1+x^2)^2}\, dx + \int_0^\infty \dfrac{2x}{(1+x^2)^2}\, dx.$

First, $\int_{-\infty}^0 \dfrac{2x}{(1+x^2)^2}\, dx = \lim_{t \to -\infty} \int_t^0 \dfrac{2x}{(1+x^2)^2}\, dx.$

For $\int_t^0 \dfrac{2x}{(1+x^2)^2}\, dx$, use the substitution $u = 1 + x^2$, $du = 2x\, dx$. At $x = 0$, $u = 1$
and at $x = t$, $u = 1 + t^2$:

$$\int_t^0 \dfrac{2x}{(1+x^2)^2}\, dx = \int_{1+t^2}^1 \dfrac{du}{u^2} = \dfrac{-1}{3u^3}\Big]_{1+t^2}^1 = \dfrac{-1}{3(1)^3} - \dfrac{-1}{3(1+t^2)^3} = \dfrac{1}{3(1+t^2)^3} - \dfrac{1}{3}.$$

Therefore, $\lim_{t \to -\infty} \int_t^0 \dfrac{2x}{(1+x^2)^2}\, dx = \lim_{t \to -\infty} \left(\dfrac{1}{3(1+t^2)^3} - \dfrac{1}{3} \right) = 0 - \dfrac{1}{3} = -\dfrac{1}{3}.$

In a similar fashion, $\int_0^\infty \dfrac{2x}{(1+x^2)^2}\, dx = \lim_{t \to \infty} \int_0^t \dfrac{2x}{(1+x^2)^2}\, dx = \lim_{t \to \infty} \left(\dfrac{-1}{3(1+t^2)^3} + \dfrac{1}{3} \right) = \dfrac{1}{3} - 0 = \dfrac{1}{3}.$

Thus, $\int_{-\infty}^\infty \dfrac{2x}{(1+x^2)^2}\, dx = \dfrac{1}{3} + \left(-\dfrac{1}{3} \right) = 0.$

d) $\int_0^1 x^{-2/3} dx$ is a Type 2 improper integral with lower limit 0. $f(x) = x^{-2/3}$ has a vertical asymptote at $x = 0$.

$$\int_0^1 x^{-2/3} dx = \lim_{t \to 0^+} \int_t^1 x^{-2/3} dx = \lim_{t \to 0^+} 3x^{1/3} \Big]_t^1 = \lim_{t \to 0^+} \left(3\sqrt[3]{1} - 3\sqrt[3]{t} \right) = 3 - 0 = 3.$$

e) $\int_0^1 \frac{1}{1-x} dx$ is a Type 2 improper integral with upper limit 1. $f(x) = \frac{1}{1-x}$ has a vertical asymptote at $x = 1$.

$$\int_0^1 \frac{1}{1-x} dx = \lim_{t \to 1^-} \int_0^t \frac{1}{1-x} dx = \lim_{t \to 1^-} -\ln(1-x) \Big]_0^t = \lim_{t \to 1^-} \left(-\ln(1-t) - (-\ln(1-0)) \right)$$

$$= \lim_{t \to 1^-} \left(-\ln(1-t) \right) = \infty.$$

The limit does not exist, so $\int_0^1 \frac{1}{1-x} dx$ diverges.

f) $\int_{-1}^1 x^{-2/3} dx$ is a Type 2 improper integral with upper and lower limits. $f(x) = x^{-2/3}$ has a vertical asymptote at $x = 0$.

$$\int_{-1}^1 x^{-2/3} dx = \int_{-1}^0 x^{-2/3} dx + \int_0^1 x^{-2/3} \text{ By example } \mathbf{d)} \text{ the second limit is 3. In a similar}$$

fashion, the first limit is also 3. Therefore $\int_{-1}^1 x^{-2/3} dx = 3 + 3 = 6$.

1) Give a definition of each:

a) $\int_0^\infty \sin 2x \, dx$

<This integral is Type 1.>

$$\lim_{t \to \infty} \int_0^t \sin 2x \, dx.$$

b) $\int_0^1 \frac{1}{\sqrt{x}} dx$

<This integral is Type 2, $x = 0$.>

$$\lim_{t \to 0^+} \int_t^1 \frac{1}{\sqrt{x}} dx.$$

c) $\displaystyle\int_{-\infty}^{\infty} e^{-x^2}\, dx$

<This integral is Type 1.>

$$\int_{-\infty}^{0} e^{-x^2}\, dx + \int_{0}^{\infty} e^{-x^2}\, dx$$

$$= \lim_{t\to-\infty}\int_{t}^{0} e^{-x^2}\, dx + \lim_{t\to\infty}\int_{0}^{t} e^{-x^2}\, dx.$$

(The choice of zero was arbitrary; any real number would do.)

d) $\displaystyle\int_{-2}^{2} \frac{1}{(x-1)^2}\, dx$

<This integral is Type 2, c = 1.>

$$\int_{-2}^{1} \frac{1}{(x-1)^2}\, dx + \int_{1}^{2} \frac{1}{(x-1)^2}\, dx$$

$$= \lim_{t\to1^-}\int_{-2}^{t} \frac{1}{(x-1)^2}\, dx + \lim_{t\to1^+}\int_{t}^{2} \frac{1}{(x-1)^2}\, dx.$$

e) $\displaystyle\int_{1}^{\infty} \frac{1}{1-x}\, dx$

<This one is both Type 1 and Type 2.>

Rewrite the integral as

$$\int_{1}^{2} \frac{1}{1-x}\, dx + \int_{2}^{\infty} \frac{1}{1-x}\, dx$$

$$= \lim_{t\to1^+}\int_{t}^{2} \frac{1}{1-x}\, dx + \lim_{t\to\infty}\int_{2}^{t} \frac{1}{1-x}\, dx.$$

(The choice of 2 was arbitrary; any real number greater than 1 would do.)

2) Make a table of values for $\displaystyle\int_{t}^{1} \frac{1}{\sqrt{x}}\, dx$ for $t = .5, .05, .005, .0005,$ and $.00005.$ Does it appear that $\displaystyle\int_{0}^{1} \frac{1}{\sqrt{x}}\, dx$ converges?

t	$\displaystyle\int_{t}^{1} \frac{1}{\sqrt{x}}\, dx$
.5	.586
.05	1.553
.005	1.859
.0005	1.955
.00005	1.986

$\displaystyle\int_{0}^{1} \frac{1}{\sqrt{x}}\, dx$ appears to converge to 2.

It does, because $\displaystyle\int_{t}^{1} \frac{1}{\sqrt{x}}\, dx = 2 - 2\sqrt{t}.$

As $t \to 0,\ 2 - 2\sqrt{t}$ approaches 2.

3) Evaluate $\int_0^{\pi/2} \sec^2 x \, dx$.

\<This integral is Type 2, $b = \dfrac{\pi}{2}$.\>

$$\int_0^{\pi/2} \sec^2 x \, dx = \lim_{t \to \pi/2^-} \int_0^t \sec^2 x \, dx$$

$$= \lim_{t \to \pi/2^-} \tan x \Big|_0^t = \lim_{t \to \pi/2^-} \tan t,$$

which does not exist (the limit is ∞). This integral diverges.

4) Evaluate $\int_1^\infty x e^{-x^2} dx$.

\<This integral is Type 1 with upper limit ∞.\>

$$\int_1^\infty x e^{-x^2} \, dx = \lim_{t \to \infty} \int_1^t x e^{-x^2} \, dx$$

$$(u = -x^2, \, du = -2x \, dx)$$

$$= \lim_{t \to \infty} \left(-\tfrac{1}{2} e^{-x^2} \Big|_1^t \right)$$

$$= \lim_{t \to \infty} \left(-\tfrac{1}{2} (e^{-t^2} - e^{-1}) \right)$$

$$= -\tfrac{1}{2} \left(0 - \tfrac{1}{e} \right) = \tfrac{1}{2e}.$$

5) a) Find the area of the shaded region.

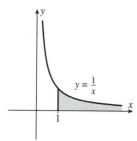

$$\int_1^\infty \frac{1}{x} \, dx = \lim_{t \to \infty} \int_1^t \frac{1}{x} \, dx = \lim_{t \to \infty} \ln x \Big|_1^t$$

$$= \lim_{t \to \infty} \ln t = \infty.$$

The improper integral does not exist; the area is infinite.

b) Evaluate

$\int_1^\infty \dfrac{\pi}{x^2} \, dx$.

$$\int_1^\infty \frac{\pi}{x^2} \, dx = \lim_{t \to \infty} \int_1^t \frac{\pi}{x^2} \, dx = \lim_{t \to \infty} \frac{-\pi}{x} \Big|_1^t$$

$$= \lim_{t \to \infty} \left(\frac{-\pi}{t} + \frac{\pi}{1} \right) = \pi.$$

c) (For fun) The region in part **a)** has an infinite area. How can it be painted with a finite amount of paint?

i) Take the region, $0 \leq y \leq \dfrac{1}{x}$, $1 \leq x \leq \infty$, and revolve it about the x-axis.

ii) The volume of the resulting solid of revolution is given by $\int_1^\infty \dfrac{\pi}{x^2} \, dx$, which by part **b)** is π.

iii) Fill the solid of revolution with π gallons of paint.

iv) Dip the region in part **a)** in the solid of revolution. The region is then painted with at most π gallons of paint!

6) Apply directly the Fundamental Theorem of Calculus to evaluate $\int_{-1}^{1} \frac{1}{x^2}\,dx$. What's wrong with this?

$$\int_{-1}^{1} \frac{1}{x^2}\,dx = \int_{-1}^{1} x^{-2}\,dx = -\frac{x^{-1}}{1}\Big]_{-1}^{1}$$

$$= -\frac{1}{x}\Big]_{-1}^{1} = -2.$$

This cannot be since $\frac{1}{x^2} > 0$ when defined. Since $\frac{1}{x^2}$ is not continuous at 0, we cannot use the FTOC.

Since the exponent $p = -2$ this integral diverges.

Page 525

B. Sometimes it can be shown that an improper integral converges without determining the exact value to which it converges. (It is useful information to know an improper integral exists before it will be approximated.) One such method to determine existence is the **Comparison Test for Improper Integrals:**

Suppose f and g are continuous functions and $f(x) \geq g(x) \geq 0$ for $x \geq a$. If $\int_{a}^{\infty} f(x)\,dx$ converges then $\int_{a}^{\infty} g(x)\,dx$ converges.

Similar statements can be made for the other forms of improper integrals. The improper integral $\int_{1}^{\infty} \frac{1}{x^p}\,dx$ is often used as a comparator.

Example: Determine whether $\int_{1}^{\infty} \cos\left(\frac{1}{x}\right)\,dx$ converges.

For $x \geq 1, 0 < \frac{1}{x} \leq 1$. Therefore, $\cos\left(\frac{1}{x}\right) \geq \cos(1) > 0.50$.

Thus, $\int_{1}^{\infty} \cos\left(\frac{1}{x}\right)\,dx \geq \int_{1}^{\infty} 0.5\,dx = \infty$.

Since $\int_{1}^{\infty} 0.5\,dx$ diverges, $\int_{1}^{\infty} \cos\left(\frac{1}{x}\right)\,dx$ diverges.

---

7) True or False:
If $f(x) \geq g(x) \geq 0$ for $x \geq a$ and f and g are continuous, then if $\int_{a}^{b} g(x)\,dx$ diverges, then $\int_{a}^{\infty} f(x)\,dx$ diverges.

True. This statement is the contrapositive of the Comparison Test for Improper Integrals.

8) True or False:

If f and g are continuous, $f(x) \geq g(x) \geq 0$ for all $x \leq b$, and if $\int_{-\infty}^{b} f(x)\, dx$ converges, then $\int_{-\infty}^{b} g(x)\, dx$ converges.

True.

9) Show that $\int_{1}^{\infty} \frac{|\sin x|}{x^2}\, dx$ converges.

For $1 \leq x \leq \infty$, $0 \leq |\sin x| \leq 1$. Thus,

$$0 \leq \frac{|\sin x|}{x^2} \leq \frac{1}{x^2}. \int_{1}^{\infty} \frac{1}{x^2}\, dx \text{ converges (to 1).}$$

Thus $\int_{1}^{\infty} \frac{|\sin x|}{x^2}\, dx$ converges (but we don't know to what value, except that it will be between 0 and 1).

Chapter 8 — Further Applications of Integration

Section 8.1   Arc Length

This section uses definite integrals to calculate the length of a curve (arc length). As with other integral applications, arc length will be the limit of the sums of approximations of the lengths of several small pieces of the curve.

Concepts to Master

A. Length of a curve

B. Arc length function and its derivative

C. Technology Plus

Summary and Focus Questions

Page 538

A. If f' is continuous on $[a, b]$, the **length of the curve** or **arc length** of $y = f(x)$ for $a \le x \le b$ is

$$L = \int_a^b \sqrt{1 + (f'(x))^2}\, dx = \int_a^b \sqrt{1 + \left(\frac{dy}{dx}\right)^2}\, dx.$$

If g' is continuous on $[c, d]$, the arc length of $x = g(y)$ for $c \le y \le d$ is

$$\int_c^d \sqrt{1 + (g'(y))^2}\, dy = \int_c^d \sqrt{1 + \left(\frac{dx}{dy}\right)^2}\, dy.$$

Because the arc length integral involves a square root, the integral will often need to be approximated.

Example: For $y = x^2$, $\dfrac{dy}{dx} = 2x$. The length of the curve $y = x^2$

for $1 \le x \le 2$ is $\displaystyle\int_1^2 \sqrt{1 + (2x)^2}\, dx = \int_1^2 \sqrt{1 + 4x^2}\, dx$, which is

approximately 3.17.

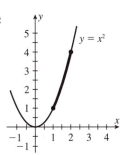

1) Find a definite integral for the length of $y = \frac{1}{x}$, $1 \le x \le 3$.

$y' = -\frac{1}{x^2}$, so

$$L = \int_1^3 \sqrt{1 + \left(\frac{-1}{x^2}\right)^2}\, dx = \int_1^3 \frac{\sqrt{x^4 + 1}}{x^2}\, dx.$$

2) Show that the circumference of a circle of radius r is $2\pi r$.

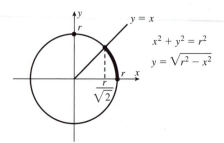

$$x^2 + y^2 = r^2$$
$$y = \sqrt{r^2 - x^2}$$

The circumference is 8 times the arc length of $y = \sqrt{r^2 - x^2}$ in the first quadrant,

$$0 \le x \le \frac{r}{\sqrt{2}}.$$

$y' = \frac{-x}{\sqrt{r^2 - x^2}}$, so the arc length is

$$\int_0^{\frac{r}{\sqrt{2}}} \sqrt{1 + \left(\frac{-x}{\sqrt{r^2 - x^2}}\right)^2}\, dx$$

$$= \int_0^{\frac{r}{\sqrt{2}}} \sqrt{1 + \frac{x^2}{r^2 - x^2}}\, dx = \int_0^{\frac{r}{\sqrt{2}}} \frac{r}{\sqrt{r^2 - x^2}}\, dx.$$

We use the substitution $x = r \sin \theta$, $dx = r \cos \theta\, d\theta$. Then

$$\int \frac{r}{\sqrt{r^2 - x^2}}\, dr = \int \frac{r}{r \cos \theta}\, r \cos \theta\, d\theta = \int r\, d\theta$$

$$= r\theta = r \sin^{-1}\left(\frac{x}{r}\right).$$

Thus, $\int_0^{\frac{r}{\sqrt{2}}} \frac{r}{\sqrt{r^2 - x^2}}\, dx = r \sin^{-1}\left(\frac{x}{r}\right)\Big]_0^{\frac{r}{\sqrt{2}}}$

$$= r\left(\sin^{-1}\left(\frac{1}{\sqrt{2}}\right) - \sin^{-1}(0)\right) = r\left(\frac{\pi}{4} - 0\right)$$

$$= \frac{r\pi}{4}.$$

The circumference of the entire circle is

$$8\left(\frac{r\pi}{4}\right) = 2\pi r.$$

3) What is the length of the arc given by
$x = 1 + y^{3/2}, 0 \le y \le 4$?

$x' = \frac{3}{2}y^{1/2}.$

$$L = \int_0^4 \sqrt{1 + \left(\frac{3}{2}y^{1/2}\right)^2}\, dy$$

$$= \int_0^4 \sqrt{1 + \frac{9}{4}y}\, dy.$$

Let $u = 1 + \frac{9}{4}y$, $du = \frac{9}{4}\, dy$

At $y = 0$, $u = 1$. At $y = 4$, $u = 10$.

$$L = \frac{4}{9}\int_1^{10}\sqrt{u}\, du = \frac{8}{27}u^{3/2}\Big]_1^{10}$$

$$= \frac{8}{27}(10\sqrt{10} - 1).$$

4) Write an expression involving definite integrals for the length of the curve
$y = \sqrt[3]{x}$ from $x = -1$ to $x = 8$.

Then approximate the value using the Midpoint Rule with $n = 3$.

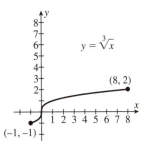

$y = \sqrt[3]{x}$

$(8, 2)$

$(-1, -1)$

Since a vertical tangent exists at $x = 0$, y' is not continuous on $[-1, 8]$. Therefore, we choose to treat x as a function of y.
From $y = \sqrt[3]{x}$, $x = y^3$ and $x' = 3y^2$.
The arc length is

$$\int_{-1}^2 \sqrt{1 + (3y^2)^2}\, dy = \int_{-1}^2 \sqrt{1 + 9y^4}\, dy.$$

Using $n = 3$ and $\Delta y = 1$, the Midpoint Rule yields

subint	$\overline{y_i}$	$\overline{x_i}$
$[-1, 0]$	1.25	-0.5
$[0, 1]$	1.25	0.5
$[1, 2]$	6.82	1.5

$$\int_{-1}^2 \sqrt{1 + 9y^4}\, dy \approx 1(1.25 + 1.25 + 6.82)$$

$$= 9.32.$$

Page 541

B. For a continuously differentiable (meaning f' is continuous) function $y = f(x)$, $a \le x \le b$, let $s(x)$ be the distance traveled along the curve from $(a, f(a))$ to $(x, f(x))$. $s(x)$ is the **arc length function**.

The function $s(x)$ may be written as

$$s(x) = \int_a^x \sqrt{1 + [f'(t)]^2} \, dt.$$

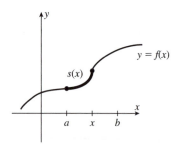

If we write the above with differentials, then the differential

$ds = \sqrt{1 + \left(\dfrac{dy}{dx}\right)^2} \, dx$ may be used to express arc length:

$$L = \int ds.$$

The expression $ds = \sqrt{1 + \left(\dfrac{dy}{dx}\right)^2}$ is useful because we can then write

$$(ds)^2 = (dx)^2 + (dy)^2.$$

This last form can be used to remember *both* arc length formulas:

$$ds = \sqrt{1 + \left(\frac{dy}{dx}\right)^2} \, dx \quad \text{and} \quad ds = \sqrt{1 + \left(\frac{dx}{dy}\right)^2} \, dy.$$

5) Find the arc length function for a particle starting at $(0, 0)$ and moving along the curve $y^2 = 4x^3$, $y \ge 0$.

$y^2 = 4x^3$, $y = 2x^{3/2}$ for $x \ge 0$ and $y \ge 0$.

$\dfrac{dy}{dx} = 3x^{1/2}$.

$s(x) = \displaystyle\int_0^x \sqrt{1 + (3t^{1/2})^2} \, dt$

$\quad = \displaystyle\int_0^x \sqrt{1 + 9t} \, dt.$

Let $u = 1 + 9t$. Then $du = 9 \, dt$ and $dt = \dfrac{1}{9} du$.

At $t = 0$, $u = 1$.
At $t = x$, $u = 1 + 9x$.

$s(x) = \dfrac{1}{9} \displaystyle\int_1^{1+9x} \sqrt{u} \, du$

$\quad = \dfrac{1}{9} \left(\dfrac{u^{3/2}}{3/2} \right) \Bigg]_1^{1+9x}$

$\quad = \dfrac{2}{27} u^{3/2} \Bigg]_1^{1+9x}$

$\quad = \dfrac{2}{27}(1 + 9x)^{3/2} - \dfrac{2}{27}.$

6) What is the length of the arc from $(0, 0)$ to $(1, 2)$ along the curve in question **5)**?

Substitute $x = 1$ in the arc length function.

$$s(1) = \frac{2}{27}(1 + 9(1))^{3/2} - \frac{2}{27}$$

$$= \frac{2}{27}(10^{3/2} - 1) \approx 2.268.$$

7)

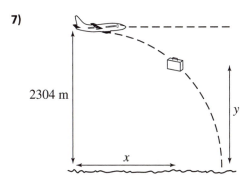

A plane is flying at an altitude of 2304 m when a suitcase falls from the cargo area. Because of air resistance, the suitcase trajectory is $y = 2304 - \frac{x^{3/2}}{6}$. Find the distance the suitcase travels along its path to the ground.

When the suitcase travels x meters horizontally, its height is $y = 2304 - \frac{x^{3/2}}{6}$.

$$y' = -\frac{3}{2}\frac{x^{1/2}}{6} = -\frac{\sqrt{x}}{4}.$$

$$s(x) = \int_0^x \sqrt{1 + \left(-\frac{\sqrt{t}}{4}\right)^2}\, dt$$

$$= \frac{1}{4}\int_0^x \sqrt{16 + t}\, dt$$

Let $u = 16 + t,\ du = dt.$
At $t = 0,\ u = 16.$
At $t = x,\ u = 16 + x.$

$$s(x) = \frac{1}{4}\int_{16}^{16+x} u^{1/2}\, du = \frac{1}{6}u^{3/2}\Big]_{16}^{16+x}$$

$$= \frac{(16 + x)^{3/2} - 64}{6}.$$

When $y = 0$,

$$2304 - \frac{x^{3/2}}{6} = 0$$

$$x^{3/2} = 13824$$

$$x = 576.$$

At $x = 576$,

$$s(576) = \frac{(16 + 576)^{3/2} - 64}{6} \approx 2389.997 \text{ m.}$$

C. Technology Plus. Use a computer algebra system or a graphing calculator to solve.

T-1) Use a calculator or a CAS to evaluate the integral for the arc length of $y = x - \cos x,\ -1 \le x \le 1.$

$y' = 1 + \sin x.$
The length of the curve is

$$\int_{-1}^{1} \sqrt{1 + (1 + \sin x)^2}\, dx \approx 2.932.$$

Section 8.2   Area of a Surface of Revolution

Objects such as a football or a round glass bowl, which are the surfaces of solids of revolution, may have their surface areas calculated by a definite integral. As with other area applications, it is helpful to draw a representative figure for a typical term in the Riemann sum that approximates the surface area.

Concepts to Master

A. Calculate the surface area of a solid of revolution

B. Technology Plus

Summary and Focus Questions

Page
545

A. If $f(x) \geq 0$ and $f'(x)$ is continuous on $[a, b]$, the **surface area** of the solid of revolution is approximated by a sum of surface areas of solids of revolution of small segments of the curve. The length of a typical segment, given by the arc length formula, is approximately $\sqrt{1 + (f'(x_i^*))^2}\, \Delta x$ and when rotated about the x-axis, produces a surface area that is approximately $2\pi f(x_i^*)\sqrt{1 + (f'(x_i^*))^2}\, \Delta x$. Thus, the surface area of the solid of revolution is

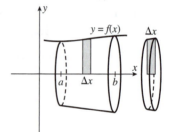

$$S = \lim_{n \to \infty} \sum_{i=1}^{n} 2\pi f(x_i^*) \sqrt{1 + (f'(x_i^*))^2}\, \Delta x$$

$$= \int_a^b 2\pi f(x)\sqrt{1 + (f'(x))^2}\, dx.$$

For curves described by $x = g(y)$, $c \leq y \leq d$, and rotated about the x-axis,

$$S = \int_c^d 2\pi y \sqrt{1 + \left(\frac{dx}{dy}\right)^2}\, dy.$$

Recalling that $(ds)^2 = (dx)^2 + (dy)^2$, we use a more compact notation to represent both the above cases of rotation about the x-axis.

$$S = \int 2\pi y\, ds, \text{ where } ds = \sqrt{1 + \left(\frac{dy}{dx}\right)^2}\, dx \text{ or } ds = \sqrt{1 + \left(\frac{dx}{dy}\right)^2}\, dy.$$

For rotation about the y-axis, $S = \int 2\pi x\, ds.$

Example: Find the surface area of the solid of revolution formed by rotating the area between $x = 0$, $x = 2$, $y = 0$, and $y = \frac{1}{3}x^3$ about the x-axis.

For $f(x) = \frac{1}{3}x^3$, $f'(x) = x^2$. The surface area is

$$S = \int_0^2 2\pi\left(\frac{1}{3}x^3\right)\sqrt{1 + (x^2)^2}\, dx$$

$$= \frac{2\pi}{3}\int_0^2 \sqrt{1 + x^4}\, x^3\, dx.$$

Let $u = 1 + x^4$, $du = 4x^3\, dx$.

At $x = 0$, $u = 1$. At $x = 2$, $u = 17$.

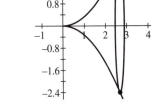

Then $S = \frac{2\pi}{3}\left(\frac{1}{4}\right)\int_1^{17}\sqrt{u}\, du = \frac{2\pi}{3}\left(\frac{1}{4}\right)\left(\frac{2}{3}\right)u^{3/2}\Big]_1^{17} = \frac{\pi}{9}(17^{3/2} - 1^{3/2}) = \frac{\pi}{9}(17^{3/2} - 1).$

1) Find the surface area of the solid obtained by revolving $y = \sin 2x$, $0 \le x \le \frac{\pi}{2}$, about the x-axis.

$$S = \int_0^{\pi/2} 2\pi \sin 2x\sqrt{1 + (2\cos 2x)^2}\, dx$$

Let $u = 2\cos 2x$. Then $du = -4\sin 2x$.
At $x = 0$, $u = 2$. At $x = \frac{\pi}{2}$, $u = -2$.

$$S = \int_0^{\pi/2}\left(-\frac{1}{4}\right)2\pi \sin 2x(-4)\sqrt{1 + (2\cos 2x)^2}\, dx$$

$$= -\frac{\pi}{2}\int_2^{-2}\sqrt{1 + u^2}\, du = \frac{\pi}{2}\int_{-2}^2 \sqrt{1 + u^2}\, du$$

(Table of Integrals, #21)

$$= \frac{\pi}{2}\left(\frac{u}{2}\sqrt{1 + u^2} + \frac{1}{2}\ln\left|u + \sqrt{1 + u^2}\right|\right)\Big]_{-2}^2$$

$$= \frac{\pi}{2}\left(\frac{2}{2}\sqrt{5} + \frac{1}{2}\ln\left|2 + \sqrt{5}\right|\right)$$
$$- \left(\frac{-2}{2}\sqrt{5} + \frac{1}{2}\ln\left|-2 + \sqrt{5}\right|\right)$$

$$= \frac{\pi}{2}\left(2\sqrt{5} + \ln\frac{\sqrt{2 + \sqrt{5}}}{\sqrt{-2 + \sqrt{5}}}\right)$$

$$\approx 9.292.$$

2) Set up two equivalent definite integrals for the surface area of the solid obtained by revolving $y = x^2, 0 \le x \le 2$, about the y-axis.

In both solutions $s = \int 2\pi x \, ds$.

i) $ds = \sqrt{1 + \left(\frac{dy}{dx}\right)^2} \, dx.$

$y = x^2$ so $\frac{dy}{dx} = 2x.$

$S = \int_0^2 2\pi x \, ds = \int_0^2 2\pi x \sqrt{1 + (2x)^2} \, dx$

$\quad = \int_0^2 2\pi x \sqrt{1 + 4x^2} \, dx$

ii) $ds = \sqrt{1 + \left(\frac{dx}{dy}\right)^2} \, dy.$

Since $y = x^2, 0 \le x \le 2,$

$x = \sqrt{y}, 0 \le y \le 4$ and $\frac{dx}{dy} = \frac{1}{2\sqrt{y}}.$

$S = \int_0^4 2\pi x \, ds$

$\quad = \int_0^4 2\pi \sqrt{y} \sqrt{1 + \left(\frac{1}{2\sqrt{y}}\right)^2} \, dy$

$\quad = \int_0^4 2\pi \sqrt{y} \sqrt{\frac{1 + 4y}{4y}} \, dy$

$\quad = \int_0^4 \pi \sqrt{1 + 4y} \, dy.$

The integral in part ii) is equal to the integral in part i) since $y = x^2, dy = 2x \, dx,$ and for $0 \le x \le 2$, we have $0 \le y \le 4$.

B. Technology Plus. Use a computer algebra system or a graphing calculator to solve.

T-1) Find the surface area of the solid obtained by rotating $y = \tan x$ about the x-axis, $0 \le x \le \frac{\pi}{3}$. Use a calculator or a CAS to evaluate the integral.

The surface area is

$\int_0^{\pi/3} 2\pi(\tan x)\sqrt{1 + (\sec^2 x)^2} \, dx \approx 10.502.$

Section 8.3   Applications to Physics and Engineering

This section gives additional applications of definite integrals, including calculating the hydrostatic pressure and the total force on an object immersed in a liquid to a given depth and finding the center of mass (balance point) of a two-dimensional object.

Concepts to Master

A. Pressure at a given depth; Force on a vertical plane due to hydrostatic pressure

B. Moments of inertia about the x-axis and y-axis; Centroid (center of mass)

C. Theorem of Pappus

Summary and Focus Questions

Page 552

A. The **pressure** on a plate suspended horizontally in a liquid with density ρ at a depth d is

$$P = \rho g d,$$

where g is the gravitational constant. The units for P are newtons per meter^2 (pascals) or in pounds per ft^2.

Suppose a plate is suspended *vertically* in a liquid with mass density ρ and the total depth of the liquid is H. Suppose the surface of the plate can be described as bounded by $x = f(y)$, $x = k(y)$, $y = c$, $y = d$ with $f(y) \geq k(y)$ for all $y \in [c, d]$. Then the force due to liquid pressure on a section of the plate is approximately

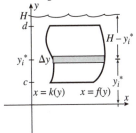

(density)(gravitational constant)(depth)(area of section)
$= \rho g(H - y_i^*)[f(y_i^*) - k(y_i^*)]\Delta y.$

The **total hydrostatic force on the surface** is $\displaystyle\int_c^d \rho g(H - y)(f(y) - k(y)) \, dy.$

1) A circular plate of radius 3 cm is suspended horizontally at a depth of 12 m in an oil having density 1500 kg/m³. What is the pressure on the plate?

Since the plate is horizontal, the pressure is uniform at all points on the plate.
$P = \rho g d = (1500 \text{ kg/m}^3)(9.8 \text{ m/s}^2)(12 \text{ m})$
$= 176{,}400 \text{ Pa (pascals)} = 176.4 \text{ kPa.}$

2) A flat isosceles triangle is suspended in water as in the figure. Water density is 1000 kg/m³. Set up a definite integral for the total hydrostatic force on the surface of the triangle.

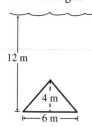

12 m

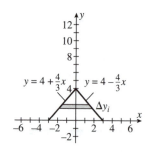

Using coordinates as shown, the sides of the triangle are

$y = 4 - \frac{4}{3}x$, so $x = -\frac{3}{4}y + 3$, and

$y = 4 + \frac{4}{3}x$, so $x = \frac{3}{4}y - 3$.

The force on the triangle is

$$\int_0^4 1000(9.8)(12 - y)\left[\left(-\frac{3}{4}y + 3\right) - \left(\frac{3}{4}y - 3\right)\right]dy$$
$$= \int_0^4 9800(12 - y)\left(6 - \frac{3}{2}y\right) dy.$$

Page 554

B. A **lamina** is a thin, flat sheet of material determined by a region R, such as in the figure, whose density is uniformly ρ, a constant.

Let $A = \int_a^b f(x)\, dx$ be the area of R. The **centroid** (or **center of mass**) of R is the point $(\overline{x}, \overline{y})$ where

$$\overline{x} = \frac{1}{A}\int_a^b x f(x)\, dx \text{ and}$$

$$\overline{y} = \frac{1}{A}\int_a^b \frac{1}{2}[f(x)]^2\, dx.$$

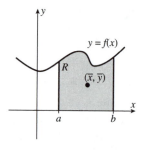

If the lamina was suspended in the air by a string attached at the centroid $(\overline{x}, \overline{y})$, then the lamina would balance. Note also that the centroid need not be located within R. A washer, for example, has its centroid in the middle of its hole.

Example: Find the centroid for the region at the right.

The area of the region is

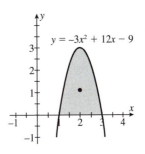

$$A = \int_1^3 (-3x^2 + 12x - 9)\, dx = (-x^3 + 6x^2 - 9x)\Big]_1^3 = 4.$$

$$\int_1^3 x(-3x^2 + 12x - 9)\, dx = \int_1^3 (-3x^3 + 12x^2 - 9x)\, dx$$

$$= \left(-\tfrac{3}{4}x^4 + 4x^3 - \tfrac{9}{2}x^2\right)\Big]_1^3 = 8.$$

$$\int_1^3 \tfrac{1}{2}(-3x^2 + 12x - 9)^2\, dx = \tfrac{1}{2}\int_1^3 (9x^4 - 72x^3 + 198x^2 - 216x + 81)\, dx$$

$$= \tfrac{1}{2}\left(\tfrac{9}{5}x^5 - 18x^4 + 66x^3 - 108x^2 + 81x\right)\Big]_1^3 \approx 4.8.$$

Therefore, $\bar{x} = \tfrac{8}{4} = 2$ and $\bar{y} = \tfrac{4.8}{4} = 1.2.$ (Note that $\bar{x} = 2$ agrees with the symmetry of the region about the line $x = 2$.)

If A is the area of the region R is bounded above by $y = f(x)$, bounded below by $y = g(x)$, and bounded on the sides by $x = a$ and $x = b$, then

$$A = \int_a^b (f(x) - g(x))\, dx,$$

$$\bar{x} = \frac{1}{A}\int_a^b x(f(x) - g(x))\, dx, \quad \text{and}$$

$$\bar{y} = \frac{1}{A}\int_1^3 \tfrac{1}{2}([f(x)]^2 - [g(x)]^2)\, dx.$$

The **moment of R about the x-axis** is

$$M_x = \rho\int_a^b \tfrac{1}{2}[f(x)]^2\, dx = \rho A\bar{y}$$

and the **moment about the y-axis** is

$$M_y = \rho\int_a^b x f(x)\, dx = \rho A\bar{x}.$$

These are measures of the tendency of the lamina to rotate about the x-axis and y-axis, respectively.

Example: The moments about the axes for the region in the example above are:

$$M_x = \rho(4.8) = 4.8\rho \quad \text{and}$$
$$M_y = \rho(8) = 8\rho.$$

3) Find the center of mass of the region bounded by $y = 4 - x^2$, $x = 0$, $y = 0$, with density ρ. Then find M_x and M_y.

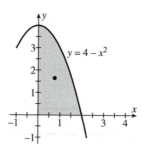

The area of the region is

$$A = \int_0^2 (4 - x^2)\, dx = 4x - \frac{x^3}{3}\Big]_0^2 = \frac{16}{3}.$$

$$\int_a^b x f(x)\, dx = \int_0^2 x(4 - x^2)\, dx$$

$$= \int_0^2 (4x - x^3)\, dx$$

$$= \left(2x^2 - \frac{x^4}{4}\right)\Big]_0^2 = 4.$$

Therefore, $\overline{x} = \dfrac{1}{\frac{16}{3}}(4) = \dfrac{3}{4}.$

$$\int_a^b \frac{1}{2}[f(x)]^2\, dx = \frac{1}{2}\int_0^2 (4 - x^2)^2\, dx$$

$$= \frac{1}{2}\int_0^2 (16 - 8x^2 + x^4)\, dx$$

$$= \frac{1}{2}\left(16x - \frac{8}{3}x^3 + \frac{x^5}{5}\right)\Big]_0^2$$

$$= \frac{128}{15}.$$

Therefore, $\overline{y} = \dfrac{1}{\frac{16}{3}}\left(\dfrac{128}{15}\right) = \dfrac{8}{5}.$

$$M_x = \rho\left(\frac{128}{15}\right) = \frac{128}{15}\rho.$$
$$M_y = \rho(4) = 4\rho.$$

Page
559

C. The Theorem of Pappus: If a region R is revolved about a line L that does not intersect R, the resulting solid has volume $V = 2\pi dA$, where A is the area of R and d is the perpendicular distance between the centroid of R and the line L. (Notice that $2\pi d$ is the distance traveled by the centroid in a circle about the line.)

Be careful that you do not interpret dA as a differential. In this case dA means the product of the distance and the area.

Example: From the first example in this subsection we found that the region at the right has area 4 and centroid $(\bar{x}, \bar{y}) = (2, 1.2)$.

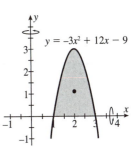

The perpendicular distance between the centroid and the y-axis ($x = 0$) is 2. If we rotate the region about the y-axis, the volume of the resulting solid is $V = 2\pi(2)(4) = 16\pi$.

If we rotate the region about the x-axis ($y = 0$), the volume of the resulting solid is $V = 2\pi(1.2)(4) = 9.6\pi$.

4) Find the volume of the solid of revolution obtained by rotating the region in question 3 about:

    a) the x-axis.

$\bar{y} = \frac{8}{5}$ is the distance from the center of mass to the axis of revolution (the x-axis). The area is $A = \frac{16}{3}$. Thus,
$$V = 2\pi\left(\frac{8}{5}\right)\left(\frac{16}{3}\right) = \frac{256\pi}{15}.$$

    b) the y-axis.

$\bar{x} = \frac{3}{4}$, so $V = 2\pi\left(\frac{3}{4}\right)\left(\frac{16}{3}\right) = 8\pi.$

Section 8.4   Applications to Economics and Biology

This section, like the previous ones, gives additional applications of definite integrals, including consumer surplus (the total amount of money that consumers save if they buy at a certain price) and various models of blood flow.

Concepts to Master

Total amount or total change of a quantity in economic and biological applications

Summary and Focus Questions

Page 563 Applications of definite integrals in this section and elsewhere all involve calculating the total amount of a quantity (that can be represented by a continuous function $y = f(x)$) within a certain range ($a \le x \le b$). If we integrate the rate of change of a quantity over an interval $[a, b]$, the result is the total change of that quantity from a to b. Here are two examples.

Example from Economics: A demand function $p(x)$ is the price necessary to sell x items. Let X be the amount actually available and P be the price (a constant) for that amount, $P = p(X)$. The **consumer's surplus** is

$$\int_0^X [p(x) - P]\,dx.$$

This is the total amount of money that could be saved by consumers if they were to buy all that is available (X) when the price is P dollars per unit.

Example from Biology: The **cardiac output** is the rate of flow of blood through the heart per unit time. We calculate this by injecting dye into the heart and measuring the concentration levels $c(t)$ at several times within some time interval $[0, T]$. If A is the amount of dye initially injected then the ratio of that initial amount of dye to the total of all the concentrations is the cardiac output:

$$\frac{A}{\int_0^T c(t)\,dt}.$$

If $c(t)$ decreases slowly (see the figure at the right) then $\int_0^T c(t)\,dt$ is large and, therefore, the cardiac output ratio $\dfrac{A}{\int_0^T c(t)\,dt}$ will be small.

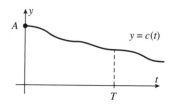

On the other hand, if, $c(t)$ decreases quickly (see the figure), then $\int_0^T c(t)\, dt$ will be small and the cardiac output ratio $\dfrac{A}{\int_0^T c(t)\, dt}$ will be large.

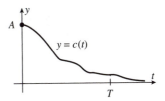

1) Find a definite integral for the consumer surplus determined when 1000 units are available and the demand function is $p(x) = 85 - 0.03x$.

At $X = 1000$,
$P = 85 - 0.03(1000) = 55.$
The consumer surplus is
$$\int_0^{1000} [85 - 0.03x - 55]\, dx$$
$$= \int_0^{1000} (30 - 0.03x)\, dx \ (= 15{,}000).$$

2) The number of black fly hatchlings as measured on the Marquette Bay during the month of June is increasing at an estimated rate of $1000 + e^{1.1t}$ per day (t is measured in days in June). How much does the fly population increase between June 5 and June 12?

If we integrate the rate of change of the population from day 5 to day 12, that will calculate the total change in the population.
$$\int_5^{12} (1000 + e^{1.1t})\, dt$$
$$= \left(1000t + \frac{e^{1.1t}}{1.1}\right)\Big]_5^{12} \approx 498{,}018 \text{ flies.}$$

3) A patient is injected with 10 mL of tracing dye to measure her cardiac output. Data is collected on the concentration of tracer left in her heart every 5 seconds for 30 seconds; see below. Estimate her cardiac output. Use the Right Endpoint Rule to estimate the total concentration after 30 seconds.

t (sec)	c(t) (mL)
0	10.00
5	8.50
10	6.30
15	4.10
20	2.20
25	0.09
30	0.01

$A = 10$ mL. $\Delta t = 5$ sec. Using the right endpoint for t_i^*:

i	t_i^*	$c(x_i^*)$	$c(x_i^*)\Delta t$
1	5	8.50	42.50
2	10	6.30	31.50
3	15	4.10	20.50
4	20	2.20	11.00
5	25	0.09	0.45
6	30	0.01	0.05
			106.00

$\int_0^{30} c(t)\, dt \approx 106$ and the cardiac output is
$$\frac{A}{\int_0^{30} c(t)\, dt} \approx \frac{10}{106} \approx .094 \text{ mL/sec.}$$

Section 8.5   Probability

The probability of an event is a number between 0 and 1 that represents the likelihood that the event will occur—the closer the probability is to one the more likely the event will occur. This section shows how calculus is used to define and calculate some basic probability concepts. We can represent probability as an area under a portion of a curve where the area under the entire curve is one. Areas, of course, are determined by definite integrals.

Concepts to Master

A. Continuous random variable, Probability density function; Probability of an event

B. Mean (average value) of a random variable

C. Normal distributions

D. Technology Plus

Summary and Focus Questions

Page 568

A. A **continuous random variable** is a variable whose values range over an entire interval. For example, the distances that an Olympic athlete can throw a javelin may be somewhere between 60 and 90 meters.

A **probability density function** for a random variable X is a non-negative function $y = f(x)$ such that the total area under the curve is equal to one. The **probability** that X could take on a value somewhere between a and b is

$$P(a \leq X \leq b) = \int_a^b f(x)\, dx.$$

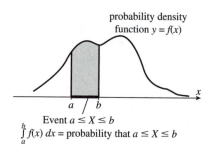

probability density function $y = f(x)$

Event $a \leq X \leq b$

$\int_a^b f(x)\, dx$ = probability that $a \leq X \leq b$

Example: a) Show that the function $f(x) = \begin{cases} 0 & \text{if } x < 1 \\ \dfrac{1}{x^2} & \text{if } x \geq 1 \end{cases}$ is a probability

density function for a random variable X.

$$\int_{-\infty}^{\infty} f(x)\, dx = \int_1^{\infty} f(x)\, dx = \lim_{t \to \infty} \int_1^t \frac{1}{x^2}\, dx = \lim_{t \to \infty} \frac{-1}{x} \Big]_1^t = \frac{-1}{t} - \frac{-1}{1} = 0 + 1 = 1.$$

Therefore, f is a probability density function.

b) Find the probability that X is between 2 and 4.

There is a 25% probability that X will take on a value between 2 and 4 because

$$\int_2^4 f(x)\,dx = \int_2^4 \frac{1}{x^2}\,dx = \frac{-1}{x}\bigg]_2^4 = \frac{-1}{4} - \frac{-1}{2} = 0.25.$$

1) For any probability density function f,

a) What is $\int_{-\infty}^{\infty} f(x)\,dx$?

1. This integral is the area under the entire curve.

b) What is $\int_a^a f(x)\,dx$?

0. This means that for any a, the chances that X will have *exactly* the value a is zero.

2) The time X it takes before a drug begins being absorbed into the bloodstream is a random variable. Suppose the probability density function for X is

$$f(x) = \begin{cases} 0 & \text{for } x < 0 \\ 2e^{-2x} & \text{for } x \geq 0. \end{cases}$$

a) Show that f is a probability density function.

$$\int_{-\infty}^{\infty} f(x)\,dx = \int_0^{\infty} f(x)\,dx = \lim_{t\to\infty} \int_0^t 2e^{-2x}\,dx$$

$$= \lim_{t\to\infty} 2\left(\frac{-e^{-2x}}{2}\right)\bigg]_0^t$$

$$= \lim_{t\to\infty} -e^{-2x}\bigg]_0^t$$

$$= \lim_{t\to\infty}\left(-e^{-2t} - (-e^{-2(0)})\right)$$

$$= \lim_{t\to\infty}\left(-e^{-2t} - (-1)\right) = 0 + 1 = 1.$$

b) What does $\int_{-\infty}^{10} f(x)\,dx$ represent?

The probability that it takes up to 10 minutes for the blood to begin absorbing the drug.

c) Find the probability that the drug takes at least 1 minute before starting to be absorbed.

$$P(1 \leq X < \infty) = \int_1^{\infty} 2e^{-2x}\,dx$$

$$= \lim_{t\to\infty} \int_1^t 2e^{-2x}\,dx$$

$$= \lim_{t\to\infty}(-e^{-2x})\bigg]_1^t$$

$$= \lim_{t\to\infty}(-e^{-2t} + e^{-2}) = e^{-2}.$$

Page
570

B. The **mean** of a random variable X with probability density $y = f(x)$ is

$$\mu = \int_{-\infty}^{\infty} x f(x)\, dx.$$

The number μ is the "weighted average" of all values of X. Because μ is the x-coordinate of the centroid of the region under f, μ is a balance point, meaning that half the area under f is to the left of the line $x = \mu$ and half the area is to the right of $x = \mu$.

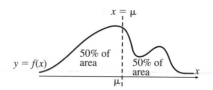

3) Show that the graph below is a probability density function and calculate the mean.

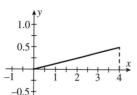

The function in the graph is $f(x) = \frac{1}{8}x$ for $0 \le x \le 4$ and $f(x) = 0$ elsewhere. The area of the region under the function is

$$\int_{0}^{4} \frac{1}{8}x\, dx = \frac{1}{16}x^2 \Big]_{0}^{4} = \frac{1}{16}(16) - 0 = 1.$$

Therefore, f is a probability density function.

The mean is

$$\int_{0}^{4} x\left(\frac{1}{8}x\right) dx = \frac{1}{8}\int_{0}^{4} x^2\, dx = \frac{x^3}{24}\Big]_{0}^{4} = \frac{8}{3}.$$

Page
572

C. A normal probability distribution with mean μ and standard deviation σ for a random variable X has a probability density function of the form

$$f(x) = \frac{1}{\sigma\sqrt{2\pi}}\, e^{-(x-\mu)^2/(2\sigma^2)}.$$

The graph is a bell-shaped curve, symmetric about the line $x = \mu$. The standard deviation σ determines how spread out the shape of the bell curve will be. The larger the value of σ, the flatter the bell curve will be and the more spread out.

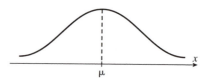

Example: A "one-hour" candle actually burns an average of 1.1 hours with a standard deviation of .05 hours. Assuming the time X that a candle burns is normally distributed, what is the probability density function for X?

$$f(x) = \frac{1}{(.05)\sqrt{2\pi}} e^{-(x-1.1)^2/(2(.05)^2)}$$

What is the probability that a candle will burn between 1.06 and 1.12 hours?

$$P(1.06 \le X \le 1.12) = \int_{1.06}^{1.12} \frac{1}{(.05)\sqrt{2\pi}} e^{-(x-1.1)^2/(2(.05)^2)} dx.$$

Using a calculator or computer, the probability is about $0.4436 = 44.36\%$.

Example: If it has been determined that the length of life of an Acme 100-watt bulb is normally distributed with mean 160 hours and standard deviation 20 hours, what is the probability that an Acme 100-watt bulb will last at least 200 hours?

Let X be the length of life of a light bulb.

$$P(200 \leq X < \infty) = \int_{200}^{\infty} \frac{1}{20\sqrt{2\pi}} e^{-(x-160)^2/(2(20)^2)} \, dx$$

$$= \frac{1}{20\sqrt{2\pi}} \int_{200}^{\infty} e^{-(x-160)^2/2(20)^2)} \, dx.$$

This improper integral will need to be approximated using an upper limit of 260 (5 standard deviations) and a calculator, computer, or one of the approximation techniques.

$$\frac{1}{20\sqrt{2\pi}} \int_{200}^{260} e^{-(x-160)^2/(2(20)^2)} \, dx \approx \frac{1}{20\sqrt{2\pi}}(1.141) \approx 0.0228.$$

About 2.28% of all Acme bulbs last at least 200 hours.

4) Both normal probability density functions below have mean 0. Which one, **a)** or **b)**, has the larger standard deviation?

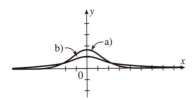

b) has the larger standard deviation because the area is more spread out.

5) Suppose that the length of time it takes to complete an income tax form is normally distributed with mean 50 minutes and standard deviation 20 minutes. Write a definite integral expression for:

a) the probability that the form can be completed in 10 minutes or less.

$$\int_{-\infty}^{10} \frac{1}{20\sqrt{2\pi}} e^{-((x-50)^2/2(20)^2)} \, dx.$$

b) the probability that the form is completed in between 30 and 60 minutes.

$$\int_{30}^{60} \frac{1}{20\sqrt{2\pi}} e^{-((x-50)^2/2(20)^2)} \, dx.$$

D. Technology Plus. Use a computer algebra system or a graphing calculator to solve.

T-1) The "Empirical Rule" for normal distributions says that the probability that the random variable lies within one standard deviation of the mean is about 68%.

a) State the Empirical Rule in terms of a definite integral.

The probability that the random variable X is within σ of the mean μ is

$$P(\mu - \sigma \le X \le \mu + \sigma)$$
$$= P(-\sigma \le X - \mu \le \sigma)$$
$$= \int_{-\sigma}^{\sigma} \frac{1}{\sigma\sqrt{2\pi}} e^{-(x-\mu)^2/2\sigma^2} \, dx \approx 68\%.$$

b) Use a calculator or a CAS to verify the Empirical Rule for $\mu = 0$ and $\sigma = 1$.

For $\mu = 0$ and $\sigma = 1$,

$$P(-1 \le X \le 1) = \int_{-1}^{1} \frac{1}{\sqrt{2\pi}} e^{-x^2/2} \, dx$$
$$\approx 0.6826, \text{ which is about 68\%.}$$

c) What does the Empirical Rule say about the probability that the random variable is within two standard deviations? Use $\mu = 0$ and $\sigma = 1$.

$$P(-2 \le X \le 2) = \int_{-2}^{2} \frac{1}{\sqrt{2\pi}} e^{-x^2/2} \, dx$$
$$\approx 0.9544 \approx 95\%.$$

Chapter 9 — Differential Equations

Section 9.1   Modeling with Differential Equations

A differential equation is an equation that relates a function and one or more of its derivatives. This section defines the basic terminology for differential equations and looks at several phenomena that are modeled by differential equations. Right now we do not know how to solve very many kinds of differential equations—that will come in later sections.

Concepts to Master

A. Differential equation; order; degree; family of solutions to a differential equation

B. Initial condition(s); solution to an initial value problem

C. Technology Plus

Summary and Focus Questions

Page 580

A. A **differential equation** is an equation involving $x, y, y', y'', \ldots, y^{(n)}$, where y is a function of x with n derivatives. The n in the highest $y^{(n)}$ is the **order** of the equation. For example, $2(y''')^4 + 5xy' + 7x = 0$ has order 3.

We shall see in later sections that differential equations can be used to model various population changes and predator-prey relationships. For example, two models for population growth/decay are $\frac{dP}{dt} = kP$ (exponential growth—the population grows ever faster with time), and $\frac{dP}{dt} = kP\left(1 - \frac{P}{K}\right)$ (logistic growth—over time the population approaches a constant population K).

A **solution** to a differential equation is a function $y = f(x)$ that satisfies the differential equation. A general solution is an expression with arbitrary constants that represents the family of all solutions.

We know how to solve differential equations of the form $y' = f(x)$ by integration, $y = \int f(x)\, dx$. In future sections we will solve several other types of differential equations.

Example: A solution to the differential equation $y' = 2x$ is $y = x^2$.
The general solution is $y = x^2 + C$.

1) The equation $x^2(y'')^3 + 4xy' - 2y + x = 0$ has order _____.

The order is 2 because of the term containing y''.

2) Is $y = e^{-2t}$ a solution to $y'' - 2y' - 8y = 0$?

Yes. $y = e^{-2t}$, $y' = -2e^{-2t}$, $y'' = 4e^{-2t}$.
Thus $y'' - 2y' - 8y$
$$= 4e^{-2t} - 2(-2e^{-2t}) - 8(e^{-2t}) = 0.$$

3) Guess a solution to $y'' = -9y$.

From your knowledge of trigonometric functions, you should see that one solution is $y = \sin 3x$. In general, $y = \sin 3x + C$.

4) Find a solution to $y' - x^3 + x = 1$.

We first solve for y':
$y' - x^3 + x = 1$
$y' = x^3 - x + 1$.
Any antiderivative of $x^3 - x + 1$ is a solution, so $y = \dfrac{x^4}{4} - \dfrac{x^2}{2} + x$. In general,
$$y = \frac{x^4}{4} - \frac{x^2}{2} + x + C.$$

5) Let $y' = \frac{x}{y}$.

   a) Could $y = f(x)$ (graphed below) be a solution to the differential equation?

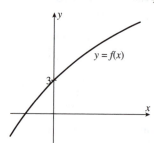

No. In the graph $y'(0)$ is positive, while $y' = \frac{x}{y}$ will be 0 or not exist when $x = 0$.

   b) Could $y = f(x)$ (graphed below) be a solution to the differential equation?

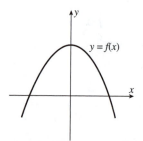

No. In the graph y' is negative when $x > 0$ and $y > 0$, but $y' = \frac{x}{y}$ will be positive when both x and y are positive.

Page
583

B. Initial conditions are specified function values that are used to determine a particular solution to a differential equation from the general solution.

Example: The equation $y' = 10x$ has general solution $y = 5x^2 + C$. If we specify an initial condition of $y(1) = 12$, then $12 = 5 + C$ implies $C = 7$ so the particular solution to $y' = 10x$ for the condition $y(1) = 12$ is $y = 5x^2 + 7$.

In general, to solve an initial value problem, find the general solution and use the initial conditions to determine the constant.

6) The general solution to $y' = 2xy$ is $y = Ce^{x^2}$ where $C > 0$. Find a solution to the initial-value problem $y' = 2xy$, $y(1) = 3$.

$y = Ce^{x^2}$. At $x = 1$, $3 = Ce^{1^2} = Ce$.

Thus $C = \frac{3}{e}$.

The particular solution is

$$y = \frac{3}{e}e^{x^2} = 3e^{x^2-1}.$$

7) Find a solution to $y' = -y$, $y(0) = 4$.

The general solution to $y' = -y$ is $y = e^{-x} + C$.
For $y(0) = 4$ we have $4 = e^{-0} + C$.
$4 = 1 + C$, so $C = 3$.
The particular solution is $y = e^{-x} + 3$.

C. Technology Plus. Use a computer algebra system or a graphing calculator to solve.

T-1) The solution to the equation $x^2y' + xy = 1$ is $y = \dfrac{\ln x + C}{x}$.
Graph, on one screen, the solutions for $C = -1, 0, 1, 2$.
Use a $[0, 5]$ by $[-5, 5]$ window.

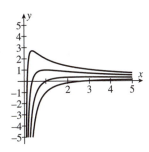

Section 9.2   Direction Fields and Euler's Method

One way to solve a differential equation is to use graphical and numeric approaches to approximate solutions. In this section we obtain the general shapes of solutions using graphs called directional fields and estimate particular solutions using Euler's method.

Concepts to Master

A. Solution curve; directional fields

B. Euler's Method

C. Technology Plus

Summary and Focus Questions

Page 586

A. The graph of a particular solution to a differential equation is called a **solution curve.** The general solution to a first order differential equation of the form $y' = F(x, y)$ is a family of functions whose graphs are related.

One way to visualize the general shape of solution curves is to draw a **direction field**—short line segments at various points (x, y) whose slope is $F(x, y)$. These line segments indicate the directions in which the family of curves proceed at each point.

The solution curves are obtained by "connecting the dots" following the directions indicated by the slopes.

Example: $y' = x + y - 1$ has this table of values for y' and corresponding direction field:

y'	-1	0	1	2	3	4
-1	-3	-2	-1	0	1	2
0	-2	-1	0	1	2	3
x 1	-1	0	1	2	3	4
2	0	1	2	3	4	5
3	1	2	3	4	5	6
4	2	3	4	5	6	7

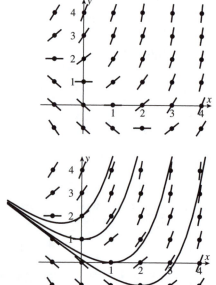

The solution curves are drawn in the figure at the right. Later we will see that the general solution is $y = -x + Ce^x$.

1) Find the directional field and sketch the general solution to $y' = xy + y$.

	y'	-1	0	1
	-2	1	0	-1
	-1	0	0	0
x	0	-1	0	1
	1	-2	0	2

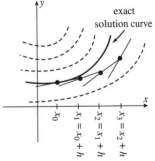

Later we will see the equation has general solution $y = Ce^{(x^2/2)+x}$.

B. Euler's method is a way to find a sequence of approximations for an initial value problem based on following tangent lines in direction fields.

For a differential equation $\dfrac{dy}{dx} = F(x, y)$ with initial condition (x_0, y_0), Euler's method constructs a table of (x, y) values that estimate the solution. The x_i values are evenly spaced out (a **step size** of h between successive x_i). Each y_{i+1} is calculated from the direction field by moving h units in the x-direction along the tangent line at (x_i, y_i):

x	y
x_0	y_0 (given)
$x_1 = x_0 + h$	$y_1 = y_0 + hF(x_0, y_0)$
$x_2 = x_1 + h$	$y_2 = y_1 + hF(x_1, y_1)$
$x_3 = x_2 + h$	$y_3 = y_2 + hF(x_2, y_2)$
$\vdots$	$\vdots$
$x_n = x_{n-1} + h$	$y_n = y_{n-1} + hF(x_{n-1}, y_{n-1})$
	$\vdots$

In general, as you step from x_0 to x_1, to x_2, and so on, the y_i become less accurate. Choosing h to be smaller will help improve accuracy.

Page 589

Example: Use Euler's method with step size 0.1 to estimate $y(1.4)$, where $y(x)$ is the solution of the initial-value problem $y' = \frac{6x^2}{y}$ and $y(1) = 2$.

We are given $x_0 = 1$, $y_0 = 2$, and $h = 0.1$. Here are four iterations of Euler's method:

x	y	$y' = \frac{6x^2}{y}$
$x_0 = 1$	$y_0 = 2$	$y' = 6(1)^2/2 = 3$
$x_1 = 1.1$	$y_1 = 2 + (0.1)3 = 2.3$	$y' = 6(1.1)^2/2.3 = 3.157$
$x_2 = 1.2$	$y_2 = 2.3 + (0.1)3.157 = 2.616$	$y' = 6(1.2)^2/2.616 = 3.303$
$x_3 = 1.3$	$y_3 = 2.616 + (0.1)3.303 = 2.946$	$y' = 6(1.3)^2/2.946 = 3.442$
$x_4 = 1.4$	$y_4 = 2.946 + (0.1)3.442 = 3.290$	

Our estimate is $y(1.4) \approx 3.290$. In Section 9.3 we shall see that $y = 2x^{3/2}$ and therefore, to three decimals, $y(1.4) = 3.313$.

2) Use Euler's method with step size 0.1 to approximate $y(1.5)$, where $y(x)$ is the solution to $y' = 2xy$, $y(1) = 3$.

$h = 0.1$.

x	y	$y' = 2xy$
1	3.000	6.000
1.1	3.600	7.920
1.2	4.392	10.541
1.3	5.446	14.160
1.4	6.862	19.214
1.5	8.783	

The estimate is $y(1.5) \approx 8.783$.

3) Repeat question **2)** with step size 0.05.

$h = 0.05$.

x	y	$y' = 2xy$
1	3.000	6.000
1.05	3.300	6.930
1.10	3.647	8.023
1.15	4.048	9.310
1.20	4.514	10.834
1.25	5.056	12.640
1.30	5.688	14.789
1.35	6.427	17.353
1.40	7.295	20.426
1.45	8.316	24.116
1.50	9.522	

The estimate is $y(1.5) \approx 9.522$.

4) The solution to $y' = 2xy$ and $y(1) = 3$ is $y = 3e^{x^2-1}$. Find the value of $y(1.5)$ to three decimals. Why is the estimate in question **3)** a better estimate than that of question **2)**?

$y(1.5) = 3e^{1.5^2-1} = 3e^{1.25} \approx 10.471$.
The estimate of 9.522 is better than 8.783 because the step size h in question **3)** is smaller than the step size in question **2)**.

C. Technology Plus. Use a computer algebra system or a graphing calculator to solve.

T-1) Use a spreadsheet, calculator, or CAS to approximate $y(2)$, where $y(x)$ is the solution to the initial value problem
$$\frac{dy}{dx} = \frac{1 - xy}{x^2}, \quad y(1) = 3.$$

a) Use $h = 0.1$.

$h = 0.100.$

i	x_i	y_i	y'
0	1.00	3.000	−2.000
1	1.10	2.800	−1.719
2	1.20	2.628	−1.496
3	1.30	2.479	−1.315
4	1.40	2.348	−1.167
5	1.50	2.231	−1.043
6	1.60	2.127	−0.938
7	1.70	2.023	−0.844
8	1.80	1.939	−0.769
9	1.90	1.862	−0.703
10	2.00	1.792	

b) Use $h = 0.05$.

$h = 0.050.$

i	x_i	y_i	y'
0	1.00	3.000	−2.000
1	1.05	2.900	−1.855
2	1.10	2.807	−1.726
3	1.15	2.721	−1.610
4	1.20	2.640	−1.506
5	1.25	2.565	−1.412
6	1.30	2.495	−1.327
7	1.35	2.428	−1.250
8	1.40	2.366	−1.180
9	1.45	2.307	−1.115
10	1.50	2.251	−1.056
11	1.55	2.198	−1.002
12	1.60	2.148	−0.952
13	1.65	2.100	−0.906
14	1.70	2.055	−0.863
15	1.75	2.012	−0.823
16	1.80	1.971	−0.786
17	1.85	1.932	−0.752
18	1.90	1.894	−0.720
19	1.95	1.858	−0.690
20	2.00	1.823	

c) The exact value is $y(2) = \dfrac{\ln 2 + 3}{2}$ ≈ 1.847. For the smaller value of h, is the approximation more accurate?

Yes.

Section 9.3 **Separable Equations**

Some differential equations can be solved explicitly. This section shows you how to solve one particular first-order type (called separable equations) where y' may be written as an expression involving only the variable x divided by an expression involving only the variable y.

Concepts to Master

Separable first-order differential equation; Solution of a separable equation

Summary and Focus Questions

Page 594

A **separable first-order differential equation** has the form $\dfrac{dy}{dx} = \dfrac{g(x)}{h(y)}$.

To solve a separable equation:

i) Rewrite the equation as $h(y)\,dy = g(x)\,dx$.

ii) Integrate to get $\displaystyle\int h(y)\,dy = \int g(x)\,dx$, an equation in x and y.

iii) Solve the equation for y.

Example: Find the general solution to the separable equation $y' = \dfrac{1}{3y^2}$.

i) $\displaystyle\int 3y^2 \, dy = \int 1 \, dx$.

ii) $y^3 = x + C$.

iii) $y = \sqrt[3]{x + C}$ is the general solution.

Some particular solutions are $y = \sqrt[3]{x + 1}$, $y = \sqrt[3]{x - 4}$, and so on. The family of solutions is pictured below:

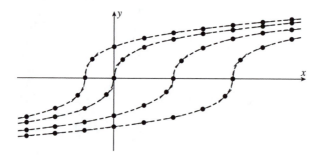

From $y' = \dfrac{1}{3y^2}$ we see that y' does not exist when $y = 0$. Therefore, each curve in the family of solutions crosses the x-axis ($y = 0$) perpendicular to the axis.

1) Is $y' = y^x$ separable?

2) Is $\dfrac{dy}{dx} = 2xy$ separable?

3) Sketch a direction field for $\dfrac{dy}{dx} = 2xy$.

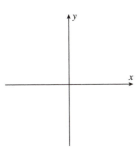

4) Solve $\dfrac{dy}{dx} = 2xy$ and draw several solution curves. (Compare your curves to the direction field in question **3**.)

No.

Yes. $\dfrac{dy}{dx} = 2xy$ may be rewritten as $\dfrac{dy}{dx} = \dfrac{2x}{y^{-1}}$.

Here is a table of some values of y':

	y				
	−2	−1	0	1	2
−2	8	4	0	−4	−8
−1	4	2	0	−2	−4
x **0**	0	0	0	0	0
1	−4	−2	0	2	4
2	−8	−4	0	4	8

Plotting these points and drawing tangent lines, the direction field is:

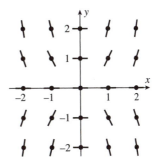

i) $\dfrac{dy}{dx} = 2xy$, so $\dfrac{dy}{y} = 2x\,dx$.

ii) $\displaystyle\int \dfrac{dy}{y} = \int 2x\,dx$.

$\ln|y| = x^2 + C$.

iii) $|y| = e^{x^2 + C} = e^{x^2}e^{C}$.

$y = Ke^{x^2}$, K a real number.

Solution curves for $K = 0.1, 0.5, 1$ and 2 are given in the upper part of the figure, while solutions for $K = -0.1, -0.5, -1$ and -2 are given in the lower part of the figure.

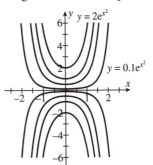

5) Solve the equation $y' = \dfrac{\sin x}{\cos y}$.

The equation is separable.

i) $\dfrac{dy}{dx} = \dfrac{\sin x}{\cos y}$

$\cos y \, dy = \sin x \, dx.$

ii) $\displaystyle\int \cos y \, dy = \int \sin x \, dx$

$\sin y = -\cos x + C.$

iii) $y = \sin^{-1}(-\cos x + C).$

6) Find the particular solution for the equation $e^{x^2}yy' + x = 0$ with initial condition $y(0) = 2$.

i) $e^{x^2}y\dfrac{dy}{dx} = -x$

$y \, dy = -xe^{-x^2} \, dx.$

ii) $\displaystyle\int y \, dy = \int -xe^{-x^2} \, dx$

$\dfrac{y^2}{2} = \dfrac{1}{2}e^{-x^2} + C.$

iii) $y^2 = e^{-x^2} + C$

$y = (e^{-x^2} + C)^{1/2} = \sqrt{e^{-x^2} + C}.$

Find the particular solution:

At $x = 0, y = 2$, so $\sqrt{e^{-0} + C} = 2.$

$1 + C = 4, C = 3.$

The particular solution is $y = \sqrt{e^{-x^2} + 3}.$

7) Solve $x\dfrac{dy}{dx} = -y^2$ $(x > 0)$ with the initial condition of $x_0 = 1, y_0 = \frac{1}{3}$.

i) $-y^{-2} \, dy = x^{-1} \, dx.$

ii) $\displaystyle\int -y^{-2} \, dy = \int x^{-1} \, dx.$

$\dfrac{1}{y} = \ln|x| + C = \ln x + C.$

iii) $y = \dfrac{1}{\ln x + C}$ is the general solution.

For $x_0 = 1, y_0 = \dfrac{1}{3}, \dfrac{1}{3} = \dfrac{1}{\ln 1 + C} = \dfrac{1}{C}$, so

$C = 3.$

The particular solution is $y = \dfrac{1}{\ln x + 3}.$

Section 9.4 Models for Population Growth

This section uses differential equations to describe changes in sizes of populations over time. Two main types of models are presented: the law of natural growth (exponential growth or decline) and the logistics model where a population increases exponentially in its early stages and then levels off at a given capacity.

Concepts to Master

A. Law of Natural Growth; Solution to $\frac{dP}{dt} = kP$

B. Logistic Differential Equation; Solution to $\frac{dP}{dt} = kP\left(1 - \frac{P}{K}\right)$

C. Technology Plus

Summary and Focus Questions

Page 605

A. If a population (P) varies over time such that its rate of change (P') is proportional to P, then we may write $\frac{P'}{P} = k$, for some constant k. The population is changing exponentially and the resulting differential equation $P' = kP$ or $\frac{dP}{dt} = kP$ is the **Law of Natural Growth.**
The solution to $\frac{dP}{dt} = kP$ is

$$P(t) = P_0 e^{kt},$$

where $P_0 = P(0)$ is the initial population.

If $k > 0$, the population is increasing; if $k < 0$, the population is decreasing.

Example: Suppose the half life of kryptonite is 2000 years. Given 50 grams of kryptonite, what is its mass after 100 years?

If $P(t)$ is the amount (population) after t years, the problem asks to find $P(100)$ given $P(0) = 50$. Thus, $P(t) = 50e^{kt}$. Since the half life is 2000 years, we know $P(2000) = 50e^{2000k} = 25$. Therefore,

$$e^{2000k} = 0.5$$
$$2000k = \ln(0.5), \text{ so } k = \frac{\ln(0.5)}{2000}.$$

The model is $P(t) = 50e^{\frac{\ln(0.5)}{2000}t} = 50(0.5)^{\frac{t}{2000}}$

and $P(100) = 50(0.5)^{\frac{100}{2000}} = 50(0.5)^{0.05} \approx 48.3$ grams.

1) True or False:
Under the law of natural growth, a
population will either continue to increase
forever or continue to decrease forever.

True.

2) A village population in the year 2000 was
800 and in 2010 it was 1000. If the village
population experiences natural growth,
what will be the size of the village in 2020?

<*Find k and P_0 for $P(t) = P_0e^{kt}$. Then find
$P(20)$.*>

Let the base year ($t = 0$) be 2000. Then
$P_0 = 800$. At $t = 10$ (the year 2010),
$$1000 = P(10) = 800e^{10k}$$
$$e^{10k} = \frac{1000}{800} = 1.25.$$

$10k = \ln(1.25)$, so $k = \frac{\ln(1.25)}{10} \approx 0.0223$.

Thus, $P(t) = 800e^{0.0223t}$ and for $t = 20$ (the
year 2020), $P(20) = 800e^{0.0223(20)} \approx 1250$.

3) Red Hawk High School had a student
population of 850 in 2008 and 950 in 2011.
To reach "Class A" status it must have a
population of 1200. Assuming the student
population follows the law of natural
growth, what year will Red Hawk High
reach Class A status?

<*Find k and P_0 for $P(t) = P_0e^{kt}$ and for what
t is $P(t) = 1200$.*>

Let the base year ($t = 0$) be 2008. Then
$P_0 = 850$. At $t = 3$ (the year 2011),
$$950 = P(3) = 850e^{3k}$$
$$e^{3k} = \frac{950}{850} = 1.118.$$

$3k = \ln(1.118)$, so $k = \frac{\ln(1.118)}{3} \approx 0.037$.

Thus, $P(t) = 850e^{0.037t}$.

$P(t) = 850e^{0.037t} = 1200$

$e^{0.037t} = \frac{1200}{850} = 1.412$

$.037t = \ln 1.412$

$t = \frac{\ln(1.412)}{.037} \approx \frac{.345}{.037} \approx 9.3$ years.

Since the base year is 2008, the first year the
population will exceed 1200 is 2018.

Page
607

B. Some populations increase rapidly for a time and then level off near some maximum capacity (K), as depicted in the graph at the right. These populations follow a logistic model of growth.

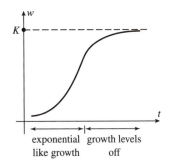

The **logistic differential equation** is

$$\frac{dP}{dt} = kP\left(1 - \frac{P}{K}\right).$$

The equation is separable and has solution

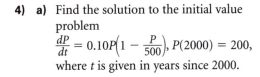

exponential   growth levels
like growth   off

$$P(t) = \frac{K}{1 + Ae^{-kt}}, \text{ where } A = \frac{K - P_0}{P_0}.$$

K is called the **carrying capacity** for the equation because $\lim\limits_{t \to \infty} P(t) = K$.

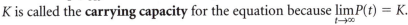

4) a) Find the solution to the initial value problem
$$\frac{dP}{dt} = 0.10P\left(1 - \frac{P}{500}\right), P(2000) = 200,$$
where t is given in years since 2000.

$P_0 = 200$ (in 2000).
$$A = \frac{K - P_0}{P_0} = \frac{500 - 200}{200} = 1.5.$$
$$P(t) = \frac{500}{1 + 1.5e^{-0.1t}}.$$

b) What is the population for 2005?

$$P(5) = \frac{500}{1 + 1.5e^{-0.1(5)}} \approx 262.$$

c) When does the population reach 450?

$$P(t) = \frac{500}{1 + 1.5e^{-0.1t}} = 450$$
$$1 + 1.5e^{-0.1t} = \frac{10}{9}$$
$$1.5e^{-0.1t} = \frac{1}{9}$$
$$e^{-0.1t} = \frac{2}{27}$$
$$-0.1t = \ln\frac{2}{27}$$
$$t = \frac{\ln\frac{2}{27}}{-0.1} = -10\ln\frac{2}{27} = 10\ln\frac{27}{2} \approx 26 \text{ years.}$$
(The year 2026).

5) A biologist stocks a shrimp farm pond with 1000 shrimp. The number of shrimp double in one year and the pond has a carrying capacity of 10,000. How long does it take the shrimp population to reach 99% of the pond's capacity?

$P_0 = 1000. A = \frac{10000 - 1000}{1000} = 9.$
So $P(t) = \frac{10000}{1 + 9e^{-kt}}.$
From $P(1) = 2000$, we have
$$2000 = \frac{10000}{1 + 9e^{-k(1)}} = \frac{10000}{1 + 9e^{-k}}$$
$$1 + 9e^{-k} = 5$$
$$9e^{-k} = 4$$
$$e^{-k} = \frac{4}{9}$$
$$-k = \ln\frac{4}{9}$$
$$k = \ln\frac{9}{4}.$$

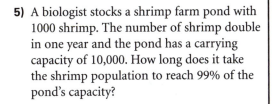

$$P(t) = \frac{10000}{1 + 9e^{-(\ln 9/4)t}} = \frac{10000}{1 + 9\left(\frac{4}{9}\right)^t}.$$

To reach 99% of capacity

$$\frac{10000}{1 + 9\left(\frac{4}{9}\right)^t} = 0.99(10000)$$

$$1 + 9\left(\frac{4}{9}\right)^t = \frac{100}{99}$$

$$9\left(\frac{4}{9}\right)^t = \frac{1}{99}$$

$$\left(\frac{4}{9}\right)^t = \frac{1}{891}$$

$$t \ln\left(\frac{4}{9}\right) = \ln\frac{1}{891}$$

$$t = \frac{\ln\frac{1}{891}}{\ln\frac{4}{9}} \approx 8.376 \text{ years.}$$

C. Technology Plus. Use a computer algebra system or a graphing calculator to solve.

T-1) An environment has a carrying capacity of 10,000 insects. The population follows a logistic model with $k = 0.02$. On the same screen, graph the solution curves for initial populations of 1000, 2000, and 3000.

For an initial population of $P_0 = 1000$,

$$A = \frac{10000 - 1000}{1000} = 9.$$

$$P(t) = \frac{10000}{1 + 9\,e^{-0.02t}}.$$

For $P_0 = 2000$,

$$A = \frac{10000 - 2000}{2000} = 4 \text{ and}$$

$$P(t) = \frac{10000}{1 + 4\,e^{-0.02t}}.$$

For $P_0 = 3000$, $A = \frac{10000 - 3000}{3000} = \frac{7}{3}$

and $P(t) = \dfrac{10000}{1 + \frac{7}{3}\,e^{-0.02t}}.$

For the window $[0, 500]$ by $[0, 10500]$ the screen looks like this.

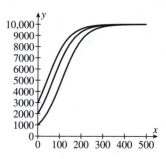

Section 9.5   Linear Equations

This section considers another type of first-order differential equation—the linear differential equation. Like separable equations, linear differential equations also have an explicit form for their solutions and have many applications modeling physical phenomena.

Concepts to Master

Linear first-order differential equations; Solutions by integrating factors

Summary and Focus Questions

Page 616

A first-order differential equation is **linear** if the equation has the form

$$\frac{dy}{dx} + P(x)y = Q(x)$$

where $P(x)$ and $Q(x)$ are continuous functions on a given interval.
To solve a linear equation:

i) Multiply both sides by the integrating factor $I(x) = e^{\int P(x)\,dx}$.

ii) Solve $I(x)y = \int I(x)Q(x)\,dx$ for y: $y = \dfrac{\int I(x)Q(x)\,dx}{I(x)}$.

Example: Solve the linear differential equation $\dfrac{\sin x}{\cos x}y' + y = \sin^2 x$, where $0 < x < \pi/2$.

Multiply both sides by $\dfrac{\cos x}{\sin x}$: $\left(\dfrac{\sin x}{\cos x}y' + y\right)\left(\dfrac{\cos x}{\sin x}\right) = (\sin^2 x)\left(\dfrac{\cos x}{\sin x}\right)$

$$y' + y\left(\frac{\cos x}{\sin x}\right) = \sin x \cos x.$$

The linear factors are $P(x) = \dfrac{\cos x}{\sin x}$ and $Q(x) = \sin x \cos x$.

$\displaystyle\int \frac{\cos x}{\sin x}\,dx = \ln|\sin x| = \ln(\sin x)$ since $0 < x < \pi/2$. Therefore, $I(x) = e^{\ln(\sin x)} = \sin x$

and $\displaystyle\int I(x)Q(x)\,dx = \int (\sin x)(\sin x \cos x)\,dx = \int \sin^2 x \cos x\,dx = \tfrac{1}{3}\sin^3 x + C.$

Thus $y = \dfrac{\frac{1}{3}\sin^3 x + C}{\sin x} = \tfrac{1}{3}\sin^2 x + \dfrac{C}{\sin x} = \tfrac{1}{3}\sin^2 x + C\csc x.$

1) Is $y' + \frac{x}{y} = x^2 + x$ linear?

No, because of the term $\frac{x}{y}$.

2) What integrating factor should be used for $y' + 7y = x^2 + x$?

$P(x) = 7.$
$I(x) = e^{\int 7\, dx} = e^{7x}.$

3) Find the general solution to $xy' + y = e^{-x}$, $x > 0$.

Put in standard form: $y' + \frac{1}{x}y = \frac{e^{-x}}{x}$.

Let $P(x) = \frac{1}{x}$ and $Q(x) = \frac{e^{-x}}{x}$.

$I(x) = e^{\int 1/x\, dx} = e^{\ln x} = x.$

$xy = \int x\frac{e^{-x}}{x}\, dx = \int e^{-x}\, dx = -e^{-x} + C.$

$y = \frac{-e^{-x} + C}{x}.$

4) Solve the initial-value problem
$y' + y = xe^x, y(0) = \frac{7}{4}.$

The equation is linear with $P(x) = 1$ and $Q(x) = xe^x$.

$I(x) = e^{\int 1\, dx} = e^x.$

$e^x y = \int e^x(xe^x)\, dx = \int xe^{2x}\, dx$

(by parts or by integral #96)

$= \frac{1}{2}xe^{2x} - \frac{1}{4}e^{2x} + C.$

Thus, $y = \frac{\frac{1}{2}xe^{2x} - \frac{1}{4}e^{2x} + C}{e^x}$

or $y = \frac{1}{2}xe^x - \frac{1}{4}e^x + \frac{C}{e^x}.$

At $x = 0, y = \frac{7}{4}$ we have

$\frac{7}{4} = \frac{1}{2}(0)e^0 - \frac{1}{4}e^0 + \frac{C}{e^0}$

$C = \frac{8}{4} = 2.$

Thus $y = \frac{1}{2}xe^x - \frac{1}{4}e^x + \frac{2}{e^x}$ is the particular solution.

5) Can a first-order differential equation be both linear and separable?

Yes, the simple equation $y' = x^2$ is both. The equation is separable because $y' = \frac{x^2}{h(y)}$, where $h(y) = 1$, and the equation is linear because $y' + (0)y = x^2$, where $P(x) = 0$.

Section 9.6 Predator-Prey Systems

This section discusses a model of a two-species predator-prey situation like that of wolf and rabbit populations. The model uses two linked differential equations. The equations are linked because change in each population is related to the number of species in both populations. If the number of rabbits increases, then the wolf population will change (increase) because there is plenty to eat and wolves to breed. Then, with more wolves around, the rabbit population will decrease.

Concepts to Master

A. Model predator-prey populations using two linked linear differential equations; Equilibrium solution

B. Phase plane

Summary and Focus Questions

Page 622

A. Let $R(t)$ be the number of prey at time t (R stands for rabbit) and let $W(t)$ be the number of predators at time t (W stands for wolf). Several reasonable assumptions are necessary in a predator-prey model:

i) the main cause of death among the prey is that they are eaten by the predators

ii) the birth and survival rate of predators depends on the amount of food available; that is, the number of prey

iii) the two species encounter each other at a rate proportional to the product of their populations (the more of either species, the more likely they will encounter each other).

The populations $R(t)$ and $W(t)$ are modeled by a pair of **predator-prey equations** (also called the **Lotka-Volterra** equations):

$$\frac{dR}{dt} = kR - aRW \quad \text{and} \quad \frac{dW}{dt} = -rW + bRW,$$

where k, r, a, and b are positive constants. It is usually impossible to find explicit solutions for R and W. This system of equations is usually analyzed by graphical methods.

An **equilibrium solution** is a solution to the system of equations for which the populations remain constant. It may be interpreted that a constant number of prey is just the right amount to support a constant number of predators. Those values of R and W are found by solving $\frac{dR}{dt} = 0$ and $\frac{dW}{dt} = 0$.

Example: In a neighborhood with cats (W) and mice (R), the populations are modeled by a system of predator-prey equations with constants $k = 0.012$, $a = 0.001$, $r = 0.05$, and $b = 0.0005$. Find the system of equations and the equilibrium point.

The equations are $\dfrac{dR}{dt} = 0.012R - 0.001RW$ and $\dfrac{dW}{dt} = -0.05W + 0.0005RW$.

The equilibrium point is where

$$0.012R - 0.001RW = 0$$
$$-0.05W + 0.0005RW = 0$$
$$R(0.012 - 0.001W) = 0$$
$$W(-0.05 + 0.0005R) = 0$$

Disregarding the solutions $R = 0$ or $W = 0$, the equations become

$$0.012 - 0.001W = 0, \text{ so } W = \frac{0.012}{0.001} = 12 \text{ cats.}$$

$$-0.05 + 0.0005R = 0, \text{ so } R = \frac{0.05}{0.0005} = 100 \text{ mice.}$$

1) A part of the Caribbean Sea contains populations of sharks and mullets governed by a system of predator-prey equations with constants $k = 0.04$, $r = 0.04$, $a = 0.008$, and $b = 0.00001$. Find the constant solutions for the populations of sharks and mullets described by the equations.

Let R refer to the mullet population and W the shark population.

At equilibrium $\dfrac{dR}{dt} = 0$ and $\dfrac{dW}{dt} = 0$.

$\dfrac{dR}{dt} = 0.04R - 0.008RW = 0$

$R(0.04 - 0.008W) = 0$

$0.04 - 0.008W = 0$

$W = 5$ sharks.

$\dfrac{dW}{dt} = -0.04W + 0.00001RW = 0$

$W(-0.04 + 0.00001R) = 0$

$-0.04 + 0.00001R = 0$

$R = 4000$ mullets.

B. The differential equation $\dfrac{dW}{dR} = \dfrac{\dfrac{dW}{dt}}{\dfrac{dR}{dt}}$ models the relationship between the populations R and W as time passes. Its solutions are graphed in the R-W plane, called the **phase plane**. Solutions to the predator-prey equations often are closed curves with the equilibrium solution(s) located inside all the closed curves.

Example: From the previous example, $\frac{dR}{dt} = 0.012R - 0.001RW$ and

$\frac{dW}{dt} = -0.05W + 0.0005RW$. A table of values for $\frac{dW}{dR} = \frac{-0.05W + 0.0005RW}{0.012R - 0.001RW}$ is:

		R			
		70	90	110	130
W	10	−1.071	−0.278	0.227	0.577
	11	−2.357	−0.611	0.500	1.269
	13	2.786	0.722	−0.591	−1.500
	14	1.500	0.389	−0.318	−0.808

For these values, the direction field is:

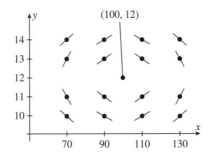

2) From question **1)** find $\frac{dW}{dR}$ and draw the direction field for these values:

		R			
		1000	3000	5000	7000
W	3.5				
	4.5				
	5.5				
	6.5				

$\frac{dW}{dR} = \frac{\frac{dW}{dt}}{\frac{dR}{dt}} = \frac{-0.04W + 0.00001RW}{0.04R - 0.008RW}.$

		R			
		1000	3000	5000	7000
W	3.5	−0.0088	−0.0010	0.0006	0.0013
	4.5	−0.0338	−0.0038	0.023	0.0048
	5.5	0.0413	0.0046	−0.0028	−0.0059
	6.5	0.0163	0.0018	−0.0011	−0.0023

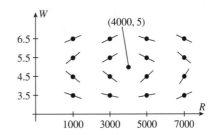

Chapter 10 — Parametric Equations and Polar Coordinates

Section 10.1   Curves Defined by Parametric Equations

As a point (x, y) moves along a curve from time $t = a$ to $t = b$, the coordinates x and y may each be described as a function of a time variable t. The idea of describing a curve with a pair of parametric equations (t is the parameter) is covered in this section. We will see that graphs of ordinary functions ($y = f(x)$) are one of many types of curves that may be defined parametrically.

Concepts to Master

A. Parameter; Parametric equations; Graphs of curves defined parametrically; Cycloid

B. Elimination of the parameter

C. Technology Plus

Summary and Focus Questions

Page 636

A. A set of **parametric equations** has the form

$$x = f(t)$$
$$y = g(t),$$

where f and g are functions of a third variable t, called a **parameter.** Each value of t determines a point (x, y) in the plane. The collection of all such points is a **parametric curve.**

A curve may be described by several different pairs of equations. Here are two sets of parametric equations for the quarter of the unit circle in the first quadrant:

$$x = t$$
$$y = \sqrt{1 - t^2}$$

for $t \in [0, 1]$.

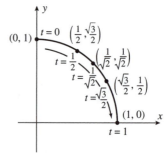

$x = \cos t$

$y = \sin t$

for $t \in \left[0, \dfrac{\pi}{2}\right].$

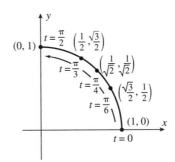

Each has the same graph (set of points). The first curve is traversed clockwise as t increases; the second graph is traversed counterclockwise.

Note that a curve defined parametrically need not be the graph of a function. The graph of

$x = 2 \sin 3t, y = 2 \cos 5t, 0 \le t \le 2\pi,$

given at the right, is clearly not the graph of a function $y = f(x).$

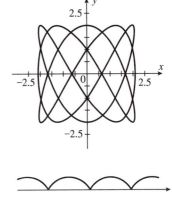

Suppose you paint a white spot on the edge of a bicycle wheel. As the bicycle rides past you the spot traces out a repeating arched curve that touches the ground once every revolution. That curve is a **cycloid**—a curve traced out by a point P on a circle as the circle rolls along a straight line. If r is the radius of the circle and the point P passes through $(0, 0)$, the parametric equations to describe the cycloid are:

$x = r(\theta - \sin \theta)$

$y = r(1 - \cos \theta).$

1) Sketch a graph of the curve given by $x = \sqrt{t}, y = t + 2.$

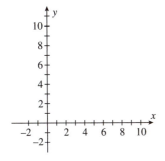

Compute some values and plot points:

t	0	1	4	9
x	0	1	2	3
y	2	3	6	11

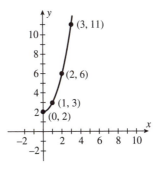

2) Describe the graph of $x = a + bt$, $y = c + dt$ where a, b, c, and d are constants.

We see the curve is half a parabola since $x = \sqrt{t}$ implies $x^2 = t$.
Thus $y = x^2 + 2$, $x \geq 0$.

If both b and d are zero, this is the single point (a, c). If $b = 0$ and $d \neq 0$ this is a vertical line through (a, c). If $b \neq 0$, then $t = \frac{x - a}{b}$ and $y = c + d\frac{(x - a)}{b}$. Thus $y = \frac{d}{b}x + c - \frac{da}{b}$, so the graph is a line through (a, c) with slope $\frac{d}{b}$.

3) Does the pair
$$x = e^t$$
$$y = e^{2t}, \quad -\infty < t < \infty$$
represent the parabola $y = x^2$?

The pair represent a portion of the parabola $y = e^{2t} = (e^t)^2 = x^2$. Since $x = e^t$, $x > 0$. The pair of equations represent that portion of $y = x^2$ to the right of the y-axis.

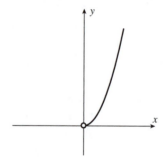

4) Use a graphing calculator to sketch
$$x = t^2 + 3t$$
$$y = 2 - t^3.$$

t	x	y
−4	4	66
−3	0	29
−2	−2	10
−1	−2	3
0	0	2
1	4	1
2	10	−6
3	18	−25
4	28	−62

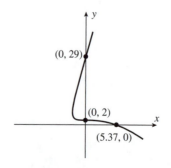

5) For a cycloid formed by a circle of radius 2:

<Use a calculator or plot several points.>

a) Draw the cycloid for $0 \le \theta \le 4\pi$.

θ	x	y
0	0	0
$\dfrac{\pi}{2}$	$\pi - 2$	2
π	2π	4
$\dfrac{3\pi}{2}$	$3\pi + 2$	2
2π	4π	0
3π	6π	4
4π	8π	0

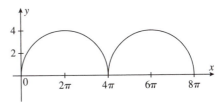

b) Where is the slope of the tangent line to the cycloid not defined for $0 \le \theta \le 4\pi$?

The tangent line does not exist at the cusps, where $\theta = 0, 2\pi, 4\pi$.

c) For what values of θ is the curve at maximum height?

By the symmetry of the cycloid, the maximum occurs midway between the cusps. The maximum is 4 and occurs at $\theta = \pi$ and $\theta = 3\pi$.

B. Sometimes parametric equations may be combined to produce a single equation not involving the parameter t; this process is called **eliminating the parameter.** The method to do so depends greatly on the nature of the parametric equations involved. Sometimes identities need to be employed, especially if trigonometric functions are involved. Other times you can solve for t in one parametric equation and substitute the result in the other equation.

Example: Eliminate the parameter in $x = 2 + 3t, y = t^2 - 2t + 1$.

Solve $x = 2 + 3t$ for t to get $t = \frac{1}{3}(x - 2)$. Substitute this in the other equation:

$$y = \left[\tfrac{1}{3}(x - 2)\right]^2 - 2\left[\tfrac{1}{3}(x - 2)\right] + 1.$$

$$y = \tfrac{1}{9}x^2 - \tfrac{10}{9}x + \tfrac{25}{9}.$$

Page 636

Example: Eliminate the parameter in $x = \sin t$, $y = \cot^2 t$.

$y = \cot^2 t = \dfrac{\cos^2 t}{\sin^2 t} = \dfrac{1 - \sin^2 t}{\sin^2 t}$. Thus, $y = \dfrac{1 - x^2}{x^2}$.

We note from $x = \sin t$, $-1 \leq x \leq 1$ and $x \neq 0$ (since $\cot t$ is undefined when $\sin t = 0$).

6) Eliminate the parameter in each

   a) $x = e^t$, $y = t^2$.

Solve for t in terms of y:
$y = t^2$, $\sqrt{y} = t$ for $t \geq 0$.
Thus, $x = e^{\sqrt{y}}$, $y \geq 0$.
A solution may also be obtained by solving
for t in terms of x:
$x = e^t$, $t = \ln x$, $y = (\ln x)^2$, $x > 0$.

   b) $x = 1 + \cos t$, $y = \sin^2 t$.

$x = 1 + \cos t$, $x - 1 = \cos t$
$\cos^2 t = (x - 1)^2$.
Since $\sin^2 t = y$, and $\cos^2 t + \sin^2 t = 1$
we have $(x - 1)^2 + y = 1$.
Thus, $y = 1 - (x - 1)^2$.
Since $x = 1 + \cos t$, $0 \leq x \leq 2$.

   c) $x = 2 \sec t$, $y = 3 \tan t$.

$\sec t = \dfrac{x}{2}$ and $\tan t = \dfrac{y}{3}$.

Hence the identity $\tan^2 t + 1 = \sec^2 t$

becomes $\left(\dfrac{y}{3}\right)^2 + 1 = \left(\dfrac{x}{2}\right)^2$ or $\dfrac{x^2}{4} - \dfrac{y^2}{9} = 1$.

7) Describe the motion of a particle moving
along a curve given by
$x = 4 \sin t$, $y = 3 \cos t$, $0 \leq t \leq \dfrac{3\pi}{2}$.

Eliminate the parameter t:

$\dfrac{x}{4} = \sin t$ and $\dfrac{y}{3} = \cos t$. Since

$\sin^2 t + \cos^2 t = 1$

$\left(\dfrac{x}{4}\right)^2 + \left(\dfrac{y}{3}\right)^2 = 1$

$\dfrac{x^2}{16} + \dfrac{y^2}{9} = 1$, which is an ellipse.

Because $0 \leq t \leq \dfrac{3\pi}{2}$, the particle starts at

$(0, 3)$ and traverses $\dfrac{3}{4}$ of the way around the
ellipse.

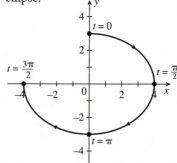

C. Technology Plus. Use a computer algebra system or a graphing calculator to solve.

T-1) Sketch the graphs of $r = \cos(2k - 1)\theta$, $0 \le \theta \le 2\pi$, for $k = 1, 2, 3$, and 4. How does the graph change as k increases?

$k = 1$

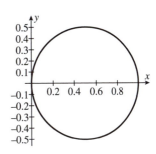

$k = 2$

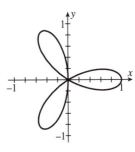

$k = 3$

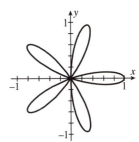

$k = 4$

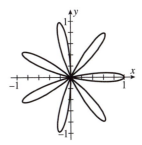

The number of leaves is $2k - 1$.

T-2) Use a graphing calculator to sketch a graph
of $x = t + 2\sin 2t$, $y = t + \sin 4t$
for $-9 \le t \le 9$.

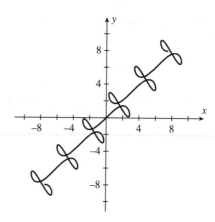

Section 10.2 Calculus with Parametric Curves

This section describes several familiar concepts for curves that are defined parametrically: slope of a tangent line to a curve, curve sketching, length of a curve, areas of regions enclosed by curves, and surfaces of solids of revolution.

Concepts to Master

A. First and second derivative of a function defined by a pair of parametric equations; applications to graphs of curves defined parametrically

B. Area of a region enclosed by curves defined parametrically

C. Length of a curve defined parametrically

D. Area of a surface of revolution for a curve defined parametrically

E. Technology Plus

Summary and Focus Questions

Page 645

A. The slope of the line tangent to a parametric curve described by $x = f(t)$, $y = g(t)$ is the first derivative of y with respect to x and is given by

$$\frac{dy}{dx} = \frac{\frac{dy}{dt}}{\frac{dx}{dt}} \quad \text{if } \frac{dx}{dt} \neq 0.$$

Likewise, the second derivative of y with respect to x is the derivative of the first derivative:

$$\frac{d^2y}{dx^2} = \frac{\frac{d}{dt}\left(\frac{dy}{dx}\right)}{\frac{dx}{dt}}.$$

The curve will have a horizontal tangent when $\frac{dy}{dt} = 0$ and $\frac{dx}{dt} \neq 0$, and will have a vertical tangent when $\frac{dx}{dt} = 0$ and $\frac{dy}{dt} \neq 0$.

1) Let $x = t^3$, $y = t^2 - 2t$.

a) Find $\frac{dy}{dx}$ and $\frac{d^2y}{dx^2}$.

$\frac{dx}{dt} = 3t^2$ and $\frac{dy}{dt} = 2t - 2$, so $\frac{dy}{dx} = \frac{2t-2}{3t^2}$.

$\frac{d}{dt}\left(\frac{dy}{dx}\right) = \frac{3t^2(2) - (2t-2)(6t)}{(3t^2)^2} = \frac{4-2t}{3t^3}.$

Thus $\frac{d^2y}{dx^2} = \frac{\frac{4-2t}{3t^3}}{3t^2} = \frac{4-2t}{9t^5}.$

b) Find the horizontal and vertical tangents for the curve.

$\frac{dy}{dx} = 0$ at $t = 1$ $(x = 1, y = -1)$.

A horizontal tangent is at $(1, -1)$.

$\frac{dy}{dx}$ does not exist at $t = 0$ $(x = 0, y = 0)$.

A vertical tangent is at $(0, 0)$.

c) Discuss the concavity of the curve.

$\dfrac{d^2y}{dx^2} > 0$ for $0 < t < 2$ $(0 < x < 8)$ and negative for $t < 0$ $(x < 0)$ and $t > 2$ $(x > 8)$. The curve is concave upward for $x \in (0, 8)$ and concave downward for $x \in (-\infty, 0)$ and $x \in (8, \infty)$.

d) Sketch the curve using the information above.

$\dfrac{d^2y}{dx^2} = \dfrac{4 - 2t}{9t^5} = 0$ at $t = 2$ $(x = 8, y = 0)$.

$(8, 0)$ and $(0, 0)$ are inflection points.

t	x	y	$\dfrac{dy}{dx}$
-3	-27	15	$-\dfrac{8}{27}$
-2	-8	8	$-\dfrac{1}{2}$
-1	-1	3	$-\dfrac{4}{3}$
0	0	0	undefined
1	1	-1	0
2	8	0	$\dfrac{1}{6}$
3	27	3	$\dfrac{4}{27}$

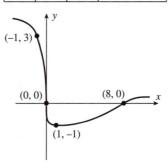

2) Find the equation of the tangent line to the curve $x = 4t^2 + 2t + 1$, $y = 7t + 2t^2$ at the point corresponding to $t = -1$.

At $t = -1$, $x = 3$ and $y = -5$.

$\dfrac{dx}{dt} = 8t + 2 = -6$ at $t = -1$.

$\dfrac{dy}{dt} = 7 + 4t = 3$ at $t = -1$.

Thus $\dfrac{dy}{dx} = \dfrac{3}{-6} = -\dfrac{1}{2}$.

The tangent line is $y + 5 = -\dfrac{1}{2}(x - 3)$.

B. Suppose the parametric equations $x = f(t)$, $y = g(t)$, $t \in [\alpha, \beta]$ define an integrable function $y = F(x) \geq 0$ over the interval $[a, b]$, where $a = f(\alpha)$, $b = f(\beta)$. The area under $y = F(x)$ is

$$\int_{\alpha}^{\beta} g(t) f'(t) \, dt \quad \text{if } (f(\alpha), g(\alpha)) \text{ is the left endpoint}$$

or

$$\int_{\beta}^{\alpha} g(t) f'(t) \, dt \quad \text{if } (f(\beta), g(\beta)) \text{ is the left endpoint.}$$

Example: Find the area bounded by the x-axis and the curve given by $x = \sqrt{t}, y = 4t - t^2, 0 \le t \le 4$.

The area is graphed at the right and is

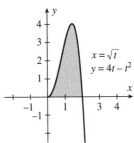

$$\int_0^4 (4t - t^2)\left(\tfrac{1}{2}t^{-1/2}\right) dt = \tfrac{1}{2} \int_0^4 \left(4t^{1/2} - t^{3/2}\right) dt$$

$$= \tfrac{1}{2}\left(4\left(\tfrac{2}{3}t^{3/2}\right) - \tfrac{2}{5}t^{5/2}\right)\Big]_0^4 = \left(\tfrac{4}{3}t^{3/2} - \tfrac{1}{5}t^{5/2}\right)\Big]_0^4$$

$$= \left(\tfrac{4}{3}(4)^{3/2} - \tfrac{1}{5}(4)^{4/2}\right) - 0 = \tfrac{32}{3} - \tfrac{32}{5} = \tfrac{64}{15}.$$

Note that we may eliminate the parameter and obtain $y = 4x^2 - x^4$,

for $0 \le x \le 2$. Thus the area is $\int_0^2 (4x^2 - x^4)\, dx = \tfrac{64}{15}$.

3) Find the area under the curve $x = t^2 + 1$, $y = e^t, 0 \le t \le 1$.

\<Set up the integral and use integration by parts to evaluate it.\>

The area is $\int_0^1 e^t\, (2t)dt$

$(u = 2t, dv = e^t\, dt; du = 2\, dt, v = e^t)$

$= 2te^t\Big]_0^1 - \int_0^1 e^t 2\, dt = (2te^t - 2e^t)\Big]_0^1$

$= (2e - 2e) - (0 - 2) = 2.$

4) Find the area inside the loop of the curve given by $x = 9 - t^2$, $y = t^3 - 3t$.

\<First draw the graph to get a sense for what area is described.\>

$$\frac{dy}{dx} = \frac{\frac{dy}{dt}}{\frac{dx}{dt}} = \frac{3t^2 - 3}{-2t}.$$

$\frac{dy}{dx} = 0$ at $t = 1, t = -1.$ $\frac{dy}{dx}$ is not defined at $t = 0$. There are horizontal tangents at $(8, 2)$ (where $t = -1$) and $(8, -2)$ (where $t = 1$). There is a vertical tangent at $(9, 0)$ (where $t = 0$).

t	x	y
-3	0	-18
-2	5	-2
-1	8	2
0	9	0
1	8	-2
2	5	2
3	0	18

The graph is

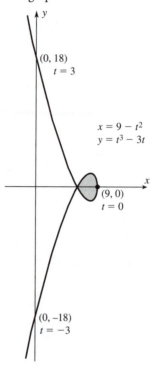

The curve crosses itself (to form a loop) at $y = 0$.

$$t^3 - 3t = 0$$
$$t(t^2 - 3) = 0$$
$$t = 0 , t = \sqrt{3}, t = -\sqrt{3}.$$

The area of the loop is twice the area under the top portion which corresponds to $t \in [-\sqrt{3}, 0]$.

The area is $2 \displaystyle\int_{-\sqrt{3}}^{0} (t^3 - 3t)(-2t)\, dt$

$$= 2 \int_{-\sqrt{3}}^{0} (-2t^4 + 6t^2)\, dt$$

$$= 2 \left(-\tfrac{2}{5}t^5 + 2t^3 \right) \Big]_{-\sqrt{3}}^{0}$$

$$= \frac{24\sqrt{3}}{5}.$$

Page
648

C. The length of a curve defined by $x = f(t)$, $y = g(t)$, $t \in [\alpha, \beta]$ with f', g' continuous and the curve traversed only once as t increases from α to β is

$$\int_{\alpha}^{\beta} \sqrt{\left(\frac{dx}{dt}\right)^2 + \left(\frac{dy}{dt}\right)^2}\, dt.$$

Recall that if s is the arc length function, then $(ds)^2 = (dx)^2 + (dy)^2$. Thus,

$$\int ds = \int \sqrt{\left(\frac{dx}{dt}\right)^2 + \left(\frac{dy}{dt}\right)^2}\, dt,$$ which is consistent with the previous formula for

arc length.

Example: Find the length of the curve given by $x = t^2 + 1$, $y = \frac{1}{3}t^3$
for $0 \le t \le 2$.

With $\dfrac{dx}{dt} = 2t$ and $\dfrac{dy}{dt} = t^2$, the length is

$$\int_0^2 \sqrt{(2t)^2 + (t^2)^2}\, dt = \int_0^2 \sqrt{4t^2 + t^4}\, dt$$

$$= \int_0^2 t\sqrt{4 + t^2}\, dt.$$

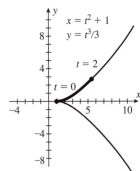

Let $u = 4 + t^2$, $du = 2t\, dt$. For $t = 0$, $u = 4$ and
for $t = 2$, $u = 8$.

The length is $\dfrac{1}{2}\displaystyle\int_4^8 u^{1/2}\, du = \dfrac{1}{2}\left(\dfrac{2}{3}u^{3/2}\right)\Big]_4^8$

$$= \tfrac{1}{3}\left(8^{3/2} - 4^{3/2}\right) = \frac{8\sqrt{8}}{3} - \frac{8}{3} \approx 4.88.$$

5) Write a definite integral for the arc length
of each:

a) the curve given by $x = \ln t$, $y = t^2$, for
$1 \le t \le 2$.

$x'(t) = \dfrac{1}{t}$ and $y'(t) = 2t$. The arc length is

$$\int_1^2 \sqrt{\left(\frac{1}{t}\right)^2 + (2t)^2}\, dt.$$

b) an ellipse given by $\dfrac{x^2}{a^2} + \dfrac{y^2}{b^2} = 1$, where
$a, b > 0$.

The ellipse may be defined parametrically
by $x = a\cos t$, $y = b\sin t$, $t \in [0, 2\pi]$. The
length of the ellipse is

$$\int_0^{2\pi} \sqrt{(-a\sin t)^2 + (b\cos t)^2}\, dt$$

$$= \int_0^{2\pi} \sqrt{a^2\sin^2 t + b^2\cos^2 t}\, dt.$$

c) The curve given by

$$x = \cos 2t,$$
$$y = \sin t,$$
$$t \in [0, 2\pi].$$

Begin with a sketch of the curve. We plot some points.

t	x	y
0	1	0
$\frac{\pi}{4}$	0	$\frac{\sqrt{2}}{2}$
$\frac{\pi}{2}$	-1	1
$\frac{3\pi}{4}$	0	$\frac{\sqrt{2}}{2}$
π	1	0
$\frac{5\pi}{4}$	0	$-\frac{\sqrt{2}}{2}$
$\frac{3\pi}{2}$	-1	-1
$\frac{7\pi}{4}$	0	$-\frac{\sqrt{2}}{2}$
2π	1	0

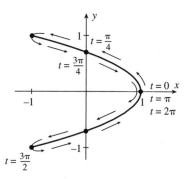

The curve is traversed once for $t \in \left[\frac{\pi}{2}, \frac{3\pi}{2}\right]$ and has length

$$\int_{\frac{\pi}{2}}^{\frac{3\pi}{2}} \sqrt{4 \sin^2 2t + \cos^2 t}\; dt.$$

D. If a curve given by $x = f(t)$, $y = g(t)$, $t \in [\alpha, \beta]$, with f', g' continuous and $g(t) \geq 0$ is rotated about the x-axis, the area of the resulting surface of revolution is

$$S = \int_{\alpha}^{\beta} 2\pi y \sqrt{\left(\frac{dx}{dt}\right)^2 + \left(\frac{dy}{dt}\right)^2}\; dt.$$

Using the ds notation this is $S = \int 2\pi y\; ds$, the same formula as in the section on areas of surfaces of revolution.

6) Find the definite integral for the area of the surface of revolution about the x-axis for each curve:

  a) $x = 2t + 1, y = t^3, t \in [0, 2]$

$$S = \int_0^2 2\pi(t^3)\sqrt{(2)^2 + (3t^2)^2}\ dt$$

$$= \int_0^2 2\pi t^3\sqrt{4 + 9t^4}\ dt.$$

  b) a "football" obtained from rotating the ellipse $\dfrac{x^2}{a^2} + \dfrac{y^2}{b^2} = 1, y \geq 0$ about the x-axis.

The top half of the ellipse is $x = a \cos t$, $y = b \sin t, t \in [0, \pi]$.

$$ds = \sqrt{(dx)^2 + (dy)^2}$$

$$= \sqrt{(-a \sin t)^2 + (b \cos t)^2}$$

$$= \sqrt{a^2 \sin^2 t + b^2 \cos^2 t}.$$

The area is

$$S = \int_0^\pi 2\pi(b \sin t)\sqrt{a^2 \sin^2 t + b^2 \cos^2 t}\ dt.$$

E. Technology Plus. Use a computer algebra system or a graphing calculator to solve.

T-1) Use a CAS to evaluate the definite integral for the arc length of

$x = t + \cos t$
$y = t + \sin t$
for $0 \leq t \leq 2\pi$.

$\dfrac{dx}{dt} = 1 - \sin t, \quad \dfrac{dy}{dt} = 1 + \cos t$

The arc length is

$$\int_0^{2\pi} \sqrt{(1 - \sin t)^2 + (1 + \cos t)^2}\ dt$$

$$\approx 10.037.$$

Section 10.3 Polar Coordinates

All our graphs have been in a rectangular coordinate system where the two coordinates are distances from perpendicular axes. In this section on polar coordinates the pair of numbers that determines a point in a plane is an angle through which to rotate from the *x*-axis together with a distance from the origin. We will see how to convert from one coordinate system to another and see that certain curves are much easier to express in polar coordinates than rectangular coordinates.

Concepts to Master

A. Points in polar coordinates; conversion to and from rectangular to polar coordinates

B. Graphs of equations in polar coordinates

C. Tangents to polar curves

Summary and Focus Questions

Page 654

A. To construct a **polar coordinate system** start with a point called the **pole** and a ray from the pole called the **polar axis.**

Pole   Polar axis

A point *P* has polar coordinates (r, θ) if

   $|r|$ = the distance from the pole to *P*, and
   θ = the measure of a directed angle with initial side the polar axis and terminal side the line through the pole and *P*.

Example: The point *P* with polar coordinates $(r, \theta) = (2, \frac{\pi}{4})$ is plotted by swinging an angle upward of $\frac{\pi}{4}$ radians (45°) and then moving out 2 units from the pole along the angled line.

In general, plotting a point *P* which has polar coordinates (r, θ) depends on the signs of *r* and θ. Here are the possibilities:

r and *θ*	How to plot *P* (*r*, *θ*)	Graph
r = 0 *θ* = any value	*P* is the pole.	
r > 0 *θ* ≥ 0	Rotate the polar axis *counterclockwise* by the angle *θ* and locate *P* on this ray a distance *r* units from the pole.	
r > 0 *θ* < 0	Rotate the polar axis *clockwise* by the angle *θ* and locate *P* on this ray a distance *r* units from the pole.	
r < 0 *θ* ≥ 0	Rotate the polar axis *counterclockwise* by the angle *θ* and then reflect a ray about the pole. *P* is located on this reflected ray a distance −*r* units from the pole.	
r < 0 *θ* < 0	Rotate the polar axis *clockwise* by the angle *θ* and then reflect a ray about the pole. *P* is located on this reflected ray a distance −*r* units from the pole.	

Example: Plot these five points in polar coordinates:

$A(0, \pi/2)$

$B(2, \pi/3)$

$C(3, -\pi/6)$

$D(-2, \pi/6)$

$E(-1, -\pi/3)$

Note that when θ is negative (points C and E) we rotate clockwise.

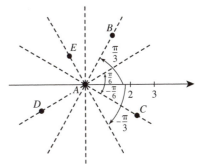

A point in a rectangular coordinate system will have a unique pair of coordinates. On the other hand, a point in a polar coordinate system will have infinitely many pairs of coordinates. In particular, the point with coordinates (r, θ) also has coordinates $(r, \theta + 2\pi)$, $(-r, \theta + \pi)$, $(-r, \theta - \pi)$, and many others.

Suppose a rectangular coordinate system is placed upon the polar coordinate system as in the figure.

To change from polar to rectangular:

$x = r \cos \theta$

$y = r \sin \theta.$

To change from rectangular to polar:

Solve $\tan \theta = \dfrac{y}{x}$ ($x \neq 0$) for θ.

Solve $r^2 = x^2 + y^2$ for r. *Note:* $r > 0$ if the terminal side of the angle θ is in the same quadrant as P; if not, then $r \leq 0$.

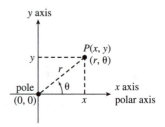

1) Plot these points in polar coordinates.

$A\left(4, \dfrac{\pi}{3}\right)$ $\qquad$ $B\left(-3, \dfrac{\pi}{4}\right)$

$C(0, 3\pi)$ $\qquad$ $D\left(2, -\dfrac{\pi}{4}\right)$

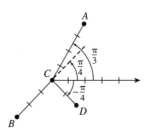

2) Find two other polar coordinates for the point with polar coordinates $\left(8, \dfrac{\pi}{3}\right)$.

There are infinitely many answers including $\left(8, \dfrac{7\pi}{3}\right), \left(-8, \dfrac{4\pi}{3}\right), \left(8, -\dfrac{5\pi}{3}\right), \ldots$

3) Sometimes, Always, or Never:

a) The polar coordinates of a point are unique.

Never.

b) $(r, \theta) = (r, \theta + 2\pi)$.

Always.

c) $(r, \theta) = (-r, \theta + \pi)$.

Always.

d) $(r, \theta) = (-r, \theta)$.

Sometimes. (True when $r = 0$.)

4) Find polar coordinates for the point P with rectangular coordinates $P: (-3, 3\sqrt{3})$.

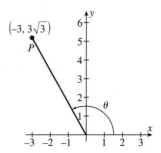

$$\tan \theta = \frac{y}{x} = \frac{3\sqrt{3}}{-3} = -\sqrt{3}.$$

A solution is $\theta = \frac{2\pi}{3}$.

$$r^2 = (-3)^2 + (3\sqrt{3})^2 = 9 + 27 = 36.$$

$$r = \pm 6.$$

Because the terminal side of $\theta = \frac{2\pi}{3}$ lies in the same quadrant as P, $r > 0$. Therefore, P has coordinates $\left(6, \frac{2\pi}{3}\right)$.

Note: $\theta = -\frac{\pi}{3}$ is another solution to $\tan \theta = -\sqrt{3}$, which results in $r = -6$ and coordinates $\left(-6, -\frac{\pi}{3}\right)$ for P.

5) Find the rectangular coordinates for the point with polar coordinates $\left(-4, \frac{5\pi}{6}\right)$.

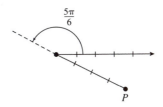

$$x = -4 \cos \frac{5\pi}{6} = (-4)\left(\frac{-\sqrt{3}}{2}\right) = 2\sqrt{3}.$$

$$y = -4 \sin \frac{5\pi}{6} = (-4)\left(\frac{1}{2}\right) = -2.$$

P has coordinates $(2\sqrt{3}, -2)$.

Page 656

B. Your first procedure for graphing a polar equation is the same as that used when first graphing functions—compute values and plot points. In some cases the graph of a polar equation is easily identified when the equation is transformed into rectangular coordinates.

Example: Graph $r = 6 \sin \theta$.

Multiply both sides of the equation by r:

$$r^2 = 6r \sin \theta$$
$$x^2 + y^2 = 6y$$
$$x^2 + y^2 - 6y = 0$$
$$x^2 + (y - 3)^2 = 9.$$

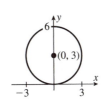

The graph is the circle of radius 3 centered at $(0, 3)$.

There are three possible symmetries that can help you draw a graph. For polar equation $r = f(\theta)$, the graph is

a) **symmetric about the polar axis** if $f(-\theta) = f(\theta)$. In other words, $r = f(\theta)$ is unchanged when θ is replaced by $-\theta$.

b) **symmetric about the pole** (unchanged when rotated 180°) if either $-f(\theta) = f(\theta)$ or $f(\theta + \pi) = f(\theta)$. In other words, $r = f(\theta)$ is unchanged when r is replaced by $-r$ or when θ is replaced by $\theta + \pi$.

c) **symmetric about the vertical line through the pole** (through the line $\theta = \frac{\pi}{2}$) if $f(\theta) = f(\pi - \theta)$. In other words, $r = f(\theta)$ is unchanged when θ is replaced by $\pi - \theta$.

Example: The graph of $r = 2 \cos 4\theta$ exhibits all three symmetries. If we use the facts that cosine has periodicity 2π, $\cos(-t) = \cos t$ for all t, and $\cos(\pi - t) = \cos t$ for all t, then

a) $2 \cos 4(-\theta) = 2 \cos(-4\theta) = 2 \cos 4\theta$.

b) $2 \cos 4(\theta + \pi) = 2 \cos(4\theta + 4\pi) = 2 \cos(4\theta + 2\pi + 2\pi) = 2 \cos 4\theta$.

c) $2 \cos 4(\pi - \theta) = 2 \cos(4\pi - 4\theta)$
$= 2 \cos(\pi - (-3\pi + 4\theta)) = 2 \cos(-3\pi + 4\theta)$
$= 2 \cos(3\pi - 4\theta) = 2 \cos(\pi - (-2\pi + 4\theta))$
$= 2 \cos(-2\pi + 4\theta) = 2 \cos(2\pi - 4\theta)$
$= 2 \cos(\pi - (-\pi + 4\theta)) = 2 \cos(-\pi + 4\theta)$
$= 2 \cos(\pi - 4\theta) = 2 \cos 4\theta$.

The graph of $r = 2 \cos 4\theta$ is an "eight-leaf rose."

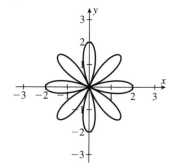

6) Sketch a graph of $r = \sin \theta - \cos \theta$.

Compute several points for values of θ.

Point	θ	r
A	0	-1
B	$\frac{\pi}{6}$	$\frac{1}{2} - \frac{\sqrt{3}}{2}$
C	$\frac{\pi}{4}$	0
D	$\frac{\pi}{2}$	1
E	$\frac{3\pi}{4}$	$\sqrt{2}$
F	π	1 (same as A)
G	$\frac{7\pi}{6}$	$\frac{-1}{2} + \frac{\sqrt{3}}{2}$ (same as B)

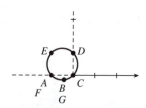

We can verify that the figure is a circle by switching to rectangular coordinates:

Multiply $r = \sin\theta - \cos\theta$ by r:

$$r^2 = r\sin\theta - r\cos\theta$$
$$x^2 + y^2 = y - x$$
$$x^2 + x + y^2 - y = 0.$$

Completing the square gives

$$\left(x + \tfrac{1}{2}\right)^2 + \left(y - \tfrac{1}{2}\right)^2 = \tfrac{1}{2},$$

the circle with center $\left(-\tfrac{1}{2}, \tfrac{1}{2}\right)$ and radius $\dfrac{1}{\sqrt{2}}$.

7) Sketch the graph of $r = \cos 3\theta$.

Since $\cos 3\theta = \cos(-3\theta)$, the curve is symmetric about the polar axis. Also $\cos 3\theta$ repeats every $\dfrac{2\pi}{3}$ units.

Point	θ	r	Point	θ	r
A	0	1	H	$\dfrac{7\pi}{12}$	$\dfrac{\sqrt{2}}{2}$
B	$\dfrac{\pi}{12}$	$\dfrac{\sqrt{2}}{2}$	I	$\dfrac{2\pi}{3}$	1
C	$\dfrac{\pi}{6}$	0	J	$\dfrac{3\pi}{4}$	$\dfrac{\sqrt{2}}{2}$
D	$\dfrac{\pi}{4}$	$-\dfrac{\sqrt{2}}{2}$	K	$\dfrac{5\pi}{6}$	0
E	$\dfrac{\pi}{3}$	-1	L	$\dfrac{11\pi}{12}$	$-\dfrac{\sqrt{2}}{2}$
F	$\dfrac{5\pi}{12}$	$-\dfrac{\sqrt{2}}{2}$	M	π	-1
G	$\dfrac{\pi}{2}$	0			

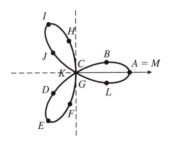

C. For polar curve $r = f(\theta)$, we can switch to rectangular coordinates $x = f(\theta)\cos\theta$, $y = f(\theta)\sin\theta$. Treating θ as a parameter,

Page 659

$$\frac{dy}{dx} = \frac{\dfrac{dy}{d\theta}}{\dfrac{dx}{d\theta}} = \frac{\dfrac{dr}{d\theta}\sin\theta + r\cos\theta}{\dfrac{dr}{d\theta}\cos\theta - r\sin\theta}.$$

This is the slope of the tangent line to $r = f(\theta)$ at (r, θ).

8) Find the slope of the tangent line to $r = e^\theta$ at $\theta = \frac{\pi}{2}$.

$x = r \cos \theta = e^\theta \cos \theta$, so

$\frac{dx}{d\theta} = e^\theta(-\sin \theta) + e^\theta(\cos \theta)$.

$y = r \sin \theta = e^\theta \sin \theta$, so

$\frac{dy}{d\theta} = e^\theta \cos \theta + e^\theta \sin \theta$.

At $\theta = \frac{\pi}{2}$,

$\frac{dx}{d\theta} = -e^{\pi/2}$ and $\frac{dy}{d\theta} = e^{\pi/2}$.

Therefore, $\frac{dy}{dx} = \frac{\frac{dy}{d\theta}}{\frac{dx}{d\theta}} = \frac{e^{\pi/2}}{-e^{\pi/2}} = -1$.

9) Find the points on $r = \sin \theta - \cos \theta$ where the tangent line is horizontal or vertical for $0 \le \theta \le \pi$.

$\frac{dy}{dx} = \frac{\frac{dy}{d\theta}}{\frac{dx}{d\theta}}$ is not defined when $\frac{dx}{d\theta} = 0$

and is zero when $\frac{dy}{d\theta} = 0$.

From $x = r \cos \theta$

$\qquad = (\sin \theta - \cos \theta) \cos \theta,$

$\frac{dx}{d\theta} = (\sin \theta - \cos \theta)(-\sin \theta)$

$\qquad + \cos \theta (\cos \theta + \sin \theta)$

$\qquad = 2 \sin \theta \cos \theta + (\cos^2 \theta - \sin^2 \theta)$

$\qquad = \sin 2\theta + \cos 2\theta.$

$\frac{dx}{d\theta} = 0$ when

$\sin 2\theta = -\cos 2\theta$

$\tan 2\theta = -1$

$2\theta = \frac{3\pi}{4}$ and $2\theta = \frac{7\pi}{4}$

$\theta = \frac{3\pi}{8}$ and $\theta = \frac{7\pi}{8}$.

At $\theta = \frac{3\pi}{8}$, $r = \sin \frac{3\pi}{8} - \cos \frac{3\pi}{8}$

$\qquad = \frac{\sqrt{\sqrt{2} + 2}}{2} - \frac{\sqrt{2 - \sqrt{2}}}{2} \approx .54.$

At $\theta = \frac{7\pi}{8}$, $r = \sin \frac{7\pi}{8} - \cos \frac{7\pi}{8}$

$\qquad = \frac{\sqrt{2 - \sqrt{2}}}{2} - \frac{-\sqrt{\sqrt{2} + 2}}{2} \approx 1.31.$

There are vertical tangents at $P\left(.54, \frac{3\pi}{8}\right)$ and $Q\left(1.31, \frac{7\pi}{8}\right)$.

From $y = r \sin \theta$

$$= (\sin \theta - \cos \theta) \sin \theta,$$

$$\frac{dy}{d\theta} = (\sin \theta - \cos \theta) \cos \theta$$
$$+ \sin \theta (\cos \theta + \sin \theta)$$
$$= 2 \sin \theta \cos \theta - (\cos^2 \theta - \sin^2 \theta)$$
$$= \sin 2\theta - \cos 2\theta = 0.$$

$$\sin 2\theta = \cos 2\theta$$

$$\tan 2\theta = 1$$

$$2\theta = \frac{\pi}{4} \text{ and } \frac{5\pi}{4}$$

$$\theta = \frac{\pi}{8} \text{ and } \frac{5\pi}{8}.$$

At $\theta = \frac{\pi}{8}, r = \sin \frac{\pi}{8} - \cos \frac{\pi}{8}$

$$= \frac{\sqrt{2 - \sqrt{2}}}{2} - \frac{\sqrt{2 + \sqrt{2}}}{2} \approx -.54.$$

At $\theta = \frac{5\pi}{8}, r = \sin \frac{5\pi}{8} - \cos \frac{5\pi}{8}$

$$= \frac{\sqrt{2 + \sqrt{2}}}{2} - \frac{-\sqrt{2 - \sqrt{2}}}{2} \approx 1.31.$$

There are horizontal tangents at $R\left(-.54, \frac{\pi}{8}\right)$ and $S\left(1.31, \frac{5\pi}{8}\right)$.

From exercise 6, the graph is a circle.

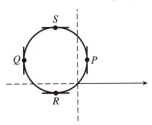

Section 10.4 Areas and Lengths in Polar Coordinates

This section develops a formula for the area of a region bounded by equations given in polar form and the arc length of a curve given by a polar equation.

Concepts to Master

A. Area of a region described by polar equations

B. Length of a curve described by a polar equation

Summary and Focus Questions

Page 665

A. Let $r = f(\theta)$ be a continuous and positive polar function for $a \le \theta \le b$, where $0 < b - a \le 2\pi$. The **area** bounded by $r = f(\theta)$, $\theta = a$ and $\theta = b$ may be approximated by partitioning the interval $[a, b]$ in n equal subintervals, each of length $\Delta\theta = \dfrac{b-a}{n}$, which correspond to n sectors. Each

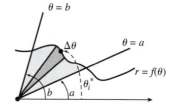

sector has a central angle of $\Delta\theta$.

The area of a sector is approximately $\frac{1}{2}\left(f(\theta_i^*)\right)^2 \Delta\theta$.

The area of the region is $\displaystyle\int_a^b \frac{1}{2}\left[f(\theta)\right]^2 \, d\theta$.

If $0 \le f(\theta) \le g(\theta)$ for all $a \le \theta \le b$ and if f and g are continuous, then the area between f and g and between $\theta = a$ and $\theta = b$ is $\displaystyle\int_a^b \frac{1}{2}\left[(g(\theta))^2 - (f(\theta))^2\right] d\theta$.

Example: Find the area in the first quadrant between the polar curves $r = 1 + \cos\theta$ and $r = 2 - \dfrac{2\theta}{\pi}$.

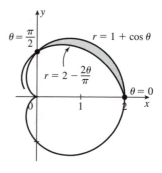

The curves are sketched at the right. The outer curve is $r = 1 + \cos\theta$ and the inner curve is $r = 2 - \dfrac{2\theta}{\pi}$. They intersect at $\theta = 0$ and $\theta = \dfrac{\pi}{2}$.

The area is $\displaystyle\int_0^{\pi/2} \frac{1}{2}\left[(1+\cos\theta)^2 - \left(2 - \frac{2\theta}{\pi}\right)^2 \right] d\theta$

$\displaystyle = \frac{1}{2}\int_0^{\pi/2} \left[(1 + 2\cos\theta + \cos^2\theta) - \left(4 - \frac{80}{\pi} + \frac{40^2}{\pi^2}\right) \right] d\theta$

$\displaystyle = \frac{1}{2}\int_0^{\pi/2} \left[\left(1 + 2\cos\theta + \frac{1}{2} + \frac{\cos 2\theta}{2}\right) - \left(4 - \frac{80}{\pi} + \frac{40^2}{\pi^2}\right) \right] d\theta$

$\displaystyle = \frac{1}{2}\int_0^{\pi/2} \left[-\frac{5}{2} + 2\cos\theta + \frac{\cos 2\theta}{2} + \frac{80}{\pi} - \frac{40^2}{\pi^2} \right] d\theta = \frac{1}{2}\left(-\frac{50}{2} + 2\sin\theta + \frac{\sin 2\theta}{4} + \frac{40^2}{\pi} - \frac{40^3}{3\pi^2} \right)\Bigg|_0^{\pi/2}$

$\displaystyle = \frac{1}{2}\left[\left(-\frac{5\pi}{4} + 2(1) + \frac{0}{4} + \pi - \frac{\pi}{6}\right) - 0 \right] = 1 - \frac{5\pi}{24} \approx 0.346.$

---

1) Find a definite integral for each region:

    a) the shaded area

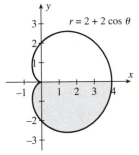

    b) The region bounded by $\theta = \frac{\pi}{3}, \theta = \frac{\pi}{2}$, $r = e^\theta$.

    c) The shaded area

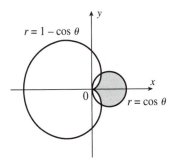

The region is bound by $\theta = \pi$, $\theta = 2\pi$, $r = 2 + 2\cos\theta$. The area is

$\displaystyle = \int_\pi^{2\pi} \frac{1}{2}(2 + 2\cos\theta)^2\, d\theta$

$\displaystyle = 2\int_\pi^{2\pi} (1 + \cos\theta)^2\, d\theta \left(= \frac{3}{2}\pi\right).$

The area is $\displaystyle\int_{\pi/3}^{\pi/2} \frac{1}{2}e^{2\theta}\, d\theta \left(= \frac{e^\pi - e^{2\pi/3}}{4}\right).$

The area is between 2 curves. First determine where the curves intersect:

$1 - \cos\theta = \cos\theta,\ 1 = 2\cos\theta,\ \cos\theta = \frac{1}{2}.$

Therefore, $\theta = \frac{\pi}{3}$ and $\theta = -\frac{\pi}{3}$.

For $-\frac{\pi}{3} \le \theta \le \frac{\pi}{3}$,

$\cos\theta \ge 1 - \cos\theta$ so the area is

$\displaystyle = \int_{-\pi/3}^{\pi/3} \frac{1}{2}[\cos^2\theta - (1 - \cos\theta)^2]\, d\theta$

$\displaystyle = \int_{-\pi/3}^{\pi/3} \left(\cos\theta - \frac{1}{2}\right) d\theta \left(= \sqrt{3} - \frac{\pi}{3}\right).$

2) Find the area above the line $r = \csc\theta$ and inside the circle $r = 2$.

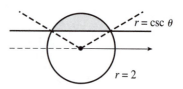

The curves intersect when

$$\csc\theta = 2$$

$$\sin\theta = \frac{1}{2}$$

$$\theta = \frac{\pi}{6} \text{ and } \frac{5\pi}{6}.$$

The area is $\displaystyle\int_{\frac{\pi}{6}}^{\frac{5\pi}{6}} \left(\frac{1}{2}(2)^2 - \frac{1}{2}\csc^2\theta \right) d\theta$

$$= \frac{1}{2}\int_{\frac{\pi}{6}}^{\frac{5\pi}{6}} (4 - \csc^2\theta)\, d\theta$$

$$= \frac{1}{2}(4\theta + \cot\theta)\Big|_{\frac{\pi}{6}}^{\frac{5\pi}{6}} = \frac{4\pi}{3} - \sqrt{3}.$$

Page 668

B. A curve given in polar coordinates by $r = f(\theta)$ for $a \le \theta \le b$ has arc length

$$L = \int_a^b \sqrt{[f(\theta)]^2 + [f'(\theta)]^2}\, d\theta.$$

3) Set up a definite integral for the length of each curve.

a) $r = e^{2\theta}$ for $0 \le \theta \le 1$.

$$\int_0^1 \sqrt{(e^{2\theta})^2 + (2e^{2\theta})^2}\, d\theta = \sqrt{5}\int_0^1 e^{2\theta}\, d\theta$$

$$\left(= \frac{\sqrt{5}}{2}(e^2 - 1) \right).$$

b) The curve sketched below.

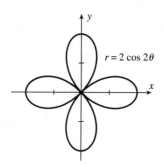

Since the curve is quite symmetric, the arc length is 8 times the length of half of one leaf:

$$8\int_0^{\frac{\pi}{4}} \sqrt{[2\cos 2\theta]^2 + [-4\sin 2\theta]^2}\, d\theta$$

$$= 8\int_0^{\frac{\pi}{4}} \sqrt{4\cos^2 2\theta + 16\sin^2 2\theta}\, d\theta$$

$$= 16\int_0^{\frac{\pi}{4}} \sqrt{1 + 3\sin^2 2\theta}\, d\theta \;(\approx 19.4).$$

Section 10.5 Conic Sections

In three dimensions, a cone and a plane may intersect in a point, a line, or a curve. Conic sections are the various types of curves that result from the intersection of a plane with a cone. This section gives geometric definitions of those curves and the equations that describe them.

Concepts to Master

Focus-directrix definition of parabolas, ellipses, and hyperbolas; Vertices; Standard form of the equations of conics

Summary and Focus Questions

Page 670

Three different types of curves may result when a cone and a plane intersect. Each may be described as the set of points in a plane satisfying a certain geometric property:

Parabola: Given a line (a **directrix**) and a point (the **focus,** F), a point P is on the parabola if the distance from P to the directrix is the same as the distance from P to F.

Ellipse: Given two points (the **foci,** F_1 and F_2), a point P is on the ellipse if the *sum* of the distances from P to F_1 and from P to F_2 is a constant.

Hyperbola: Given two points (the **foci,** F_1 and F_2), a point P is on the hyperbola if the *difference* of the distances from P to F_1 and from P to F_2 is a constant.

Parabola: $d_1 = d_2$

Ellipse: $d_1 + d_2 =$ constant
Hyperbola: $d_1 - d_2 =$ constant

The equations of the conics and their graphs in rectangular coordinates are given below.

Conic Section	Equation	Properties	Graphs
Parabola	$x^2 = 4py$	$p > 0$ focus: $(0, p)$ vertex: $(0, 0)$ directrix: $y = -p$	
	$x^2 = 4py$	$p < 0$ focus: $(0, p)$ vertex: $(0, 0)$ directrix: $y = -p$	

Parabola $y^2 = 4px$

$p > 0$
focus: $(p, 0)$
vertex: $(0, 0)$
directrix: $x = -p$

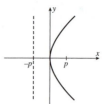

$y^2 = 4px$

$p < 0$
focus: $(p, 0)$
vertex: $(0, 0)$
directrix: $x = -p$

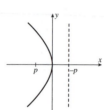

Ellipse $\dfrac{x^2}{a^2} + \dfrac{y^2}{b^2} = 1$

$a \geq b > 0$
$c^2 = a^2 - b^2$
foci: $(c, 0)$, $(-c, 0)$
vertices: $(a, 0)$, $(-a, 0)$
constant sum $= 2a$
center: $(0, 0)$
major axis: line segment
from $(-a, 0)$ to $(a, 0)$

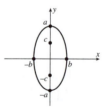

Ellipse $\dfrac{x^2}{b^2} + \dfrac{y^2}{a^2} = 1$

$a \geq b > 0$
$c^2 = a^2 - b^2$
foci: $(0, c)$, $(0, -c)$
vertices: $(0, a)$, $(0, -a)$
constant sum $= 2a$
center: $(0, 0)$
major axis: line segment
from $(0, -a)$ to $(0, a)$

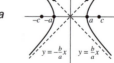

Hyperbola $\dfrac{x^2}{a^2} - \dfrac{y^2}{b^2} = 1$

$c^2 = a^2 + b^2$
foci: $(c, 0)$, $(-c, 0)$
vertices: $(a, 0)$, $(-a, 0)$
constant difference $= 2a$
center $(0, 0)$
asymptotes: $y = \dfrac{b}{a}x$,
$\qquad\qquad y = -\dfrac{b}{a}x$

Hyperbola $\dfrac{y^2}{a^2} - \dfrac{x^2}{b^2} = 1$

$c^2 = a^2 + b^2$
foci: $(0, c)$, $(0, -c)$
vertices: $(0, a)$, $(0, -a)$
constant difference $= 2a$
center: $(0, 0)$
asymptotes: $y = \dfrac{a}{b}x$,
$\qquad\qquad y = -\dfrac{a}{b}x$

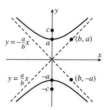

A second degree equation in x and y (with no xy term) represents a conic section whose center or vertex may be shifted from $(0, 0)$. To determine the type of conic, complete the square for x and y as in this example.

Example: What type of conic is given by the equation $x^2 + 6x + 4y^2 - 8y = 3$?
Complete the square by adding 9 and $4(1)$ to both sides:

$$x^2 + 6x \qquad + 4y^2 - 8y \qquad = 3$$
$$x^2 + 6x + 9 + 4(y^2 - 2y + 1) = 3 + 9 + 4(1).$$
$$(x + 3)^2 + 4(y - 1)^2 = 16$$
$$\frac{(x + 3)^2}{16} + \frac{(y - 1)^2}{4} = 1.$$

This is an ellipse shifted 3 units left and one unit upward.
Its center is $(-3, 1)$, $a = 4$ and $b = 2$.

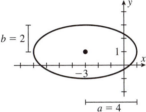

1) Find the vertices, foci, directrix (if a parabola), and asymptotes (if a hyperbola). Sketch the graph of each:

a) $\frac{x^2}{144} = 1 + \frac{y^2}{25}.$

Rewrite in standard form: $\frac{x^2}{144} - \frac{y^2}{25} = 1.$

This is a hyperbola with $a = 12$, $b = 5$.
$c^2 = 12^2 + 5^2 = 169$, $c = 13$.
Foci: $(13, 0)$, $(-13, 0)$.
Vertices: $(12, 0)$, $(-12, 0)$.
Asymptotes: $y = \frac{5}{12}x$, $y = -\frac{5}{12}x.$

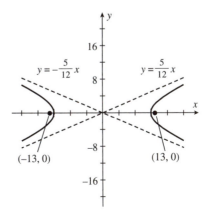

b) $x^2 = 4x + 8y - 4$.

$x^2 = 4x + 8y - 4$
$x^2 - 4x + 4 = 8y$
$(x - 2)^2 = 4(2)y$
This is a parabola, shifted two units to the right. The vertex is $(2, 0)$.
$p = 2$; the directrix is $y = -2$.
The focus is $(2, 2)$, also shifted two units.

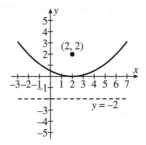

2) What is the conic given by each?

  a) $2x(6 - x) = y(8 + y)$

<Put each in standard form.>

$2x(6 - x) = y(8 + y)$
$-2x^2 + 12x = y^2 + 8y$
$2x^2 - 12x + y^2 + 8y = 0$
$2(x^2 - 6x + 9) + y^2 + 8y + 16 = 18 + 16$
$2(x - 3)^2 + (y + 4)^2 = 34$
$\dfrac{(x - 3)^2}{17} + \dfrac{(y + 4)^2}{34} = 1$.
This is an ellipse, $a = \sqrt{34}$, $b = \sqrt{17}$.

  b) $x(2 + x) - 24y = x^2 + 4y^2$

$x(2 + x) - 24y = x^2 + 4y^2$
$2x + x^2 - 24y = x^2 + 4y^2$
$2x = 4y^2 + 24y$
$x = 2y^2 + 12y = 2(y^2 + 6y)$
$x + 18 = 2(y^2 + 6y + 9) = 2(y + 3)^2$
$(y + 3)^2 = \tfrac{1}{2}(x + 18)$
This is a parabola opening to the right with vertex $(-18, -3)$.

Section 10.6   Conic Sections in Polar Coordinates

This section gives a polar coordinate definition for each conic section using just one focus and one directrix. This approach leads to simple forms for the equations of parabolas, ellipses, and hyperbolas in polar coordinates.

Concepts to Master

A. Eccentricity; Eccentric definition of conic sections; Equations of conics in polar form

B. Technology Plus

Summary and Focus Questions

Page 678

A. Another way to define conic sections is to specify a fixed point F (the focus), a fixed line l (the directrix), and a positive constant e called the **eccentricity***. Then the set of all points P such that

$$\frac{\text{distance from } P \text{ to } F}{\text{distance from } P \text{ to } l} = e$$

is a conic section. The value of e determines whether the conic is a parabola, ellipse, or hyperbola:

Eccentricity	Type	Graph	Ratio
$e = 1$	Parabola		$\dfrac{\lvert PF \rvert}{\lvert Pl \rvert} = 1$
$e < 1$	Ellipse		$\dfrac{\lvert PF \rvert}{\lvert Pl \rvert} = e < 1$
$e > 1$	Hyperbola		$\dfrac{\lvert PF \rvert}{\lvert Pl \rvert} = e > 1$

* This use of *e* for eccentricity is not to be confused with the use of *e* as the symbol for the base of natural logarithms.

Suppose a conic with eccentricity $e > 0$ is drawn in a rectangular coordinate system with focus F at $(0, 0)$, and one of the lines $x = d$, $x = -d$, $y = d$, or $y = -d$, where $d > 0$, is the directrix l. By superimposing a polar coordinate system the polar equation of the conic has one of these forms:

Conic	**Polar Form of Equation**			
Equation:	$r = \dfrac{ed}{1 + e\cos\theta}$	$r = \dfrac{ed}{1 - e\cos\theta}$	$r = \dfrac{ed}{1 + e\sin\theta}$	$r = \dfrac{ed}{1 - e\sin\theta}$
Directrix:	$x = d$	$x = -d$	$y = d$	$y = -d$
Parabola ($e = 1$)	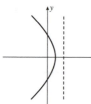	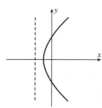	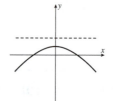	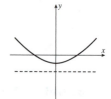
Ellipse ($0 < e < 1$)	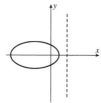	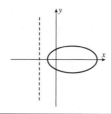	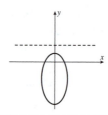	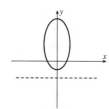
Hyperbola ($e > 1$)	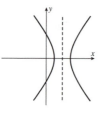	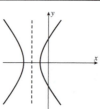	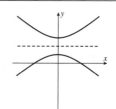	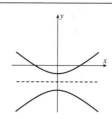

1) The sketch below shows one point P on a conic and its distances from the focus and directrix. What type of conic is it?

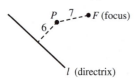

$e = \dfrac{|PF|}{|Pl|} = \dfrac{7}{6} > 1$. The conic is a hyperbola.

2) What is the polar equation of the conic with:

   a) eccentricity 2, directrix $y = 3$.

$e = 2$ and $d = 3$, so $r = \dfrac{6}{1 + 2\sin\theta}$.

   b) eccentricity $\frac{1}{2}$, directrix $x = -3$.

$e = \frac{1}{2}$ and $d = 3$, so $r = \dfrac{\frac{3}{2}}{1 - \frac{1}{2}\cos\theta}$,

$r = \dfrac{3}{2 - \cos\theta}$.

   c) directrix $x = 4$ and is a parabola.

$e = 1$ and $d = 4$, so $r = \dfrac{4}{1 + \cos\theta}$.

3) What polar form does the equation of the graph below have?

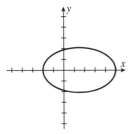

The graph is that of an ellipse. The directrix has the form $x = -d$ so the equation is

$r = \dfrac{ed}{1 - e\cos\theta}$ with $0 < e < 1$.

4) Find the eccentricity and directrix and identify the conic given by $r = \dfrac{3}{4 - 5\cos\theta}$.

Divide numerator and denominator by 4.

$r = \dfrac{\frac{3}{4}}{1 - \frac{5}{4}\cos\theta}$. We see that $e = \frac{5}{4}$.

Since $ed = \frac{3}{4}$, $d = \frac{3}{4} \cdot \frac{1}{e} = \frac{3}{4} \cdot \frac{4}{5} = \frac{3}{5}$.

The trigonometric term is $-\cos\theta$ so the directrix is $x = -\frac{3}{5}$. Since $e > 1$, the conic is a hyperbola.

B. Technology Plus. Use a computer algebra system or a graphing calculator to solve.

T-1) Sketch a graph of $r = \dfrac{1}{1 - \frac{1}{2}\cos\theta}$ on a graphing calculator. What type of conic is it?

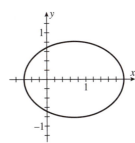

The graph is an ellipse with eccentricity $e = \frac{1}{2}$.

Chapter 11 — Infinite Sequences and Series

Section 11.1   Sequences

The main topic of this chapter is the representation of a function as a series. A series (introduced in the next section) is a sum of an infinite list of numbers—that is, the sum of a sequence of numbers. This section describes the basic concepts for sequences.

Concepts to Master

A. Sequences; Limit of a sequence; Convergence and divergence

B. Monotone sequence (increasing, decreasing); Bounded sequence; Monotonic Sequence Theorem

C. Sequences defined recursively

D. Technology Plus

Summary and Focus Questions

Page 690

A. A **sequence** is an infinite list of numbers given in a specific order:

$$\{a_n\} = \{a_1, a_2, a_3, \ldots, a_n, a_{n+1}, \ldots\}$$

a_n is called the nth **term** of the sequence.

Examples: a) The sequence $a_n = \dfrac{1}{n^2}$ has first few terms $1, \frac{1}{4}, \frac{1}{9}, \frac{1}{16}, \frac{1}{25}, \ldots$. For $n = 10$, $a_{10} = \dfrac{1}{100}$.

b) The sequence $b_n = (-1)^n$ has alternating terms: $-1, 1, -1, 1, -1, 1, \ldots$. Here $b_{109} = (-1)^{109} = -1$.

The sequence $\{a_n\}$ **converges** to a number L, written $\lim\limits_{n \to \infty} a_n = L$, means that the values of a_n get closer and closer to L as n grows larger. If $\lim\limits_{n \to \infty} a_n$ does not exist, we say $\{a_n\}$ **diverges.**

Sometimes writing out the first several terms of the sequence helps reveal whether it converges.

Examples: For the sequences above,

a) $a_n = \dfrac{1}{n^2}$ converges. The terms of a_n get smaller as n increases. Thus, $\lim\limits_{n \to \infty} \dfrac{1}{n^2} = 0$.

346

b) $b_n = (-1)^n$ diverges because the terms of b_n do not settle in near one particular number.

The formal definition of **convergence** of $\{a_n\}$ is:

$\lim\limits_{n\to\infty} a_n = L$ means for all $\epsilon > 0$ there exists a positive integer N such that if $n > N$, then $|a_n - L| < \epsilon$.

$\lim\limits_{n\to\infty} a_n = \infty$ means the a_n terms grow without bound.

One way to evaluate $\lim\limits_{n\to\infty} a_n$ is to find a real function $f(x)$ such that $f(n) = a_n$ for all n. If $\lim\limits_{n\to\infty} f(x) = L$, then $\lim\limits_{n\to\infty} a_n = L$. The converse is false.

Example: To find $\lim\limits_{n\to\infty} \dfrac{\ln n}{n^2}$, we let $f(x) = \ln \dfrac{x}{x^2}$ and evaluate

$$\lim_{x\to\infty} \frac{\ln x}{x^2} \left(\text{form } \frac{\infty}{\infty}\right) = \lim_{x\to\infty} \frac{1/x}{2x} = \lim_{x\to\infty} \frac{1}{2x^2} = 0. \text{ Therefore, } \frac{\ln n}{n^2} \text{ also converges to 0.}$$

Versions of the limit laws at infinity for functions are valid for convergent sequences. Thus, for example, $\lim\limits_{n\to\infty} ca_n = c \lim\limits_{n\to\infty} a_n$, where c is a constant. In addition,

if $\lim\limits_{n\to\infty} |a_n| = 0$, then $\lim\limits_{n\to\infty} a_n = 0$.

There is a version of the Squeeze Theorem for sequences:

if $a_n \le b_n \le c_n$ for all $n \ge N$ and $\lim\limits_{n\to\infty} a_n = \lim\limits_{n\to\infty} c_n = L$, then $\lim\limits_{n\to\infty} b_n = L$.

For a sequence whose nth term is a rational expression, such as $a_n = \dfrac{2n^2 - 4n + 7}{3n^2 + n + 10}$,

dividing both numerator and denominator by the highest power of n will yield an expression that may be easier to evaluate.

Example: $\lim\limits_{n\to\infty} \dfrac{2n^2 - 4n + 7}{3n^2 + n + 10} = \lim\limits_{n\to\infty} \dfrac{\frac{2n^2}{n^2} - \frac{4n}{n^2} + \frac{7}{n^2}}{\frac{3n^2}{n^2} + \frac{n}{n^2} + \frac{10}{n^2}} = \lim\limits_{n\to\infty} \dfrac{2 - \frac{4}{n} + \frac{7}{n^2}}{3 + \frac{1}{n} + \frac{10}{n^2}} = \lim\limits_{n\to\infty} \dfrac{2 - 0 + 0}{3 + 0 + 0} = \dfrac{2}{3}.$

Example: By the limit laws, $\lim\limits_{n\to\infty} \left(\dfrac{1}{n} + 3\dfrac{n^2 - 1}{n^2 + 1}\right) = \lim\limits_{n\to\infty} \dfrac{1}{n} + 3 \lim\limits_{n\to\infty} \dfrac{n^2 - 1}{n^2 + 1} = 0 + 3(1) = 3.$

1) Find the fourth term of the sequence $a_n = \dfrac{(-1)^n}{n^2}$.

$a_4 = \dfrac{(-1)^4}{4^2} = \dfrac{1}{16}.$

2) $\lim\limits_{n\to\infty} x_n = K$ means for all ____ there exists ____ such that if ____ then ____.

$\epsilon > 0$
a positive integer N
$n > N$
$|x_n - K| < \epsilon.$

3) Determine whether each converges.

a) Let c_n be the decimal expansion of π to n places. For example, $c_5 = 3.14159$ (5 places after the decimal).

Converges. $\lim\limits_{n \to \infty} a_n = \pi$.

b) $x_n = \dfrac{n}{n^2 + 1}$.

$\lim\limits_{n \to \infty} \dfrac{n}{n^2 + 1}$ (divide numerator and denominator by n^2)

$$= \lim_{n \to \infty} \frac{\frac{1}{n}}{1 + \frac{1}{n^2}} = \frac{0}{1 + 0} = 0.$$

Thus, $\{x_n\}$ converges to 0.

c) $b_n = \dfrac{n^2 + 1}{2n}$.

b_n grows without bound, so $\{b_n\}$ diverges. In this case we may write

$$\lim_{n \to \infty} \frac{n^2 + 1}{2n} = \infty.$$

d) $s_n = \dfrac{(-1)^n n}{n + 1}$.

The sequence $\dfrac{n}{n+1}$ converges to 1, so $s_n = \dfrac{(-1)^n n}{n + 1}$ alternately takes on values near 1 and -1. Thus, $\{s_n\}$ diverges.

e) $d_n = \dfrac{n}{n + 1} + \dfrac{2}{n^2}$.

Converges.

$$\lim_{n \to \infty} d_n = \lim_{n \to \infty} \frac{n}{n + 1} + \lim_{n \to \infty} \frac{2}{n^2} = 1 + 0 = 1.$$

4) Find $\lim\limits_{n \to \infty} \dfrac{\cos n}{n}$.

Since $-1 \leq \cos n \leq 1$ for all n,

$$-\frac{1}{n} \leq \frac{\cos n}{n} \leq \frac{1}{n} \text{ for all } n.$$

Because both $\left\{-\dfrac{1}{n}\right\}$ and $\left\{\dfrac{1}{n}\right\}$ converge to 0,

$$\lim_{n \to \infty} \frac{\cos n}{n} = 0.$$

5) Find $\lim\limits_{n \to \infty} \dfrac{\sin x}{n}$.

Since $\sin x$ is a constant as far as n is concerned, $\lim\limits_{n \to \infty} \dfrac{\sin x}{n} = (\sin x) \lim\limits_{n \to \infty} \dfrac{1}{n}$

$= (\sin x)\,(0) = 0.$

Silliness: Don't conclude the limit is 6 with this "computation":

$$\lim_{n \to \infty} \frac{\sin x}{n} = \lim_{n \to \infty} \frac{\sin x}{n} = \lim_{n \to \infty} \text{six} = 6.$$

6) Find $\displaystyle\lim_{n\to\infty}\frac{1-\cos\left(\frac{1}{n}\right)}{\sin\left(\frac{1}{n}\right)}$.

Let $f(x) = \dfrac{1-\cos\frac{1}{x}}{\sin\frac{1}{x}}$. Then, by

l'Hospital's Rule,

$$\lim_{x\to\infty}\frac{1-\cos\frac{1}{x}}{\sin\frac{1}{x}}\left(\text{form }\frac{0}{0}\right) = \lim_{x\to\infty}\frac{0+\left(\sin\frac{1}{x}\right)\left(\frac{-1}{x^2}\right)}{\left(\cos\frac{1}{x}\right)\left(\frac{-1}{x^2}\right)}$$

$$= \lim_{x\to\infty}\frac{\sin\frac{1}{x}}{\cos\frac{1}{x}} = \frac{0}{1} = 0.\text{ Therefore }\lim_{n\to\infty}\frac{1-\cos\left(\frac{1}{n}\right)}{\sin\left(\frac{1}{n}\right)} = 0.$$

7) True, False:

   a) If $\{a_n\}$ and $\{b_n\}$ converge, then $\{a_n + b_n\}$ converges.

True.

   b) If $\{a_n\}$ and $\{b_n\}$ diverge, then $\{a_n + b_n\}$ diverges.

False. For example, $a_n = n^2$ and $b_n = -n^2$ diverge, but $\{a_n + b_n\}$ is the constant sequence of all zeros that converges to zero.

   c) If $a_n \le b_n \le c_n$ for all n, and $\{a_n\}$ and $\{c_n\}$ converge, then $\{b_n\}$ converges.

False. Let $\{c_n\}$ be the sequence in Exercise **3a)**. Let $a_n = -c_n$ and $b_n = (-1)^n$ for all n. Then $\{c_n\}$ converges to π, $\{a_n\}$ converges to $-\pi$, but we have seen that $\{b_n\}$ diverges.

8) If $\displaystyle\lim_{n\to\infty} s_n = 4$ and $\displaystyle\lim_{n\to\infty} t_n = 2$, then

   a) $\displaystyle\lim_{n\to\infty}(8s_n - 2t_n) = \underline{\hspace{1cm}}$.

$8(4) - 2(2) = 28.$

   b) $\displaystyle\lim_{n\to\infty}\frac{3s_n}{t_n} = \underline{\hspace{1cm}}$.

$\dfrac{3(4)}{2} = 6.$

B. Let $\{a_n\}$ be a sequence.

$\{a_n\}$ is **increasing** means $a_{n+1} > a_n$ for all n.
$\{a_n\}$ is **decreasing** means $a_{n+1} < a_n$ for all n.
$\{a_n\}$ is **monotonic** if it is either increasing or decreasing.
$\{a_n\}$ is **bounded above** means $a_n \le M$ for some M and all n.
$\{a_n\}$ is **bounded below** means $a_n \ge m$ for some m and all n.
$\{a_n\}$ is **bounded** if it is both bounded above and bounded below.

Examples: Let $a_n = n^2 + 1$, $b_n = \frac{1}{n}$, and $c_n = \dfrac{(-1)^n}{n^2}$. Then

$\{a_n\}$ is increasing and not bounded above; $\{a_n\}$ is bounded below by 2.
$\{b_n\}$ is decreasing, bounded above by 1 and below by 0.
$\{c_n\}$ is neither increasing nor decreasing, and is bounded above by $\frac{1}{4}$ and below by -1.

Example: $t_n = 1 + \dfrac{1}{e^n}$ is a decreasing sequence because $e > 1$ and

$$t_n = 1 + \frac{1}{e^n} > 1 + \frac{1}{e^n}\cdot\frac{1}{e} = 1 + \frac{1}{e^{n+1}} = t_{n+1}.$$

Page 696

Every convergent sequence is bounded. For example, the sequence $b_n = \dfrac{3600}{n^2}$ has several large first terms ($b_1 = 3600$, $b_2 = 900$, $b_3 = 400, \ldots$), but after the 60th term, $0 < b_n < 1$. Therefore, $\{b_n\}$ is bounded below by 0 and bounded above by 3600.

Monotonic Sequence Theorem:

If $\{a_n\}$ is bounded and monotonic, then $\{a_n\}$ converges.

Example: $t_n = 1 + \dfrac{1}{e^n}$ is a monotonic (decreasing) sequence and is bounded

($1 < t_n < 2$ for all n). Therefore, $\displaystyle\lim_{n \to \infty} 1 + \dfrac{1}{e^n}$ exists. (The limit is 1.)

The converse of the Monotonic Sequence Theorem is false—a convergent sequence need not be monotonic. For example, $\displaystyle\lim_{n \to \infty} \dfrac{(-1)^n}{n^2} = 0$ but $\dfrac{(-1)^n}{n^2}$ is neither increasing nor decreasing.

9) Is $c_n = \dfrac{1}{3n}$ bounded above? bounded below?

$\left\{\dfrac{1}{3n}\right\}$ is bounded above by $\dfrac{1}{3}$ and bounded below by 0. (There are many other bounds.)

10) Is $s_n = \dfrac{n}{n+1}$ an increasing sequence?

Yes, because
$$s_n = \frac{n}{n+1}$$
$$< \frac{n}{n+1} + \frac{1}{(n+1)(n+2)}$$
$$= \frac{n(n+2)+1}{(n+1)(n+2)}$$
$$= \frac{n^2 + 2n + 1}{(n+1)(n+2)}$$
$$= \frac{(n+1)^2}{(n+1)(n+2)}$$
$$= \frac{n+1}{n+2} = s_{n+1}.$$

11) True or False:

a) If $\{a_n\}$ is not bounded below then $\{a_n\}$ diverges.

True.

b) If $\{a_n\}$ is decreasing and $a_n \geq 0$ for all n then $\displaystyle\lim_{n \to \infty} a_n$ exists.

True.

c) If $\{a_n\}$ is bounded, then $\{a_n\}$ converges.

False. For example, $a_n = (-1)^n$.

C. A sequence $\{a_n\}$ is **defined by a recurrence relation** (defined **recursively**) means:

   i) a_1 is defined.

   ii) a_{n+1}, for $n = 1, 2, 3, \ldots$, is defined in terms of previous a_i. (Often a_{n+1} is defined using only a_n.)

For example, the sequence $\{a_n\}$ given by

$$a_1 = \frac{1}{2}$$
$$a_{n+1} = \frac{a_n}{2}, n = 1, 2, 3, \ldots$$

is the sequence $\frac{1}{2}, \frac{1}{4}, \frac{1}{8}, \frac{1}{16}, \ldots$. This is a recursive definition of $a_n = \frac{1}{2^n}$.

12) Define the sequence $\frac{1}{2}, \frac{3}{4}, \frac{7}{8}, \frac{15}{16}, \frac{31}{32}, \ldots$ with a recurrence relation.

$a_1 = \frac{1}{2}$, $a_2 = \frac{3}{4} = \frac{1}{2} + \frac{1}{4} = a_1 + \frac{1}{4}$,

$a_3 = \frac{7}{8} = \frac{3}{4} + \frac{1}{8} = a_2 + \frac{1}{8}$.

The pattern shows that

$a_{n+1} = a_n + \frac{1}{2^{n+1}}$, for $n = 1, 2, 3, \ldots$

(Note: a non-recursive answer is

$a_n = \frac{2^n - 1}{2^n}$ for $n = 1, 2, 3, \ldots$)

13) Let $a_1 = 2$ and $a_{n+1} = \frac{2a_n}{3}$. Find $\lim\limits_{n \to \infty} a_n$.

$a_1 = 2$, $a_2 = \frac{2(2)}{3} = \frac{4}{3}$,

$a_3 = \frac{2\left(\frac{4}{3}\right)}{3} = \frac{8}{9}$, $a_4 = \frac{2\left(\frac{8}{9}\right)}{3} = \frac{16}{27}$,

$a_5 = \frac{2\left(\frac{16}{27}\right)}{3} = \frac{32}{81}$.

Since the denominator is growing faster than the numerator, it seems that $\lim\limits_{n \to \infty} a_n = 0$. This is so since a non-recursive formula for a_n is $\frac{2^n}{3^{n-1}} = 2\left(\frac{2}{3}\right)^{n-1}$.

14) Suppose $\{a_n\}$ is defined as $a_1 = 2$, $a_2 = 1$ and $a_{n+2} = a_{n+1} - a_n$ for $n = 1, 2, 3, \ldots$. Find $\lim\limits_{n \to \infty} a_n$.

Calculate a few terms to understand the pattern.

$a_1 = 2$, $a_2 = 1$
$a_3 = 1 - 2 = -1$
$a_4 = -1 - 1 = -2$
$a_5 = -2 - (-1) = -1$
$a_6 = -1 - (-2) = 1$
$a_7 = 1 - (-1) = 2$
$a_8 = 2 - 1 = 1$
$a_9 = 1 - 2 = -1$
$\vdots$

The terms repeat the pattern $1, -1, 2, -1, 1, 2$ every six terms. $\lim\limits_{n \to \infty} a_n$ does not exist.

Page 699

D. Technology Plus. Use a computer algebra system or a graphing calculator to solve.

T-1) Using a spreadsheet or calculator, find the first 20 partial sums of $\sum_{n=1}^{\infty} \dfrac{1}{n^4 + 1}$. Estimate the sum of the series.

n	$\dfrac{1}{n^4 + 1}$	$\sum_{k=1}^{n} \dfrac{1}{k^4 + 1}$
1	0.50000	0.50000
2	0.05882	0.55882
3	0.01220	0.57102
4	0.00389	0.57491
5	0.00160	0.57651
6	0.00077	0.57728
7	0.00042	0.57770
8	0.00024	0.57794
9	0.00015	0.57809
10	0.00010	0.57819
11	0.00007	0.57826
12	0.00005	0.57831
13	0.00004	0.57835
14	0.00003	0.57838
15	0.00002	0.57840
16	0.00002	0.57842
17	0.00001	0.57843
18	0.00001	0.57844
19	0.00001	0.57845
20	0.00001	0.57846

$$\sum_{n=1}^{\infty} \frac{1}{n^4 + 1} \approx 0.57846.$$

Section 11.2 Series

A series is the sum of all the terms of an infinite sequence. To determine whether (and if so, to what) a series sums, we build a second sequence consisting of the first term, the sum of the first two terms, the sum of the first three terms, the sum of the first four terms, etc. The sum of the series exists (that is, the series converges) if the limit of this sequence of partial sums exists. In this section we define these concepts precisely and look at some specific series which converge and others which do not.

Concepts to Master

A. Infinite series, Partial sums; Convergent and divergent series; Convergent series laws

B. Geometric series; Value of a converging geometric series; Harmonic series

Summary and Focus Questions

Page 703

A. Adding up all the terms of a sequence $\{a_n\}$ is an (**infinite**) **series:**

$$\sum_{k=1}^{\infty} a_k = a_1 + a_2 + a_3 + \ldots + a_n + \ldots$$

If we add up just the first n terms, we have the **nth partial sum** of the series:

$$s_n = \sum_{k=1}^{n} a_k = a_1 + a_2 + a_3 + \ldots + a_n.$$

A series **converges** if the limit of its sequence of partial sums exists. In other words $\sum_{n=1}^{\infty} a_n$ converges (to a number s) means $\lim_{n\to\infty} s_n$ exists (and is s).

If $\lim_{n\to\infty} s_n$ does not exist, then $\sum_{n=1}^{\infty} a_n$ **diverges.**

Example: Let $a_n = \frac{1}{2^n}$. Then $s_1 = a_1 = \frac{1}{2}$, $s_2 = a_1 + a_2 = \frac{1}{2} + \frac{1}{4} = \frac{3}{4}$,

$s_3 = a_1 + a_2 + a_3 = \frac{1}{2} + \frac{1}{4} + \frac{1}{8} = \frac{7}{8}$. In general, s_n is $\dfrac{2^n - 1}{2^n}$, which is a sequence that

converges to 1. Thus, $\sum_{n=1}^{\infty} \frac{1}{2^n} = 1$.

Test for Divergence:

If $\lim_{n\to\infty} a_n$ does not exist or $\lim_{n\to\infty} a_n \neq 0$, then $\sum_{n=1}^{\infty} a_n$ **diverges.**

Example: The series $\sum_{n=1}^{\infty} 2^n$ diverges since $\lim_{n\to\infty} 2^n \neq 0$.

Important: Just because the terms a_n approach zero is not enough to conclude that $\sum_{n=1}^{\infty} a_n$ converges. We will see an example in part B.

If $\sum\limits_{n=1}^{\infty} a_n$ converges to L and $\sum\limits_{n=1}^{\infty} b_n$ converges to M then:

$$\sum_{n=1}^{\infty} (a_n + b_n) \text{ converges to } L + M.$$

$$\sum_{n=1}^{\infty} (a_n - b_n) \text{ converges to } L - M.$$

For any constant c, $\sum\limits_{n=1}^{\infty} ca_n = c\sum\limits_{n=1}^{\infty} a_n$ converges to cL.

---

1) A series converges if the _____ of the sequence of _____ exists.

> limit
>
> partial sums

2) Find the first six partial sums of $\sum\limits_{n=1}^{\infty} \dfrac{1}{n^2}$.

> $s_1 = \dfrac{1}{1^2} = 1.$
>
> $s_2 = \dfrac{1}{1^2} + \dfrac{1}{2^2} = 1 + \dfrac{1}{4} = \dfrac{5}{4} = 1.25$
>
> $s_3 = \dfrac{1}{1^2} + \dfrac{1}{2^2} + \dfrac{1}{3^2} = \dfrac{5}{4} + \dfrac{1}{9} = \dfrac{49}{36} \approx 1.3611$
>
> $s_4 = \dfrac{49}{36} + \dfrac{1}{16} = \dfrac{205}{144} \approx 1.4236$
>
> $s_5 = \dfrac{205}{144} + \dfrac{1}{5^2} = \dfrac{5269}{3600} \approx 1.4636.$
>
> $s_6 = \dfrac{5269}{3600} + \dfrac{1}{6^2} = \dfrac{5369}{3600} \approx 1.4914.$

3) From the sequence of partial sums in question 2), estimate $\sum\limits_{n=1}^{\infty} \dfrac{1}{n^2}$.

> This is very difficult to guess. Perhaps $\sum\limits_{n=1}^{\infty} \dfrac{1}{n^2}$ is near 1.5. The famous mathematician Euler showed that $\sum\limits_{n=1}^{\infty} \dfrac{1}{n^2} = \dfrac{\pi^2}{6} \approx 1.6449.$
>
> (The series converges very slowly.) Later we will see methods to help with estimations.

4) Sometimes, Always, or Never:

   a) If $\lim\limits_{n\to\infty} a_n = 0$, $\sum\limits_{n=1}^{\infty} a_n$ converges.

> Sometimes.

   b) If $\lim\limits_{n\to\infty} a_n \neq 0$, $\sum\limits_{n=1}^{\infty} a_n$ diverges.

> Always.

5) Suppose $\sum\limits_{n=1}^{\infty} a_n = 3$ and $\sum\limits_{n=1}^{\infty} b_n = 4$.

   Evaluate each of the following:

   a) $\sum\limits_{n=1}^{\infty} (a_n + 2b_n).$

> $3 + 2(4) = 11.$

   b) $\sum\limits_{n=1}^{\infty} \dfrac{a_n}{5}.$

> $\dfrac{3}{5}.$

c) $\displaystyle\sum_{n=1}^{\infty} \frac{1}{a_n}$.

This series diverges. Because $\displaystyle\sum_{n=1}^{\infty} a_n = 3$,

$\displaystyle\lim_{n\to\infty} a_n = 0$. Thus, $\displaystyle\lim_{n\to\infty} \frac{1}{a_n} \neq 0$, so $\displaystyle\sum_{n=1}^{\infty} \frac{1}{a_n}$

diverges.

6) Sometimes, Always, or Never:

a) $\displaystyle\sum_{n=1}^{\infty} a_n$ converges, but $\displaystyle\lim_{n\to\infty} a_n$ does not exist.

Never. If $\displaystyle\sum_{n=1}^{\infty} a_n$ converges, then $\displaystyle\lim_{n\to\infty} a_n$ exists and is 0.

b) $\displaystyle\sum_{n=1}^{\infty} a_n$ converges, $\displaystyle\sum_{n=1}^{\infty} (a_n + b_n)$

converges, but $\displaystyle\sum_{n=1}^{\infty} b_n$ diverges.

Never. $\displaystyle\sum_{n=1}^{\infty} b_n = \sum_{n=1}^{\infty} (a_n + b_n) - \sum_{n=1}^{\infty} a_n$ is the difference of two convergent series and must converge.

7) Does $\displaystyle\sum_{n=1}^{\infty} \cos\left(\frac{1}{n}\right)$ converge?

No, the nth term does not approach 0.

Page 706

B. We need to build up a library of series that are known to converge or diverge. Here are the first two.

A **geometric series** has the form

$$\sum_{n=1}^{\infty} ar^{n-1} = a + ar + ar^2 + ar^3 + \dots$$

and converges (for $a \neq 0$) to $\dfrac{a}{1-r}$ if and only if $-1 < r < 1$.

Geometric series are important because not only do we always know whether the series converges, when it does converge we know exactly what it converges to.

Example: The series $\displaystyle\sum_{n=1}^{\infty} 4\left(\frac{1}{3}\right)^{n-1}$ is geometric with $a = 4$ and $r = \frac{1}{3}$. Since $\frac{1}{3} < 1$,

$\displaystyle\sum_{n=1}^{\infty} 4\left(\frac{1}{3}\right)^{n-1}$ converges and $\displaystyle\sum_{n=1}^{\infty} 4\left(\frac{1}{3}\right)^{n-1} = \dfrac{4}{1 - \frac{1}{3}} = \dfrac{4}{\frac{2}{3}} = 6.$

The **harmonic series** has the form

$$\sum_{n=1}^{\infty} \frac{1}{n} = 1 + \frac{1}{2} + \frac{1}{3} + \dots + \frac{1}{n} + \dots$$

This series diverges and is an example where $\displaystyle\lim_{n\to\infty} a_n = 0$ but $\displaystyle\sum_{n=1}^{\infty} a_n$ diverges.

8) a) $\displaystyle\sum_{n=1}^{\infty} \frac{2^n}{100}$ is a geometric series in which

$a = $ _____ and $r = $ _____.

$\displaystyle\sum_{n=1}^{\infty} \frac{2^n}{100} = \frac{2}{100} + \frac{4}{100} + \frac{8}{100} + \dots$

$a = \dfrac{2}{100}$ and $r = 2$.

b) Does the series converge?

No, since $r \geq 1$.

9) $\displaystyle\sum_{n=1}^{\infty} \left(\frac{-2}{9}\right)^{n} =$ _____.

This is a geometric series with $a = r = -\frac{2}{9}$

which converges to $\dfrac{-\frac{2}{9}}{1 - \left(-\frac{2}{9}\right)} = -\frac{2}{11}$.

10) Does $\displaystyle\sum_{n=1}^{\infty} \frac{6}{n}$ converge?

The series diverges, since it is a multiple of the harmonic series:

$$\sum_{n=1}^{\infty} \frac{6}{n} = 6 \sum_{n=1}^{\infty} \frac{1}{n}.$$

11) $\displaystyle\sum_{n=1}^{\infty} \frac{2^{n}}{3^{n+1}} =$ _____.

$\displaystyle\sum_{n=1}^{\infty} \frac{2^{n}}{3^{n+1}} = \sum_{n=1}^{\infty} \frac{2}{9}\left(\frac{2}{3}\right)^{n-1}$, which is a

geometric series with $a = \frac{2}{9}$, $r = \frac{2}{3}$ that

converges to $\dfrac{\frac{2}{9}}{1 - \frac{2}{3}} = \frac{2}{3}$.

12) In a race between Achilles and a tortoise Achilles gives the tortoise a 100 m head start. If Achilles runs at 5 m/s and the tortoise moves at $\frac{1}{2}$ m/s how far has the tortoise traveled by the time Achilles catches him?

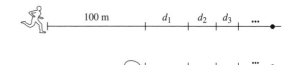

Let d_1 = distance tortoise traveled while Achilles was running the 100 m to the tortoise's starting point.

Since Achilles' velocity is 5 m/s,

$$d_1 = \left(\frac{100 \text{ m}}{5 \text{ m/s}}\right)\left(\frac{1}{2} \text{ m/s}\right) = 10 \text{ m}.$$

Let d_2 = distance tortoise traveled while Achilles was running the distance d_1.

Since $d_1 = 10$, $d_2 = \left(\frac{10 \text{ m}}{5 \text{ m/s}}\right)\left(\frac{1}{2} \text{ m/s}\right) = 1 \text{ m}$.

In general, for each n,

$$d_n = \left(\frac{d_{n-1}}{5}\right)\left(\frac{1}{2}\right) = \frac{d_{n-1}}{10}.$$

The total distance traveled by the tortoise is

$\displaystyle\sum_{n=1}^{\infty} d_n = 10 + 1 + \frac{1}{10}$. This is a geometric

series with $a = 10$, $r = \frac{1}{10}$.

This series converges to $\dfrac{10}{1 - \frac{1}{10}} = \frac{100}{9}$ m.

Section 11.3   The Integral Test and Estimates of Sums

This section and the next three sections provide tests to determine whether a series converges without explicitly finding the sum. (It is a good first step to determine whether something exists before trying to calculate or estimate it.) The Integral Test in this section is a natural first choice, for it relates infinite series $\left(\sum\limits_{n=1}^{\infty} ... \right)$ to improper integrals $\left(\int_{1}^{\infty} ... \, dx \right)$. However, the test only applies to certain series of positive terms. The test, when applicable, also permits us to estimate the sum of the convergent series.

Concepts to Master

A. Integral Test for convergence; *p*-series

B. Estimate of the sum of a convergent series using the Integral Test

Summary and Focus Questions

Page 714

A. Integral Test: Let $\{a_n\}$ be a sequence of positive, decreasing terms ($0 < a_{n+1} < a_n$ for all n). Suppose $f(x)$ is a positive, continuous, and decreasing function on $[1, \infty)$ such that $a_n = f(n)$ for all n. Then

$$\sum_{n=1}^{\infty} a_n \text{ converges if and only if } \int_{1}^{\infty} f(x)dx \text{ converges.}$$

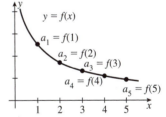

The Integral Test works well when a_n has the form of a function whose antiderivative is easily found. It is one of the few general tests that gives necessary and sufficient conditions for convergence.

Note that the Integral Test does not say $\sum\limits_{n=1}^{\infty} a_n = \int_{1}^{\infty} f(x)dx$. When these converge, they may be different numbers.

Example: To determine whether $\sum\limits_{n=1}^{\infty} \dfrac{2n}{n^2 + 1}$ converges, let $f(x) = \dfrac{2x}{x^2 + 1}$. Then

$$\int_{1}^{\infty} \frac{2x}{x^2 + 1} dx = \lim_{t \to \infty} \int_{1}^{t} \frac{2x}{x^2 + 1} dx = \lim_{t \to \infty} \ln(x^2 + 1) \Big]_{1}^{t} = \lim_{t \to \infty} \left(\ln(t^2 + 1) - \ln 2 \right) = \infty.$$

Since $\int_{1}^{\infty} \dfrac{2x}{x^2 + 1} dx$ diverges, $\sum\limits_{n=1}^{\infty} \dfrac{2n}{n^2 + 1}$ diverges by the Integral Test.

A *p*-series is a series of the form $\sum\limits_{n=1}^{\infty} \dfrac{1}{n^p}$, where p is a constant.

If $p > 1$, then $\sum\limits_{n=1}^{\infty} \dfrac{1}{n^p}$ converges.

If $p \leq 1$, then $\sum\limits_{n=1}^{\infty} \dfrac{1}{n^p}$ diverges.

Examples:

The harmonic series $\sum\limits_{n=1}^{\infty} \dfrac{1}{n}$ diverges because it is a *p*-series ($p = 1$).

The series $\sum\limits_{n=1}^{\infty} \dfrac{1}{n^4}$ converges because it is a *p*-series ($p = 4$).

1) True or False:
The Integral Test will determine to what value a series converges.

False. The test only indicates whether a series converges.

2) Test $\sum\limits_{n=1}^{\infty} \dfrac{n}{e^n}$ for convergence.

$\dfrac{n}{e^n}$ suggests the function $f(x) = xe^{-x}$.
On the interval $[1, \infty]$, f is continuous and decreasing, and $f(n) = \dfrac{n}{e^n}$ for $n = 1, 2, 3, \ldots$

$$\int_1^{\infty} xe^{-x} = \lim_{t \to \infty} \int_1^t xe^{-x}\,dx.$$

Using integration by parts,
($u = x$, $dv = e^{-x}\,dx$)

$$\int_1^t xe^{-x}\,dx = \left(-xe^{-x} - e^{-x}\right)\Big]_1^t = \dfrac{2}{e} - \dfrac{t+1}{e^t}.$$

$$\lim_{t \to \infty} \left(\dfrac{2}{e} - \dfrac{t+1}{e^t}\right) = \dfrac{2}{e} - 0 = \dfrac{2}{e}.$$

(By l'Hospital's Rule, $\dfrac{t+1}{e^t} \to 0$.)

Thus $\int_1^{\infty} xe^{-x}\,dx = \dfrac{2}{e}$, so $\sum\limits_{n=1}^{\infty} \dfrac{n}{e^n}$ converges.

Note: We can *not* conclude that $\sum\limits_{n=1}^{\infty} \dfrac{n}{e^n} = \dfrac{2}{e}$.

3) Which of these converge?

a) $\sum\limits_{n=1}^{\infty} \dfrac{1}{n^2}$.

Converges (*p*-series with $p = 2$).

b) $\sum\limits_{n=1}^{\infty} \dfrac{1}{\sqrt{n}}$.

Diverges $\left(\text{*p*-series with } p = \dfrac{1}{2}\right)$.

c) $\displaystyle\sum_{n=1}^{\infty} \frac{3}{2n^3}.$

Converges. $\displaystyle\sum_{n=1}^{\infty} \frac{3}{2n^3} = \frac{3}{2}\sum_{n=1}^{\infty} \frac{1}{n^3}$ is a multiple of a p-series with $p = 3$.

d) $\displaystyle\sum_{n=1}^{\infty} \left(\frac{1}{n^3} + \frac{1}{8^n}\right).$

Converges.

$$\sum_{n=1}^{\infty} \left(\frac{1}{n^3} + \frac{1}{8^n}\right) = \sum_{n=1}^{\infty} \frac{1}{n^3} + \sum_{n=1}^{\infty} \frac{1}{8^n},$$

the sum of a convergent p-series ($p = 3$) and a convergent geometric series $\left(r = \frac{1}{8}\right)$.

e) $\displaystyle\sum_{n=1}^{\infty} n^{-e}.$

Converges (p-series with $p = e$).

f) $\displaystyle\sum_{n=1}^{\infty} e^{-n}.$

Converges (This is not a p-series. It is a geometric series with $a = r = \frac{1}{e}$.)

Page 718

B. If $\displaystyle\sum_{n=1}^{\infty} a_n = s$ has been found to converge by the Integral Test using $f(x)$, then the **error** $R_n = s - s_n$ between the series value and the nth partial sum satisfies

$$\int_{n+1}^{\infty} f(x)\,dx \le R_n \le \int_{n}^{\infty} f(x)dx.$$

Example: The p-series $\displaystyle\sum_{n=1}^{\infty} \frac{1}{n^4}$ converges. (Remarkably, Euler showed that this series converges to $\dfrac{\pi^4}{90} \approx 1.08232$.) Let $s = \displaystyle\sum_{n=1}^{\infty} \frac{1}{n^4}$ and $f(x) = \dfrac{1}{x^4}$. For $n = 5$,

$$\int_{6}^{\infty} \frac{1}{x^4}dx = \lim_{t\to\infty}\int_{6}^{t} \frac{1}{x^4}dx = \lim_{t\to\infty} \frac{-1}{3x^3}\bigg]_{6}^{t} = \lim_{t\to\infty}\left(\frac{-1}{3t^3} - \frac{-1}{3(6)^3}\right) = \lim_{t\to\infty}\left(\frac{1}{648} - \frac{1}{3t^3}\right) = \frac{1}{648} \approx 0.00154.$$

Likewise, $\displaystyle\int_{5}^{\infty} \frac{1}{x^4}dx = \frac{1}{375} \approx 0.00267.$ Therefore, R_5, the difference between the infinite series sum s and the fifth partial sum s_5, satisfies:

$$0.00154 \le R_5 \le 0.00267.$$

To 5 decimals, the fifth partial sum is

$$s_5 = \sum_{n=1}^{5} \frac{1}{n^4} = \frac{1}{1^4} + \frac{1}{2^4} + \frac{1}{3^4} + \frac{1}{4^4} + \frac{1}{5^4} = \frac{1}{1} + \frac{1}{16} + \frac{1}{81} + \frac{1}{256} + \frac{1}{625} = \frac{14001361}{12960000} \approx 1.08035.$$

The actual value of $R_5 = s - s_5$ is $1.08232 - 1.08035 = 0.00197.$

4) In exercise **2** of Section 11.2, we found that the sixth partial sum of $\sum\limits_{n=1}^{\infty} \frac{1}{n^2}$ is

$$s_6 = \frac{1}{1} + \frac{1}{4} + \frac{1}{9} + \frac{1}{16} + \frac{1}{25} + \frac{1}{36}$$

$$= \frac{5369}{3600} \approx 1.4914.$$

a) Estimate the difference R_6 between s_6 and the exact value of

$$s = \sum_{n=1}^{\infty} \frac{1}{n^2}.$$

We know that $\sum\limits_{n=1}^{\infty} \frac{1}{n^2}$ converges (p-series, $p = 2$). Let $f(x) = \frac{1}{x^2}$.

$$\int_7^{\infty} \frac{1}{x^2}\, dx \leq R_6 \leq \int_6^{\infty} \frac{1}{x^2}\, dx.$$

$$\int_7^{\infty} \frac{1}{x^2}\, dx = \lim_{t\to\infty} \int_7^t x^{-2}\, dx$$

$$= \lim_{t\to\infty} \left(\frac{1}{7} - \frac{1}{t}\right) = \frac{1}{7}.$$

Likewise $\int_6^{\infty} \frac{1}{x^2}\, dx = \frac{1}{6}.$

Therefore, $\frac{1}{7} \leq R_6 \leq \frac{1}{6}.$

b) Estimate $s = \sum\limits_{n=1}^{\infty} \frac{1}{n^2}$ using the results above.

From part **a)** $\frac{1}{7} \leq s - s_6 \leq \frac{1}{6}$

$$\frac{1}{7} + s_6 \leq s \leq \frac{1}{6} + s_6$$

$$\frac{1}{7} + \frac{5369}{3600} \leq s \leq \frac{1}{6} + \frac{5369}{3600}$$

$$\frac{41183}{25200} \leq s \leq \frac{5969}{3600}$$

$$1.6342 \leq s \leq 1.6581.$$

It turns out that $s = \frac{\pi^2}{6} \approx 1.6449.$

c) Calculate R_6 and compare it to the error bounds $\frac{1}{7}$ and $\frac{1}{6}$.

$R_6 = s - s_6 = 1.6449 - 1.4914 = 0.1535.$

R_6 is between $\frac{1}{7} \approx 0.1429$ and $\frac{1}{6} \approx 0.1667.$

5) How many terms of $\sum\limits_{n=1}^{\infty} \frac{1}{n^2+1}$ are necessary to find the sum to within 0.01?

Let $f(x) = \frac{1}{x^2 + 1}.$

$\sum\limits_{n=1}^{\infty} \frac{1}{n^2 + 1}$ converges because $\int_1^{\infty} \frac{1}{x^2 + 1}\, dx$

$$= \lim_{t\to\infty} \int_1^t \frac{1}{x^2 + 1}\, dx$$

$$= \lim_{t\to\infty} \left(\tan^{-1}(t) - \tan^{-1}(1)\right) = \frac{\pi}{2} - \frac{\pi}{4} = \frac{\pi}{4}.$$

Since $s - s_n \leq \int_n^{\infty} f(x)\, dx$, it is sufficient to find n such that $\int_n^{\infty} f(x)\, dx \leq 0.01.$

From above, we see that

$$\int_n^{\infty} \frac{1}{x^2 + 1}\, dx = \frac{\pi}{2} - \tan^{-1}(n).$$

$$\frac{\pi}{2} - \tan^{-1}(n) < 0.01$$

$$\tan^{-1}(n) > \frac{\pi}{2} - 0.01$$

$$n > \tan\left(\frac{\pi}{2} - 0.01\right) \approx 99.997.$$

The first 100 terms are sufficient.

Section 11.4   The Comparison Tests

This section deals only with series of positive terms. If a series with positive terms converges, then any other positive series that is term for term smaller than the convergent series will also converge. This section makes precise this notion of comparing series of positive terms and discusses estimating limits.

Concepts to Master

A. Comparison Test; Limit Comparison Test

B. Estimate the sum of a convergent series

Summary and Focus Questions

Page 722

A. Let $\sum_{n=1}^{\infty} a_n$ and $\sum_{n=1}^{\infty} b_n$ be series whose terms are all positive.

Comparison Test:

i) If $a_n \leq b_n$ for all n and $\sum_{n=1}^{\infty} b_n$ converges, then $\sum_{n=1}^{\infty} a_n$ converges.

ii) If $a_n \geq b_n$ for all n and $\sum_{n=1}^{\infty} b_n$ diverges, then $\sum_{n=1}^{\infty} a_n$ diverges.

Limit Comparison Test:

Let c be a real number. If $\lim_{n \to \infty} \dfrac{a_n}{b_n} = c > 0$, then

$\sum_{n=1}^{\infty} a_n$ and $\sum_{n=1}^{\infty} b_n$ either both converge or both diverge.

When $\lim_{n \to \infty} \dfrac{a_n}{b_n} = 0$, if $\sum_{n=1}^{\infty} b_n$ converges, then $\sum_{n=1}^{\infty} a_n$ converges.

When $\lim_{n \to \infty} \dfrac{a_n}{b_n} = \infty$, if $\sum_{n=1}^{\infty} b_n$ diverges, then $\sum_{n=1}^{\infty} a_n$ diverges.

For a given series $\sum_{n=1}^{\infty} a_n$, success using either comparison test to determine whether $\sum_{n=1}^{\infty} a_n$ converges depends on coming up with another series $\sum_{n=1}^{\infty} b_n$ (geometric, p-series, ...) whose convergence is known and for which a_n and b_n may be compared.

Example: Determine whether $\displaystyle\sum_{n=1}^{\infty} \frac{1}{\sqrt{n}+4}$ converges by the Comparison Test.

To start, you need a hunch whether $\displaystyle\sum_{n=1}^{\infty} \frac{1}{\sqrt{n}+4}$ converges. Because

$\dfrac{1}{\sqrt{n}+4}$ is "like" $\dfrac{1}{\sqrt{n}}$ and $\displaystyle\sum_{n=1}^{\infty} \frac{1}{\sqrt{n}}$ is a divergent p-series $\left(p = \frac{1}{2}\right)$, we have

reason to believe that $\displaystyle\sum_{n=1}^{\infty} \frac{1}{\sqrt{n}+4}$ diverges. We also have an idea of the type

of b_n to look for—one for which $a_n \geq b_n$:

Let $b_n = \dfrac{1}{3\sqrt{n}}$. Then for all n, $\dfrac{1}{\sqrt{n}+4} \geq \dfrac{1}{3\sqrt{n}}$. We see that $\displaystyle\sum_{n=1}^{\infty} \frac{1}{3\sqrt{n}} = \frac{1}{3}\sum_{n=1}^{\infty} n^{-\frac{1}{2}}$

is a divergent p-series. Therefore, $\displaystyle\sum_{n=1}^{\infty} \frac{1}{\sqrt{n}+4}$ diverges.

Example: Determine whether $\displaystyle\sum_{n=1}^{\infty} \frac{1}{\sqrt{n}+4}$ converges by the Limit Comparison Test.

Let $a_n = \dfrac{1}{\sqrt{n}+4}$. We choose $b_n = \dfrac{1}{\sqrt{n}}$ for all n because it is similar to a_n. We know that

$\displaystyle\sum_{n=1}^{\infty} \frac{1}{\sqrt{n}}$ diverges (a p-series with $p = \frac{1}{2}$). Then

$$\lim_{n\to\infty} \frac{a_n}{b_n} = \lim_{n\to\infty} \frac{\frac{1}{\sqrt{n}+4}}{\frac{1}{\sqrt{n}}} = \lim_{n\to\infty} \frac{\sqrt{n}}{\sqrt{n}+4} = \lim_{n\to\infty} \sqrt{\frac{n}{n+4}} = 1.$$

Since $1 > 0$ and $\displaystyle\sum_{n=1}^{\infty} \frac{1}{\sqrt{n}}$ diverges, $\displaystyle\sum_{n=1}^{\infty} \frac{1}{\sqrt{n}+4}$ also diverges.

1) Determine whether each of the following
converge. Find a series $\displaystyle\sum_{n=1}^{\infty} b_n$ to use
with the Comparison Test or Limit
Comparison Test.

a) $\displaystyle\sum_{n=1}^{\infty} \frac{1}{n^2+2n}$, $b_n = $ _____.

Because $\dfrac{1}{n^2+2n}$ is "like" $\dfrac{1}{n^2}$ for large n, and

$\displaystyle\sum_{n=1}^{\infty} \frac{1}{n^2}$ is a convergent p-series, we suspect

that the given series converges.

Use $b_n = \dfrac{1}{n^2}$. Since $\dfrac{1}{n^2+2n} < \dfrac{1}{n^2}$ and

$\displaystyle\sum_{n=1}^{\infty} \frac{1}{n^2}$ converges, $\displaystyle\sum_{n=1}^{\infty} \frac{1}{n^2+2n}$ converges by

the Comparison Test.

b) $\displaystyle\sum_{n=1}^{\infty} \frac{\sqrt[3]{n}}{n+4}$, $b_n =$ _____.

Since $\dfrac{\sqrt[3]{n}}{n+4}$ is "like" $\dfrac{\sqrt[3]{n}}{n} = \dfrac{1}{\sqrt[3]{n^2}}$,

use $b_n = \dfrac{1}{\sqrt[3]{n^2}}$.

Then $\displaystyle\lim_{n\to\infty} \frac{a_n}{b_n} = \lim_{n\to\infty} \frac{\dfrac{\sqrt[3]{n}}{n+4}}{\dfrac{1}{\sqrt[3]{n^2}}} = \lim_{n\to\infty} \frac{n}{n+4} = 1.$

Since $\displaystyle\sum_{n=1}^{\infty} \frac{1}{\sqrt[3]{n^2}}$ diverges $\left(\text{a } p\text{-series with}\right.$

$\left. p = \frac{2}{3}\right)$, $\displaystyle\sum_{n=1}^{\infty} \frac{\sqrt[3]{n}}{n+4}$ diverges by the Limit

Comparison Test.

c) $\displaystyle\sum_{n=1}^{\infty} \frac{1}{n+2^n}$, $b_n =$ _____.

Use $b_n = \dfrac{1}{2^n}$. Then for $n \ge 1$, $\dfrac{1}{n+2^n} \le \dfrac{1}{2^n}$.

Since $\displaystyle\sum_{n=1}^{\infty} \frac{1}{2^n}$ converges $\left(\text{a geometric series}\right.$

with $\left. r = \frac{1}{2}\right)$, $\displaystyle\sum_{n=1}^{\infty} \frac{1}{n+2^n}$ converges by the

Comparison Test.

2) Suppose $\{a_n\}$ and $\{b_n\}$ are positive
sequences. Sometimes, Always, Never:

a) If $\displaystyle\lim_{n\to\infty} \frac{a_n}{b_n} = 0$ and $\displaystyle\sum_{n=1}^{\infty} a_n$ converges,

then $\displaystyle\sum_{n=1}^{\infty} b_n$ converges.

Sometimes. True for $a_n = \dfrac{1}{n^3}$ and $b_n = \dfrac{1}{n^2}$.

False for $a_n = \dfrac{1}{n^2}$ and $b_n = \dfrac{1}{n}$.

b) If $\displaystyle\lim_{n\to\infty} \frac{a_n}{b_n} = \infty$ and $\displaystyle\sum_{n=1}^{\infty} b_n$ diverges,

then $\displaystyle\sum_{n=1}^{\infty} a_n$ diverges.

Always.

Page 725

B. If $\displaystyle\sum_{n=1}^{\infty} a_n = s$ converges by the Comparison Test using $\displaystyle\sum_{n=1}^{\infty} b_n = t$, then

$$s - s_n \le t - t_n, \text{ for } n = 1, 2, 3, \dots.$$

This means that $T_n = t - t_n$ (which may be easier to calculate) is an upper
bound for an estimate of $R_n = s - s_n$.

Example: Show that $s = \displaystyle\sum_{n=1}^{\infty} \frac{5+n}{n \cdot 4^n}$ converges by the Comparison Test and find an

upper bound for the difference between $\displaystyle\sum_{n=1}^{\infty} \frac{5+n}{n \cdot 4^n}$ and its fifth partial sum s_5.

We see that $0 \le \dfrac{5+n}{n \cdot 4^n} \le \dfrac{6n}{n \cdot 4^n} = \dfrac{6}{4^n}$ and $\displaystyle\sum_{n=1}^{\infty} \frac{6}{4^n}$ converges to $t = \dfrac{\frac{6}{4}}{1 - \frac{1}{4}} = \dfrac{\frac{6}{4}}{\frac{3}{4}} = 2$

because it is a geometric series (with $a = \frac{6}{4}$ and $r = \frac{1}{4}$). Therefore $\sum_{n=1}^{\infty} \frac{5+n}{n \cdot 4^n}$ converges.

The fifth partial sum is $t_5 = \frac{6}{4} + \frac{6}{4^2} + \frac{6}{4^3} + \frac{6}{4^4} + \frac{6}{4^5} \approx 1.99805$. Therefore

$$T_5 = 2.00000 - 1.99805 = 0.00195$$

is an upper bound for $R_5 = s - s_5$.

3) In part **c)** of question **1)** we saw that
$\sum_{n=1}^{\infty} \frac{1}{n + 2^n}$ converges.

a) Find s_4, the fourth partial sum.

$$s_4 = \frac{1}{1+2} + \frac{1}{2+4} + \frac{1}{3+8} + \frac{1}{4+16}$$
$$= \frac{1}{3} + \frac{1}{6} + \frac{1}{11} + \frac{1}{20} = \frac{141}{220}.$$

b) Estimate the difference between the sum of the series and its fourth partial sum.

From **1c)**, $b_n = \frac{1}{2^n}$ may be used to show
$\sum \frac{1}{n+2^n}$ converges. Let $s = \sum_{n=1}^{\infty} \frac{1}{n+2^n}$ and
$t = \sum_{n=1}^{\infty} \frac{1}{2^n}$. Then $s - s_4 \le t - t_4$.

Since t is the sum of a geometric series we can calculate it:

$$t = \frac{\frac{1}{2}}{1 - \frac{1}{2}} = 1.$$

$$t_4 = \frac{1}{2} + \frac{1}{4} + \frac{1}{8} + \frac{1}{16} = \frac{15}{16}.$$

Thus, $t - t_4 = 1 - \frac{15}{16} = \frac{1}{16}$ and

$s - s_4 \le \frac{1}{16}$. Therefore, we know $s_4 = \frac{141}{220}$
is within $\frac{1}{16}$ of the value of s.

Section 11.5   **Alternating Series**

An alternating series is one for which consecutive terms have opposite signs. This section describes a test for convergence of these kinds of series and a simple method to estimate the sum of a convergent alternating series.

Concepts to Master

A. Alternating Series Test
B. Estimating the sum of a convergent alternating series
C. Technology Plus

Summary and Focus Questions

Page 727

A. An **alternating series** has successive terms of opposite sign—that is, it has one of these forms:

$$\sum_{n=1}^{\infty} (-1)^{n-1} a_n \text{ or } \sum_{n=1}^{\infty} (-1)^n a_n, \text{ where } a_n > 0.$$

Here are two examples of alternating series:

$$\sum_{n=1}^{\infty} \frac{(-1)^n}{n} = -\frac{1}{1} + \frac{1}{2} - \frac{1}{3} + \frac{1}{4} - \frac{1}{5} + \cdots$$

$$\text{and } \sum_{n=1}^{\infty} (-1)^{n-1} n! = 1 - 2! + 3! - 4! + 5! - \dots.$$

The Alternating Series Test:

If $\{a_n\}$ is a decreasing sequence with $a_n > 0$ for all n and $\lim_{n \to \infty} a_n = 0$,

then $\sum_{n=1}^{\infty} (-1)^{n-1} a_n$ converges.

Example: The series $\sum_{n=1}^{\infty} \frac{(-1)^n}{2^n}$ is an alternating series whose nth term is

$(-1)^n \frac{1}{2^n}$. Since $\lim_{n \to \infty} \frac{1}{2^n} = 0$, the series converges.

Note: The Alternating Series Test is another "existence test"—it confirms convergence but does not identify the actual sum of the series.

1) Is $\sum_{n=1}^{\infty} \frac{\sin n}{n}$ an alternating series?

No, although some terms are positive and others negative.

2) Use the Alternating Series Test to determine whether each converge:

a) $\displaystyle\sum_{n=1}^{\infty} \frac{(-1)^{n-1}}{e^n}$.

This is an alternating series with $a_n = \frac{1}{e^n}$.

$\frac{1}{e^{n+1}} \leq \frac{1}{e^n}$, so the terms decrease.

Since $\displaystyle\lim_{n\to\infty} \frac{1}{e^n} = 0$, $\displaystyle\sum_{n=1}^{\infty} \frac{(-1)^{n-1}}{e^n}$ converges by the Alternating Series Test.

b) $\displaystyle\sum_{n=1}^{\infty} \frac{(-1)^n n}{n+1}$.

This is an alternating series but the other conditions for the test do not hold.

Since $\displaystyle\lim_{n\to\infty} \frac{n}{n+1} \neq 0$, $\displaystyle\sum_{n=1}^{\infty} \frac{(-1)^n n}{n+1}$ diverges by the Divergence Test.

c) $\displaystyle\sum_{n=3}^{\infty} \frac{(-1)^{n-1} \ln n}{n}$. (Note that the series starts at $n = 3$.)

It may not be obvious that $a_n = \frac{\ln n}{n}$ decreases. Let $f(x) = \frac{\ln x}{x}$.

Then $f'(x) = \dfrac{x\left(\frac{1}{x}\right) - \ln x\,(1)}{x^2} = \dfrac{1 - \ln x}{x^2}$.

Since $n \geq 3$, $x \geq 3 > e$. Thus, $f'(x) < 0$ and f is decreasing. Therefore, $\{a_n\}$, $n \geq 3$, is decreasing. Finally,

$$\lim_{n\to\infty} \frac{\ln n}{n} = \lim_{n\to\infty} \frac{\frac{1}{n}}{1} = 0 \text{ by l'Hospital's Rule.}$$

Therefore, $\displaystyle\sum_{n=3}^{\infty} \frac{(-1)^{n-1} \ln n}{n}$ converges.

d) $\displaystyle\sum_{n=1}^{\infty} (-1)^{n-1} n!$

This series diverges. It is an alternating series but $\displaystyle\lim_{n\to\infty} (-1)^{n-1} n!$ does not exist.

(Remember that for a series $\displaystyle\sum_{n=1}^{\infty} a_n$ to converge we must have $\displaystyle\lim_{n\to\infty} a_n = 0$.)

B. Alternation Series Estimation Theorem:

If $\displaystyle\sum_{n=1}^{\infty} (-1)^{n-1} a_n$ is an alternating series with $a_{n+1} \leq a_n$ and $\displaystyle\lim_{n\to\infty} a_n = 0$ then $|s - s_n| \leq a_{n+1}$.

If $R_n = s - s_n$, the theorem says s_n may be used to approximate s to within a_{n+1}. In other words, $|R_n| \leq a_{n+1}$ for all n.

3) a) Show that the series $\sum_{n=1}^{\infty} \dfrac{(-1)^n}{\sqrt{n+1}}$ converges.

Since the series alternates and $\dfrac{1}{\sqrt{n+1}}$ decreases to 0, the series converges.

b) Estimate the error between the sum of the series and its fifteenth partial sum.

The error $|s - s_{15}|$ does not exceed

$$a_{16} = \frac{1}{\sqrt{16+1}} = 0.2.$$

4) For what value of n is s_n, the nth partial sum, within 0.001 of $s = \sum_{n=1}^{\infty} \dfrac{(-1)^{n+1}}{2n+5}$?

Let $a_n = \dfrac{1}{2n+5}$.

Since $|s - s_n| \leq a_{n+1}$ we need to find n such that $a_{n+1} < 0.001$.

$$a_{n+1} = \frac{1}{2(n+1)+5} = \frac{1}{2n+7}.$$

$$\frac{1}{2n+7} \leq 0.001$$

$$2n + 7 \geq 1000$$

$$2n \geq 993$$

$$n \geq 496.5.$$

Let $n = 497$. Then s_{497} is within 0.001 of s.

5) Approximate $\sum_{n=1}^{\infty} \dfrac{(-1)^{n+1}}{n^3}$ to within 0.005.

Let $a_n = \dfrac{1}{n^3}$. Then

$$|s - s_n| \leq a_{n+1} = \frac{1}{(n+1)^3} \leq 0.005 = \frac{1}{200}.$$

$$(n+1)^3 \geq 200$$

$$n + 1 \geq \sqrt[3]{200} \approx 5.84$$

$$n \geq 4.84. \text{ Let } n = 5.$$

Therefore, s_5 is within 0.005 of s.

$$s_5 = \frac{1}{1^3} - \frac{1}{2^3} + \frac{1}{3^3} - \frac{1}{4^3} + \frac{1}{5^3}$$

$$= 1 - \frac{1}{8} + \frac{1}{27} - \frac{1}{64} + \frac{1}{125} \approx 0.9044.$$

We do not know to what number the series converges, but we do know that 0.9044 is within 0.005 of the series sum.

C. Technology Plus. Use a computer algebra system or a graphing calculator to solve.

T-1) Use the 15th partial sum to estimate
$$\sum_{n=1}^{\infty} \frac{(-1)^{n-1}}{n^3}.$$ How accurate is the estimate?

n	$\frac{(-1)^{n-1}}{n^3}$	$\sum_{k=1}^{n} \frac{(-1)^{k-1}}{k^3}$
1	1.00000	1.00000
2	−0.12500	0.87500
3	0.03704	0.91204
4	−0.01563	0.89641
5	0.00800	0.90441
6	−0.00463	0.89978
7	0.00292	0.90270
8	−0.00195	0.90075
9	0.00137	0.90212
10	−0.00100	0.90112
11	0.00075	0.90187
12	−0.00058	0.90129
13	0.00046	0.90175
14	−0.00036	0.90139
15	0.00030	0.90169

The 15th partial sum is

$$s_{15} = \sum_{n=1}^{15} \frac{(-1)^{n-1}}{n^3} \approx 0.90169.$$

The absolute value of the 16th term is

$$\left| \frac{(-1)^{16-1}}{16^3} \right| \approx 0.00024.$$

The 15th partial sum is within approximately 0.00024 of the sum of the series.

Section 11.6 Absolute Convergence and the Ratio and Root Tests

The concept of convergence of an infinite series may be split into two separate subconcepts: absolute convergence and conditional convergence. This section gives a precise definition of each. It also describes the Root and Ratio Tests, which may be used to check for absolute convergence.

Concepts to Master

A. Absolute convergence; Conditional convergence

B. Ratio Test; Root Test

C. Rearrangement of terms of a series

Summary and Focus Questions

Page 732

A. For any series $\sum_{n=1}^{\infty} a_n$ (with a_n not necessarily positive or alternating) two types of convergence may be defined:

$\sum_{n=1}^{\infty} a_n$ **converges absolutely** means that $\sum_{n=1}^{\infty} |a_n|$ converges.

$\sum_{n=1}^{\infty} a_n$ **converges conditionally** means that $\sum_{n=1}^{\infty} a_n$ converges but $\sum_{n=1}^{\infty} |a_n|$ diverges.

Examples: a) $\sum_{n=1}^{\infty} \dfrac{(-1)^{n-1}}{n^3}$ is absolutely convergent because the series

$\sum_{n=1}^{\infty} \left| \dfrac{(-1)^{n-1}}{n^3} \right| = \sum_{n=1}^{\infty} \dfrac{1}{n^3}$ converges (p-series, $p = 3$).

b) $\sum_{n=1}^{\infty} \dfrac{(-1)^{n-1}}{\sqrt[3]{n}}$ is conditionally convergent because it converges (by the

Alternating Series Test) but $\sum_{n=1}^{\infty} \left| \dfrac{(-1)^{n-1}}{\sqrt[3]{n}} \right| = \sum_{n=1}^{\infty} \dfrac{1}{\sqrt[3]{n}}$ diverges

$\left(p\text{-series}, p = \dfrac{1}{3} \right)$.

Either one of absolute convergence or conditional convergence implies (ordinary) convergence. Conversely, convergence implies either absolute or conditional convergence (but not both!).

Thus *every* series must behave in exactly one of these three ways: diverge, converge absolutely, or converge conditionally.

1) Sometimes, Always, or Never:

a) If $\sum_{n=1}^{\infty} |a_n|$ diverges, then $\sum_{n=1}^{\infty} a_n$ diverges.

Sometimes. True for $a_n = n$ but false for $a_n = \dfrac{(-1)^n}{n}$.

b) If $\sum_{n=1}^{\infty} a_n$ diverges, then $\sum_{n=1}^{\infty} |a_n|$ diverges.

Always.

c) If $a_n \geq 0$ for all n, then $\sum_{n=1}^{\infty} a_n$ is not conditionally convergent.

Always. Since $|a_n| = a_n$ for all n, we cannot have $\sum_{n=1}^{\infty} a_n$ converges and $\sum_{n=1}^{\infty} |a_n|$ diverges.

2) Determine whether each converges conditionally, converges absolutely, or diverges.

a) $\sum_{n=1}^{\infty} \dfrac{(-1)^n}{\sqrt{n}}$.

The series is alternating with $\dfrac{1}{\sqrt{n}}$ decreasing to 0 so it converges. It remains to check for absolute convergence:

$$\sum_{n=1}^{\infty} \left| \frac{(-1)^n}{\sqrt{n}} \right| = \sum_{n=1}^{\infty} \frac{1}{\sqrt{n}} \text{ diverges } \left(p\text{-series} \right.$$

with $p = \frac{1}{2}$). Thus, $\sum_{n=1}^{\infty} \dfrac{(-1)^n}{\sqrt{n}}$ converges conditionally.

b) $\sum_{n=1}^{\infty} \dfrac{\sin n + \cos n}{n^3}$.

Check for absolute convergence first:

$$\sum_{n=1}^{\infty} \left| \frac{\sin n + \cos n}{n^3} \right| = \sum_{n=1}^{\infty} \frac{|\sin n + \cos n|}{n^3}.$$

Since $|\sin n + \cos n| \leq 2$ and $\sum_{n=1}^{\infty} \dfrac{2}{n^3}$ converges (it is a p-series with $p = 3$), $\sum_{n=1}^{\infty} \dfrac{|\sin n + \cos n|}{n^3}$ converges by the Comparison Test. Thus, $\sum_{n=1}^{\infty} \dfrac{\sin n + \cos n}{n^3}$ converges absolutely.

3) Does $\displaystyle\sum_{n=1}^{\infty} \frac{\sin e^n}{e^n}$ converge?

Yes. The series is not alternating but does contain both positive and negative terms. Check for absolute convergence:

$$\sum_{n=1}^{\infty} \left|\frac{\sin e^n}{e^n}\right| = \sum_{n=1}^{\infty} \frac{|\sin e^n|}{e^n}.$$

Since $|\sin e^n| \le 1$, $\dfrac{|\sin e^n|}{e^n} \le \dfrac{1}{e^n}$.

$\displaystyle\sum_{n=1}^{\infty} \frac{1}{e^n}$ converges $\left(\text{geometric series with}\right.$

$r = \dfrac{1}{e}\bigg)$ so by the Comparison Test

$\displaystyle\sum_{n=1}^{\infty} \left|\frac{\sin e^n}{e^n}\right|$ converges. Therefore, $\displaystyle\sum_{n=1}^{\infty} \frac{\sin e^n}{e^n}$

converges absolutely and hence converges.

Page 734

B. The following tests are very useful for determining whether a series converges absolutely. Both tests decide whether $|a_n|$ approaches 0 fast enough so that $\displaystyle\sum_{n=1}^{\infty} a_n$ converges. Neither test involves another series or function, which makes them relatively easy to use. However, each has cases where it fails to provide any information.

The Ratio Test works well when the nth term, a_n, contains exponentials or factorials or when a_n is defined recursively. It will fail when a_n is a rational function of n.

Ratio Test: Let a_n be a sequence of nonzero terms.

i) If $\displaystyle\lim_{n\to\infty} \left|\frac{a_{n+1}}{a_n}\right| = L < 1$, then $\displaystyle\sum_{n=1}^{\infty} a_n$ converges absolutely.

ii) If $\displaystyle\lim_{n\to\infty} \left|\frac{a_{n+1}}{a_n}\right| = L > 1$ or $\displaystyle\lim_{n\to\infty} \left|\frac{a_{n+1}}{a_n}\right| = \infty$, then $\displaystyle\sum_{n=1}^{\infty} a_n$ diverges.

iii) If $\displaystyle\lim_{n\to\infty} \left|\frac{a_{n+1}}{a_n}\right| = 1$, the Ratio Test fails: $\displaystyle\sum_{n=1}^{\infty} a_n$ may converge or diverge.

The Root Test works well when a_n contains an expression to the nth power.

Root Test:

i) If $\displaystyle\lim_{n\to\infty} \sqrt[n]{|a_n|} = L < 1$, then $\displaystyle\sum_{n=1}^{\infty} a_n$ converges absolutely.

ii) If $\displaystyle\lim_{n\to\infty} \sqrt[n]{|a_n|} = L > 1$ or $\displaystyle\lim_{n\to\infty} \sqrt[n]{|a_n|} = \infty$, $\displaystyle\sum_{n=1}^{\infty} a_n$ diverges.

iii) If $\displaystyle\lim_{n\to\infty} \sqrt[n]{|a_n|} = 1$, the Root Test fails: $\displaystyle\sum_{n=1}^{\infty} a_n$ may converge or diverge.

The limit $\displaystyle\lim_{n\to\infty} \sqrt[n]{n} = 1$ is sometimes useful in applying the Root Test.

Examples: a) $\displaystyle\sum_{n=1}^{\infty} \frac{n^2}{n!}$ converges absolutely by the Ratio Test:

$$\lim_{n\to\infty} \left|\frac{a_{n+1}}{a_n}\right| = \lim_{n\to\infty} \left|\frac{\frac{(n+1)^2}{(n+1)!}}{\frac{n^2}{n!}}\right| = \lim_{n\to\infty} \frac{\frac{(n+1)^2}{(n+1)}}{n^2} = \lim_{n\to\infty} \frac{n+1}{n^2} = 0. \text{ Since this limit is less}$$

than 1 the series converges absolutely.

b) $\displaystyle\sum_{n=1}^{\infty} \frac{n}{e^{2n}}$ converges absolutely by the Root Test:

$$\lim_{n\to\infty} \sqrt[n]{|a_n|} = \lim_{n\to\infty} \sqrt[n]{\left|\frac{n}{e^{2n}}\right|} = \lim_{n\to\infty} \frac{\sqrt[n]{n}}{\sqrt[n]{e^{2n}}} = \lim_{n\to\infty} \frac{\sqrt[n]{n}}{e^2} = \frac{1}{e^2}. \text{ Since this limit is less}$$

than 1 the series converges absolutely.

4) Determine whether each converges by the Ratio Test.

a) $\displaystyle\sum_{n=1}^{\infty} \frac{n}{4^n}$

$$\lim_{n\to\infty} \left|\frac{a_{n+1}}{a_n}\right| = \lim_{n\to\infty} \frac{\frac{n+1}{4^{n+1}}}{\frac{n}{4^n}} = \lim_{n\to\infty} \frac{n+1}{4n} = \frac{1}{4}.$$

Thus, $\displaystyle\sum_{n=1}^{\infty} \frac{n}{4^n}$ converges absolutely and therefore converges.

b) $\displaystyle\sum_{n=1}^{\infty} \frac{(-4)^n}{n!}$

$$\lim_{n\to\infty} \left|\frac{a_{n+1}}{a_n}\right| = \lim_{n\to\infty} \left|\frac{\frac{(-4)^{n+1}}{(n+1)!}}{\frac{(-4)^n}{n!}}\right| = \lim_{n\to\infty} \frac{4}{n+1} = 0.$$

Thus, $\displaystyle\sum_{n=1}^{\infty} \frac{(-4)^n}{n!}$ converges absolutely and therefore converges.

c) $\displaystyle\sum_{n=1}^{\infty} \frac{(-1)^n}{\sqrt[3]{n}}$

$$\lim_{n\to\infty} \left|\frac{a_{n+1}}{a_n}\right| = \lim_{n\to\infty} \left|\frac{\frac{(-1)^{n+1}}{\sqrt[3]{n+1}}}{\frac{(-1)^n}{\sqrt[3]{n}}}\right|$$

$$= \lim_{n\to\infty} \sqrt[3]{\frac{n}{n+1}} = 1. \text{ The Ratio Test fails.}$$

The Alternating Series Test shows that this series converges. It converges conditionally because $\displaystyle\sum_{n=1}^{\infty} \frac{1}{\sqrt[3]{n}}$ diverges (p-series, $p = \frac{1}{3}$.)

d) $\sum_{n=1}^{\infty} a_n$, where $a_1 = 4$ and

$$a_{n+1} = \frac{3a_n}{2n+1}.$$

$$\lim_{n\to\infty} \left| \frac{a_{n+1}}{a_n} \right| = \lim_{n\to\infty} \frac{\frac{3a_n}{2n+1}}{a_n} = \lim_{n\to\infty} \frac{3}{2n+1} = 0.$$

Thus, $\sum_{n=1}^{\infty} a_n$ converges absolutely and

therefore converges.

5) Determine whether each converges by the Root Test:

a) $\sum_{n=1}^{\infty} \frac{1}{(n+1)^n}$

$$\lim_{n\to\infty} \sqrt[n]{|a_n|} = \lim_{n\to\infty} \sqrt[n]{\frac{1}{(n+1)^n}} = \lim_{n\to\infty} \frac{1}{n+1} = 0.$$

Thus, $\sum_{n=1}^{\infty} \frac{1}{(n+1)^n}$ converges absolutely and

therefore converges.

b) $\sum_{n=1}^{\infty} \frac{3^n}{n^3}$

$$\lim_{n\to\infty} \sqrt[n]{|a_n|} = \lim_{n\to\infty} \sqrt[n]{\frac{3^n}{n^3}} = \lim_{n\to\infty} \frac{3}{n^{3/n}} = \frac{3}{1} = 3.$$

Therefore by the Root Test $\sum_{n=1}^{\infty} \frac{3^n}{n^3}$ diverges.

To show $\lim_{n\to\infty} n^{3/n} = 1$ let $y = \lim_{n\to\infty} n^{3/n}$.

Then $\ln y = \lim_{n\to\infty} \ln n^{3/n} = \lim_{n\to\infty} \frac{3 \ln n}{n}$

$$\left(\text{form } \frac{\infty}{\infty}\right) = \lim_{n\to\infty} \frac{\frac{3}{n}}{1} = 0.$$

Thus, $y = e^0 = 1$.

c) $\sum_{n=1}^{\infty} \frac{(-1)^n}{n}$

$$\lim_{n\to\infty} \sqrt[n]{\left|\frac{(-1)^n}{n}\right|} = \lim_{n\to\infty} \frac{1}{\sqrt[n]{n}} = 1.$$

The Root Test fails, but we know this series converges because it is the alternating harmonic series.

C. A **rearrangement** of an infinite series is another series obtained from the series by changing the order of the terms. For example, $\frac{1}{2} + 1 + \frac{1}{8} + \frac{1}{4} + \frac{1}{32} + \frac{1}{16} + \cdots$ is a rearrangement of the geometric series $1 + \frac{1}{2} + \frac{1}{4} + \frac{1}{8} + \frac{1}{16} + \frac{1}{32} + \cdots$.

If $\sum_{n=1}^{\infty} a_n$ converges absolutely to s, then all rearrangements converge to the same sum s.

If $\sum_{n=1}^{\infty} a_n$ converges conditionally, its terms can be rearranged so that their sum is a different number.

Page 737

6) Do all rearrangements of $\sum_{n=1}^{\infty} \frac{(-1)^{n-1}}{n^4}$

converge to the same value?

Yes. The series is absolutely convergent since

$\sum_{n=1}^{\infty} \frac{1}{n^4}$ is a p-series ($p = 4$).

Section 11.7 Strategy for Testing Series

There are no new techniques in this section—just a summary of the tests to use for determining whether a given series converges.

Concepts to Master

Apply the various tests of convergence of a series.

Summary and Focus Questions

Page 739

To decide whether a given series $\sum_{n=1}^{\infty} a_n$ converges you should determine the form of the nth term—whether it matches a certain type (p-series, geometric or alternating), resembles a known form (comparison tests or Integral Test), or is amenable using the Ratio or Root Tests. Here is a brief summary of each test together with a representative example of its use:

Test	Form/Conditions	Conclusion(s)	Example
Divergence	$\lim\limits_{n\to\infty} a_n \neq 0$	$\sum a_n$ diverges	$\sum_{n=1}^{\infty} \dfrac{n}{n+1}$ diverges. $\lim\limits_{n\to\infty} \dfrac{n}{n+1} = 1 \neq 0.$
Integral	Find $f(x)$, continuous, decreasing with $f(n) = a_n \geq 0$.	$\sum a_n$ converges if and only if $\int_1^{\infty} f(x)\,dx$ converges.	$\sum_{n=1}^{\infty} \dfrac{\ln n}{n}$ diverges. $\int_1^{\infty} \dfrac{\ln x}{x}\,dx$ does not exist.
p-series	$a_n = \dfrac{1}{n^p}$	$\sum \dfrac{1}{n^p}$ converges if and only if $p > 1$.	$\sum_{n=1}^{\infty} \dfrac{1}{n^3}$ converges. $p = 3.$
Geometric Series	$a_n = ar^{n-1}$	$\sum ar^{n-1}$ converges if and only if $|r| < 1$.	$\sum_{n=1}^{\infty} \dfrac{2}{3^n}$ converges. $r = \dfrac{1}{3}.$
Alternating Series	$b_n = (-1)^n a_n$ or $b_n = (-1)^{n-1} a_n,$ where $a_n > 0$.	If $\lim\limits_{n\to\infty} a_n = 0$ and a_n is decreasing, $\sum b_n$ converges.	$\sum_{n=1}^{\infty} \dfrac{(-1)^n}{n}$ converges. $a_n = \dfrac{1}{n}.$

Test	Form/Conditions	Conclusion(s)	Example
Comparison	$0 \le a_n \le b_n$	If $\sum b_n$ converges then $\sum a_n$ converges. If $\sum a_n$ diverges then $\sum b_n$ diverges.	$\displaystyle\sum_{n=1}^{\infty} \frac{1}{n^3 + 1}$ converges. $b_n = \dfrac{1}{n^3}$.
Limit Comparison	$\displaystyle\lim_{n\to\infty} \frac{a_n}{b_n} = c$	If $0 < c < \infty$, $\sum a_n$ converges if and only if $\sum b_n$ converges. If $c = 0$, then $\sum b_n$ converges implies $\sum a_n$ converges. If $c = \infty$, then $\sum a_n$ diverges implies $\sum b_n$ diverges.	$\displaystyle\sum_{n=1}^{\infty} \frac{1}{2n^3 + 1}$ converges. $b_n = \dfrac{1}{n^3}$. $\displaystyle\lim_{n\to\infty} \frac{a_n}{b_n} = \frac{1}{2}$.
Ratio	$\displaystyle\lim_{n\to\infty} \left| \frac{a_{n+1}}{a_n} \right| = L$	If $L > 1$, $\sum a_n$ diverges. If $L < 1$, $\sum a_n$ converges absolutely.	$\displaystyle\sum_{n=1}^{\infty} \frac{n^2}{3^n}$ converges absolutely. $\displaystyle\lim_{n\to\infty} \left| \frac{a_{n-1}}{a_n} \right| = \frac{1}{3}$.
Root	$\displaystyle\lim_{n\to\infty} \sqrt[n]{|a_n|} = L$	If $L > 1$, $\sum a_n$ diverges. If $L < 1$, $\sum a_n$ converges absolutely.	$\displaystyle\sum_{n=1}^{\infty} \left(\frac{2n+1}{3n+4} \right)^n$ converges absolutely. $\displaystyle\lim_{n\to\infty} \sqrt[n]{|a_n|} = \frac{2}{3}$.

For each series, what is an appropriate test to apply for each?

1) $\displaystyle\sum_{n=1}^{\infty} \frac{(n+1)^n}{4^n}$.

Root Test, because the $(n+1)^n$ and 4^n terms have n in the exponent.
$$\lim_{n\to\infty} \sqrt[n]{|a_n|} = \lim_{n\to\infty} \frac{n+1}{4} = \infty.$$
This series diverges.

2) $\displaystyle\sum_{n=1}^{\infty} \frac{5}{n^2 + 6n + 3}$.

Comparison Test, compare to $\sum \dfrac{1}{n^2}$.
$$\frac{5}{n^2 + 6n + 3} \le 5 \cdot \frac{1}{n^2} \quad (p\text{-series}, p = 2).$$
This series converges.

3) $\displaystyle\sum_{n=1}^{\infty} \frac{(-1)^n n}{n^6 + 1}$.

Alternating Series Test for convergence.

$\displaystyle\lim_{n\to\infty} \frac{n}{n^6 + 1} = 0$ so the series converges.

Comparison Test (compare to $\sum \frac{1}{n^5}$) for absolute convergence.

$$\left| \frac{(-1)^n n}{n^6 + 1} \right| = \frac{n}{n^6 + 1} < \frac{1}{n^5}.$$

$\displaystyle\sum_{n=1}^{\infty} \frac{1}{n^5}$ converges (p-series, $p = 5$).

Therefore $\displaystyle\sum_{n=1}^{\infty} \frac{n}{n^6 + 1}$ converges.

This series converges absolutely.

4) $\displaystyle\sum_{n=1}^{\infty} \frac{2^n n^3}{n!}$.

Ratio Test, because of the $n!$ term.

$$\lim_{n\to\infty} \left| \frac{a_{n+1}}{a_n} \right| = \lim_{n\to\infty} \frac{2(n + 1)^2}{n^3} = 0.$$

This series converges absolutely.

5) $\displaystyle\sum_{n=1}^{\infty} \frac{n}{e^{n^2}}$.

The exponential e^{n^2} suggests the Ratio Test (which will work) but the function

$f(x) = \frac{x}{e^{x^2}} = xe^{-x^2}$ may be used in the

Integral Test.

$$\int_1^{\infty} xe^{-x^2}\, dx = \frac{1}{2e}.$$

This series converges.

6) $\displaystyle\sum_{n=1}^{\infty} \frac{n!}{2^n}$.

Divergence Test.

$\displaystyle\lim_{n\to\infty} \frac{n!}{2^n} = \infty$. Since the nth term does not go to zero, this series diverges.

Section 11.8 **Power Series**

This section introduces a special type of function called a power series—a function f whose functional values are infinite series. As such, there may be values for its variable x for which the power series $f(x)$ is defined (for that x, the resulting series converges) and others for which $f(x)$ does not exist (resulting series diverges). The tests of convergence will help determine the domain of a power series.

Concepts to Master

Power series; Interval of Convergence; Radius of convergence

Summary and Focus Questions

Page 741

A **power series in $(x - a)$** is an expression of the form

$$\sum_{n=0}^{\infty} c_n (x - a)^n, \text{ where } c_n \text{ and } a \text{ are constants.}$$

When a particular value of x is specified, the power series becomes an infinite series. Thus, for some values of x the expression converges and for other values of x it may diverge.

The domain of a power series in $(x - a)$ is all values of x for which $\sum_{n=0}^{\infty} c_n (x - a)^n$ converges and is called the **interval of convergence.** It consists of all real numbers or all real numbers from $a - R$ to $a + R$ for some number R, where $R \geq 0$. The power series diverges for all x outside the interval of convergence. The number a is called the **center** and R is the **radius** of convergence. When $R = 0$, the interval is the single point $\{a\}$. We use the convention $R = \infty$ to mean the interval of convergence is all real numbers.

The value of R is found by applying the Ratio Test to $\sum_{n=0}^{\infty} c_n (x - a)^n$ and solving for $|x - a|$ in the resulting inequality:

$$\lim_{n \to \infty} \left| \frac{c_{n+1}(x - a)^{n+1}}{c_n(x - a)^n} \right| = \lim_{n \to \infty} \left| \frac{c_{n+1}}{c_n} \right| |x - a| < 1.$$

(Sometimes the Root Test may be used.)

When R is a positive number, the two endpoints $a - R$ and $a + R$ pose special problems because the Ratio Test fails. Each endpoint may or may not be in the interval of convergence. You must use tests other than the Ratio Test to individually check the power series for convergence when $x = a - R$ and when $x = a + R$.

Example: The power series $\sum\limits_{n=0}^{\infty} \dfrac{(x-4)^n}{n+3} = \sum\limits_{n=0}^{\infty} \dfrac{1}{n+3}(x-4)^n$ has nth coefficient $\dfrac{1}{n+3}$.

The center of the interval of convergence is $a = 4$. To find the radius (by the Ratio Test) we solve $\lim\limits_{n\to\infty} \left|\dfrac{c_{n+1}}{c_n}\right| |x-4| < 1.$

$$\lim_{n\to\infty} \left|\frac{c_{n+1}}{c_n}\right| |x-4| = \lim_{n\to\infty} \left|\frac{\frac{1}{n+4}}{\frac{1}{n+3}}\right| |x-4| = \lim_{n\to\infty} \frac{n+3}{n+4}|x-4| = |x-4| < 1.$$

Thus, $-1 < x - 4 < 1$, or $3 < x < 5$. The radius of convergence is 1.

For $x = 3$, $\sum\limits_{n=0}^{\infty} \dfrac{(3-4)^n}{n+3} = \sum\limits_{n=0}^{\infty} \dfrac{(-1)^n}{n+3}$, which converges by the Alternating Series Test.

For $x = 5$, $\sum\limits_{n=0}^{\infty} \dfrac{(5-4)^n}{n+3} = \sum\limits_{n=0}^{\infty} \dfrac{1}{n+3}$, which diverges (it is all but the first three terms of the harmonic series).

The interval of convergence is $[3, 5)$.

Think of a power series as an infinitely long polynomial. (In fact every polynomial is a power series in which all coefficients except the first few are zero!) Just as polynomial function values are easy to calculate, we have seen that power series values are sometimes easy to approximate within a given degree of accuracy.

1) True or False:

A power series is an infinite series.

False. A power series is an expression for a function $f(x)$. For any x in the domain of f, $f(x)$ is the sum of a convergent series.

2) For what value of x does $\sum\limits_{n=0}^{\infty} c_n(x-a)^n$ always converge?

At $x = a$, $\sum\limits_{n=0}^{\infty} c_n(x-a)^n = c_0$.

The power series always converges when x is the center a.

3) Compute the value of

a) $\sum\limits_{n=0}^{\infty} n^2(x-3)^n$ at $x = 4$.

When $x = 4$ the power series is

$\sum\limits_{n=0}^{\infty} n^2 1^n = \sum\limits_{n=0}^{\infty} n^2$. This infinite series

diverges so there is no value for $x = 4$.

b) $\displaystyle\sum_{n=0}^{\infty} 2^n(x-1)^n$ at $x = \frac{5}{6}$.

4) Find the center, radius, and interval of convergence for:

a) $\displaystyle\sum_{n=0}^{\infty} (2n)!(x-1)^n$.

b) $\displaystyle\sum_{n=1}^{\infty} \frac{(x-4)^n}{2^n n}$. (Although this series starts at $n = 1$, we proceed in the same way.)

When $x = \frac{5}{6}$ the power series is

$$\sum_{n=0}^{\infty} 2^n\left(-\frac{1}{6}\right)^n = \sum_{n=0}^{\infty}\left(-\frac{1}{3}\right)^n$$

$$= 1 - \frac{1}{3} + \frac{1}{9} - \frac{1}{27} + \dots$$

This is a geometric series with $a = 1$, $r = -\frac{1}{3}$. It converges to $\dfrac{1}{1-\left(-\frac{1}{3}\right)} = \dfrac{3}{4}$.

Use the Ratio Test:

$$\lim_{n\to\infty}\left|\frac{(2(n+1))!(x-1)^{n+1}}{(2n)!(x-1)^n}\right|$$

$$= \left(\lim_{n\to\infty}(2n+1)(2n+2)\right)|x-1| = \infty,$$

for all x except when $x = 1$. Thus, the center is 1, the radius is 0, and the interval of convergence is $\{1\}$.

Use the Ratio Test:

$$\lim_{n\to\infty}\left|\frac{(x-4)^{n+1}}{2^{n+1}(n+1)} \cdot \frac{2^n n}{(x-4)^n}\right|$$

$$= \left(\lim_{n\to\infty}\frac{n}{2(n+1)}\right)|x-4| = \frac{1}{2}|x-4|.$$

From $\frac{1}{2}|x-4| < 1$ we conclude $|x-4| < 2$. The center is 4 and radius is 2.
When $|x-4| = 2$, $x = 2$ or $x = 6$.
At $x = 2$, the power series becomes

$$\sum_{n=1}^{\infty} \frac{(-1)^n}{n}$$ which converges by the

Alternating Series Test.
At $x = 6$, the power series is the divergent

harmonic series $\displaystyle\sum_{n=1}^{\infty} \frac{1}{n}$.

The interval of convergence is $[2, 6)$.

Note that the Root Test could have been used to start the solution:

$$\lim_{n\to\infty}\sqrt[n]{\left|\frac{(x-4)^n}{2^n n}\right|} = \left(\lim_{n\to\infty}\sqrt[n]{\frac{1}{n}}\right)\frac{1}{2}|x-4|$$

$$= 1 \cdot \frac{1}{2}|x-4|$$

$$= \frac{1}{2}|x-4| < 1.$$

c) $\displaystyle\sum_{n=1}^{\infty} \frac{x^n}{n!}.$

Use the Ratio Test:

$$\lim_{n\to\infty} \left| \frac{\frac{x^{n+1}}{(n+1)!}}{\frac{x^n}{n!}} \right| = \lim_{n\to\infty} \frac{|x|}{n+1} = 0$$

for all values of x. Thus, the interval of convergence is all real numbers, $(-\infty, \infty)$.

5) Suppose $\displaystyle\sum_{n=0}^{\infty} c_n (x-5)^n$ converges for $x = 8$. For what other values must it converge?

The interval of convergence is $5 - R$ to $5 + R$.

Because 8 is in this interval $8 \leq 5 + R$. Thus, $3 \leq R$ which means the interval at least contains all numbers between $2(= 5 - 3)$ and $8(= 5 + 3)$. We do not have enough information about the endpoint 2. The power series converges for at least all x such that $2 < x \leq 8$.

6) True, False: The two power series

$$\sum_{n=0}^{\infty} c_n(x-a)^n \text{ and } \sum_{n=0}^{\infty} 2c_n (x-a)^n \text{ have}$$

exactly the same interval of convergence.

True.

Section 11.9   Representations of Functions as Power Series

In the last section we saw that a power series is an expression that may be used to define a function whose domain is either all real numbers or includes the interval $(a - R, a + R)$ and possibly either or both endpoints (where $R \geq 0$). In this section we turn things around and begin with a function f and show that it is possible to rewrite $f(x)$ as a power series for at least some of the elements in the domain of f.

Concepts to Master

A. Obtaining a power series expression for a function by manipulating known series

B. Differentiation and integration of power series

C. Power series expansions by differentiation and integration

Summary and Focus Questions

Page 746

A. The geometric series $\displaystyle\sum_{n=1}^{\infty} ar^{n-1}$ converges to $\dfrac{a}{1 - r}$ for $a \neq 0$ and $-1 < r < 1$.

If we use $a = 1$, $r = x$ and simplify $\displaystyle\sum_{n=1}^{\infty} 1(x^{n-1})$ by rewriting it as $\displaystyle\sum_{n=0}^{\infty} x^n$ (starting at $n = 0$), then

$$\frac{1}{1 - x} = \sum_{n=0}^{\infty} x^n.$$

Even though the domain of $f(x) = \dfrac{1}{1 - x}$ is all real numbers except $x = 1$, this expression is valid only for $-1 < x < 1$. Nevertheless, we have written the function f as a power series for some of the values in its domain.

In this section, to write a given function as a power series, we start with a known power series expression for a similar function (such as $f(x) = \dfrac{1}{1 - x}$) and modify it by substitutions (such as x^2 for x) and multiplication by constants or powers of x. Here are two examples:

Example: Find a power series for $f(x) = \dfrac{3}{1 - 8x^3}$.

We start with the similar function $\dfrac{1}{1 - x} = \displaystyle\sum_{n=0}^{\infty} x^n$. Substitute $8x^3$ for x:

$$\frac{1}{1 - 8x^3} = \frac{1}{1 - (8x^3)} = \sum_{n=0}^{\infty} (8x^3)^n = \sum_{n=0}^{\infty} 8^n x^{3n}.$$

Then multiply by 3:

$$\frac{3}{1 - 8x^3} = 3\sum_{n=0}^{\infty} 8^n x^{3n} = \sum_{n=0}^{\infty} (3)8^n x^{3n}.$$

Since the interval of convergence for the original power series is $|x| < 1, |8x^3| < 1.$ Thus, $|x^3| < \frac{1}{8}$, so $|x| < \frac{1}{2}.$ The interval of convergence for f is $\left(-\frac{1}{2}, \frac{1}{2}\right).$

Example: Find a power series for $f(x) = \frac{2x}{1 + x^2}.$

We start with $\frac{1}{1 - x} = \sum_{n=0}^{\infty} x^n$ with interval of convergence $(-1, 1).$

Then substitute $-x^2$ for x:

$$\frac{1}{1 + x^2} = \sum_{n=0}^{\infty} (-x^2)^n = \sum_{n=0}^{\infty} (-1)^n x^{2n}.$$

Then multiply by $2x$:

$$\frac{2x}{1 + x^2} = \sum_{n=0}^{\infty} (-1)^n (2x)x^{2n} = \sum_{n=0}^{\infty} (-1)^n 2x^{2n+1}.$$

Since the interval of convergence for $\frac{1}{1 - x} = \sum_{n=0}^{\infty} x^n$ is $(-1, 1), |x| < 1.$ The substitution $-x^2$ for x yields $\left|-x^2\right| < 1$, which is still $|x| < 1.$ Therefore, the interval of convergence for $\frac{2x}{1 + x^2} = \sum_{n=0}^{\infty} (-1)^n 2x^{2n+1}$ is $(-1, 1).$

1) Given $\frac{1}{1 - x} = \sum_{n=0}^{\infty} x^n, |x| < 1$, find a power series expression for:

a) $\frac{1}{2 - x}.$

$$\frac{1}{2 - x} = \frac{1}{2\left(1 - \frac{x}{2}\right)} = \frac{1}{2}\left(\frac{1}{1 - \frac{x}{2}}\right).$$

Thus, $\frac{1}{2 - x} = \frac{1}{2}\sum_{n=0}^{\infty} \left(\frac{x}{2}\right)^n = \sum_{n=0}^{\infty} \frac{1}{2} \cdot \frac{x^n}{2^n}$

$$= \sum_{n=0}^{\infty} \frac{x^n}{2^{n+1}}.$$

From $\left|\frac{x}{2}\right| < 1, |x| < 2$, so $(-2, 2)$ is the interval of convergence.

b) $\dfrac{x}{x^4 - 1}$.

We note that $\dfrac{x}{x^4 - 1} = -\dfrac{x}{1 - x^4}$.

First substitute x^4 for x:

$$\frac{1}{1 - x^4} = \sum_{n=0}^{\infty} (x^4)^n = \sum_{n=0}^{\infty} x^{4n}.$$

Now multiply by $-x$:

$$\frac{-x}{1 - x^4} = -x \sum_{n=0}^{\infty} x^{4n} = -\sum_{n=0}^{\infty} x^{4n+1}.$$

Since $|x| < 1, |x^4| < 1$. Thus, the interval of convergence is still $(-1, 1)$.

Page 748

B. If $f(x) = \displaystyle\sum_{n=0}^{\infty} c_n (x - a)^n$ has interval of convergence with radius $R \geq 0$ then:

i) f is continuous on $(a - R, a + R)$.

ii) For all $x \in (a - R, a + R)$, f may be differentiated term by term:

$$f'(x) = \sum_{n=1}^{\infty} nc_n (x - a)^{n-1}.$$

Note that when the series for f starts at $n = 0$, we may write the series for f' starting at $n = 1$ since the derivative of c_0 is zero.

iii) For all $x \in (a - R, a + R)$, f may be integrated term by term:

$$\int f(x)dx = C + \sum_{n=0}^{\infty} \frac{c_n(x - a)^{n+1}}{n + 1} \text{ where } C \text{ is the constant of integration.}$$

Both $f'(x)$ and $\displaystyle\int f(x)dx$ have radius of convergence R but the endpoints $a - R$ and $a + R$ must be checked individually.

2) Find $f'(x)$ for

$$f(x) = \sum_{n=0}^{\infty} 2^n(x - 1)^n.$$

$$f'(x) = \sum_{n=0}^{\infty} 2^n n(x - 1)^{n-1}$$

$$= \sum_{n=1}^{\infty} 2^n n(x - 1)^{n-1}.$$

3) Find $\int f(x)\,dx$ where

$$f(x) = \sum_{n=1}^{\infty} \frac{(x-3)^n}{n!}.$$

$$\int f(x)\,dx = C + \sum_{n=0}^{\infty} \frac{1}{n!} \frac{(x-3)^{n+1}}{n+1}$$

$$= C + \sum_{n=0}^{\infty} \frac{(x-3)^{n+1}}{(n+1)!}.$$

4) Is $f(x) = \sum_{n=0}^{\infty} \frac{x^n}{4^n}$ continuous at $x = 2$?

Yes. 2 is in $(-4, 4)$, the interval of convergence for $f(x)$.

Page
748

C. Term by term integration and differentiation is another technique to help find power series for functions.

Example: To find a power series for $\ln(1 + x^2)$ we first note

that $\ln(1 + x^2) = \int \frac{2x}{1 + x^2}\,dx$. Recall from part **A** that

$$\frac{2x}{1+x^2} = \sum_{n=0}^{\infty} (-1)^n 2x^{2n+1}.$$

Thus $\ln(1 + x^2) = \int \frac{2x}{1+x^2}\,dx = \int \sum_{n=0}^{\infty} (-1)^n 2x^{2n+1}\,dx$

$$= \sum_{n=0}^{\infty} (-1)^n 2 \int x^{2n+1}\,dx = C + \sum_{n=0}^{\infty} (-1)^n 2 \frac{x^{2n+2}}{2n+2}$$

$$= C + \sum_{n=0}^{\infty} (-1)^n \frac{x^{2n+2}}{n+1}.$$

To determine C, let $x = 0$. Then $0 = \ln(1 + 0) = C + \sum_{n=0}^{\infty} (-1)^n \frac{0^{2n+2}}{n+1} =$

$C + 0$. Thus $C = 0$ and $\ln(1 + x^2) = \sum_{n=0}^{\infty} (-1)^n \frac{x^{2n+2}}{n+1}.$

This power series has the same interval of convergence $(-1, 1)$ as the power

series for $\frac{2x}{1+x^2}$ but we need to check the endpoints. At both $x = 1$ and

$x = -1$, $x^{2n+2} = 1$ and the resulting series $\sum_{n=0}^{\infty} \frac{(-1)^n}{n+1}$ converges. Therefore,

the interval of convergence for $\ln(1 + x^2) = \sum_{n=0}^{\infty} (-1)^n \frac{x^{2n+2}}{n+1}$ is $[-1, 1]$.

5) Given $\dfrac{1}{1-x} = \displaystyle\sum_{n=0}^{\infty} x^n, |x| < 1$, find a power series for $\dfrac{1}{(1+x)^2}$.

Substitute $-x$ for x:

$$\frac{1}{1+x} = \sum_{n=0}^{\infty} (-x)^n = \sum_{n=0}^{\infty} (-1)^n x^n.$$

Differentiate term by term:

$$\frac{-1}{(1+x)^2} = \sum_{n=1}^{\infty} (-1)^n n x^{n-1}.$$

$$\frac{1}{(1+x)^2} = (-1) \sum_{n=1}^{\infty} (-1)^n n x^{n-1}$$

$$= \sum_{n=1}^{\infty} (-1)^{n+1} n x^{n-1}.$$

Since $|-x| < 1$ is the same as $|x| < 1$, the radius of convergence is $R = 1$.

At $x = 1$ we have $\displaystyle\sum_{n=1}^{\infty} (-1)^{n+1} n$.

At $x = -1$ we have

$$\sum_{n=1}^{\infty} (-1)^{n+1} n (-1)^{n-1} = \sum_{n=1}^{\infty} n.$$

Both series diverge so the interval of convergence is $(-1, 1)$.

6) Find $\displaystyle\int \dfrac{1}{1+x^5}\, dx$ as a power series.

From $\dfrac{1}{1-x} = \displaystyle\sum_{n=0}^{\infty} x^n$, substitute $-x^5$ for x:

$$\frac{1}{1+x} = \sum_{n=0}^{\infty} (-x^5)^n$$

$$= \sum_{n=0}^{\infty} (-1)^n x^{5n}.$$

Integrate term by term:

$$\int \frac{1}{1+x^5}\, dx = \sum_{n=0}^{\infty} \int (-1)^n x^{5n}\, dx$$

$$= \sum_{n=0}^{\infty} \frac{(-1)^n}{5n+1} x^{5n+1} + C.$$

For the power series for $\dfrac{1}{1-x}$ the interval of convergence is $|x| < 1$. Substituting $-x^5$ for x, $|-x^5| < 1$ is still $|x| < 1$.
We individually check the endpoints:
At $x = 1$ the power series becomes

$$\sum_{n=0}^{\infty} \frac{(-1)^n}{5n+1} 1^{5n+1} + C = \sum_{n=0}^{\infty} \frac{(-1)^n}{5n+1} + C,$$

which is a converging alternating series.

At $x = -1$ the power series becomes

$$\sum_{n=0}^{\infty} \frac{(-1)^n}{5n + 1}(-1)^{5n+1} + C$$

$$= \sum_{n=0}^{\infty} \frac{(-1)^{6n+1}}{5n + 1} + C,$$

which is also a converging alternating series. The interval of convergence is $[-1, 1]$.

7) From $\dfrac{1}{1 - x} = \displaystyle\sum_{n=0}^{\infty} x^n$, find the sum of the series $\displaystyle\sum_{n=1}^{\infty} n\left(\frac{1}{3}\right)^{n-1}$.

By differentiation, $\left(\dfrac{1}{1 - x}\right)' = \displaystyle\sum_{n=1}^{\infty} nx^{n-1}$.

Since $\left(\dfrac{1}{1 - x}\right)' = \dfrac{1}{(1 - x)^2}$,

$$\sum_{n=1}^{\infty} nx^{n-1} = \frac{1}{(1 - x)^2}.$$

At $x = \dfrac{1}{3}$, $\displaystyle\sum_{n=1}^{\infty} n\left(\frac{1}{3}\right)^{n-1} = \dfrac{1}{\left(1 - \frac{1}{3}\right)^2} = \dfrac{9}{4}$.

Section 11.10   Taylor and Maclaurin Series

This section describes which functions may be written as power series and how to find such representations. The second partial sum of this representation will turn out to be the equation of the tangent line. Partial sums of the series for a function will be useful for approximating values of the function.

Concepts to Master

A. Taylor series; Maclaurin series

B. Taylor polynomials of degree n

C. Remainder of a Taylor series; Taylor's Inequality

D. Binomial coefficient; Binomial series for $(1 + x)^k$

E. Multiplication and division of power series

F. Technology Plus

Summary and Focus Questions

Page
753

A. Not every function can be written as a power series. However, if a function $y = f(x)$ can be expressed as a power series in $(x - a)$, then that series must have the form:

$$f(x) = \sum_{n=0}^{\infty} \frac{f^{(n)}(a)}{n!}(x - a)^n$$

This is called the **Taylor series for f at a.** In the special case of $a = 0$, it is called the **Maclaurin series for f.**

To find a Taylor series for a given function $y = f(x)$, you first find a formula for $f^{(n)}(a)$, usually in terms of n. Computing $f(a)$ and the first few derivatives, $f'(a), f''(a), f'''(a), f^{(4)}(a), \ldots$ often helps to see what the general term $f^{(n)}(a)$ looks like.

Example: Find the Maclaurin series for $f(x) = 3^x$.

First find the expression for the nth derivative at $x = 0$.

$$
\begin{array}{ll}
f(x) = 3^x & f(0) = 1 \\
f'(x) = 3^x \,(\ln 3) & f'(0) = \ln 3 \\
f''(x) = 3^x \,(\ln 3)^2 & f''(0) = (\ln 3)^2 \\
f'''(x) = 3^x \,(\ln 3)^3 & f'''(0) = (\ln 3)^3
\end{array}
$$

In general, $f^{(n)}(0) = (\ln 3)^n$. Therefore, the Maclaurin series is

$$3^x = \sum_{n=0}^{\infty} \frac{(\ln 3)^n}{n!} x^n.$$

By the Ratio Test, $\lim\limits_{n\to\infty} \left| \dfrac{\frac{(\ln 3)^{n+1}}{(n+1)!}}{\frac{(\ln 3)^n}{n!}} \right| = \lim\limits_{n\to\infty} \left| \dfrac{\ln 3}{n+1} \right| = 0$. Thus, the radius of

convergence is $R = \infty$, which means $3^x = \sum\limits_{n=0}^{\infty} \dfrac{(\ln 3)^n}{n!} x^n$ for all x.

Here are some basic Maclaurin series:

Maclaurin Series	**Domain**
$\dfrac{1}{1-x} = \sum\limits_{n=0}^{\infty} x^n = 1 + x + x^2 + x^3 + \dots$	$(-1, 1)$
$e^x = \sum\limits_{n=0}^{\infty} \dfrac{x^n}{n!} = 1 + \dfrac{x}{1!} + \dfrac{x^2}{2!} + \dfrac{x^3}{3!} + \dots$	all real numbers
$\ln(1 + x) = \sum\limits_{n=1}^{\infty} \dfrac{(-1)^{n-1}}{n} x^n = x - \dfrac{x^2}{2} + \dfrac{x^3}{3} - \dfrac{x^4}{4} + \dots$	$(-1, 1]$
$\sin x = \sum\limits_{n=0}^{\infty} \dfrac{(-1)^n}{(2n+1)!} x^{2n+1} = x - \dfrac{x^3}{3!} + \dfrac{x^5}{5!} - \dfrac{x^7}{7!} + \dots$	all real numbers
$\cos x = \sum\limits_{n=0}^{\infty} \dfrac{(-1)^n}{(2n)!} x^{2n} = 1 - \dfrac{x^2}{2!} + \dfrac{x^4}{4!} - \dfrac{x^6}{6!} + \dots$	all real numbers
$\tan^{-1} x = \sum\limits_{n=0}^{\infty} (-1)^n \dfrac{x^{2n+1}}{2n+1} = x - \dfrac{x^3}{3} + \dfrac{x^5}{5} - \dfrac{x^7}{7} + \dots$	$[-1, 1]$

$(1 + x)^k = \sum\limits_{n=0}^{\infty} \binom{k}{n} x^n = 1 + kx + \dfrac{k(k-1)}{2!} x^2 + \dots + \binom{k}{n} x^n + \dots$ Depends on k.

See part **D**.

For some functions f it is easier to find the Taylor series for $f'(x)$ or $\int f(x)\,dx$ then integrate or differentiate term by term to obtain the series for f. In some other cases, substitutions in the basic Taylor series and algebra can be used to find the Taylor series for f.

1) Find the Taylor series for $f(x) = \frac{1}{x}$ at $a = 1$.

a) Directly from the definition.

First find the general form of $f^{(n)}(1)$:

$$f(x) = x^{-1} \qquad\qquad f(1) = 1$$
$$f'(x) = -x^{-2} \qquad\quad f'(1) = -1$$
$$f''(x) = 2x^{-3} \qquad\quad f''(1) = 2$$
$$f'''(x) = -6x^{-4} \qquad f'''(1) = -6$$
$$f^{(4)}(x) = 24x^{-5} \qquad f^{(4)}(1) = 24.$$

In general, $f^{(n)}(1) = (-1)^n n!$.

Thus, the Taylor series is

$$\frac{1}{x} = \sum_{n=0}^{\infty} \frac{(-1)^n n!}{n!} (x-1)^n$$

$$= \sum_{n=0}^{\infty} (-1)^n (x-1)^n = \sum_{n=0}^{\infty} (1-x)^n.$$

b) Using substitution in a geometric series $\left(\text{Hint: } \frac{1}{x} = \frac{1}{1-(1-x)}\right)$.

In the form $\frac{1}{1-(1-x)}$, this is the value of a geometric series with $a = 1$, $r = 1 - x$.

$$\frac{1}{x} = \sum_{n=1}^{\infty} 1(1-x)^{n-1} = \sum_{n=0}^{\infty} (1-x)^n.$$

c) Using the Maclaurin series for $\ln(1+x)$ and then substituting $x - 1$ for x.

Differentiate $\ln(1+x) = \sum_{n=1}^{\infty} \frac{(-1)^{n-1}}{n} x^n$:

$$\frac{1}{1+x} = \sum_{n=1}^{\infty} \frac{(-1)^{n-1} n}{n} x^{n-1} = \sum_{n=1}^{\infty} (-1)^{n-1} x^{n-1}$$

(Note the derivative also starts at $n = 1$.)

$$= \sum_{n=1}^{\infty} (-x)^{n-1} = \sum_{n=0}^{\infty} (-x)^n.$$

(Change the sum to start at $n = 0$.)

Now substitute $x - 1$ for x:

$$\frac{1}{1+(x-1)} = \sum_{n=0}^{\infty} (-(x-1))^n = \sum_{n=0}^{\infty} (1-x)^n.$$

Therefore, $\dfrac{1}{x} = \displaystyle\sum_{n=0}^{\infty} (1-x)^n.$

d) What is the interval of convergence for your answer to part **a)**?

Apply the Ratio Test:

$$\lim_{n\to\infty} \left| \frac{(1-x)^{n+1}}{(1-x)^n} \right| = \lim_{n\to\infty} |1-x| = |1-x|.$$

$|1-x| < 1$ is equivalent to $0 < x < 2$.

At $x = 0$ and $x = 2$ the terms of $\displaystyle\sum_{n=0}^{\infty} (1-x)^n$ do no approach zero so both those series diverge. The interval of convergence is $(0, 2)$.

2) Find directly the Maclaurin series for $f(x) = e^{4x}$.

$$f(x) = e^{4x} \qquad f(0) = 1$$
$$f'(x) = 4e^{4x} \qquad f'(0) = 4$$
$$f''(x) = 16e^{4x} \qquad f''(0) = 16.$$

In general $f^{(n)}(0) = 4^n$.

$$e^{4x} = \sum_{n=0}^{\infty} \frac{4^n}{n!}x^n.$$

(Using the basic series we can write

$$e^{4x} = \sum_{n=0}^{\infty} \frac{(4x)^n}{n!} = \sum_{n=0}^{\infty} \frac{4^n}{n!}x^n.)$$

By the Ratio Test,

$$\lim_{n\to\infty} \left| \frac{\frac{4^{n+1}}{(n+1)!}x^{n+1}}{\frac{4^n}{n!}x^n} \right| = \lim_{n\to\infty} \frac{4}{n+1}|x| = 0.$$

The interval of convergence is $(-\infty, \infty)$.

3) Obtain the Maclaurin series for $\dfrac{x}{1+x^2}$ from the series for $\dfrac{1}{1-x} = \displaystyle\sum_{n=0}^{\infty} x^n$.

$$\frac{1}{1-x} = 1 + x + x^2 + \dots x^n + \dots$$

Substitute $-x^2$ for x:

$$\frac{1}{1+x^2} = 1 - x^2 + x^4 - x^6 + \dots$$
$$+ (-1)^n x^{2n} + \dots$$

Now multiply by x:

$$\frac{x}{1+x^2} = x - x^3 + x^5 - x^7 + \dots$$
$$+ (-1)^n x^{2n+1} + \dots$$
$$= \sum_{n=0}^{\infty} (-1)^n x^{2n+1}.$$

4) Using $e^x = \displaystyle\sum_{n=0}^{\infty} \frac{x^n}{n!}$ find the Maclaurin series for $\sinh x$.

$\left(\text{Remember, } \sinh x = \dfrac{e^x - e^{-x}}{2}.\right)$

$$e^x = 1 + x + \frac{x^2}{2!} + \dots \frac{x^n}{n!} + \dots$$
$$e^{-x} = 1 - x + \frac{x^2}{2!} + \dots \frac{(-1)^n x^n}{n!} + \dots$$

Subtracting term by term (the even terms cancel): $e^x - e^{-x}$

$$= 2x + \frac{2x^3}{3!} + \dots + \frac{2x^{2n+1}}{(2n+1)!} + \dots$$

Now divide by 2:

$$\sinh x = \frac{e^x - e^{-x}}{2}$$
$$= x + \frac{x^3}{3!} + \dots + \frac{x^{2n+1}}{(2n+1)!} + \dots$$
$$= \sum_{n=0}^{\infty} \frac{x^{2n+1}}{(2n+1)!}.$$

5) Evaluate $\displaystyle\int \frac{1+e^x}{x}\,dx$ using a series for the integrand.

Since $\displaystyle e^x = \sum_{n=0}^{\infty} \frac{x^n}{n!} = 1 + \frac{x}{1!} + \frac{x^2}{2!} + \frac{x^3}{3!} + \cdots,$

$$1 + e^x = 2 + \frac{x}{1!} + \frac{x^2}{2!} + \frac{x^3}{3!} + \cdots \text{ and}$$

$$\frac{1+e^x}{x} = \frac{2}{x} + \frac{1}{1!} + \frac{x}{2!} + \frac{x^2}{3!} + \cdots.$$

Therefore,

$$\int \frac{1+e^x}{x}\,dx = \int \left(\frac{2}{x} + \frac{1}{1!} + \frac{x}{2!} + \frac{x}{3!} + \cdots \right) dx$$

$$= 2\ln x + \frac{x}{1!} + \frac{x^2}{2\cdot 2!} + \frac{x^3}{3\cdot 3!} + \cdots + C$$

$$= \ln x^2 + \sum_{n=1}^{\infty} \frac{x^n}{n\cdot n!} + C.$$

6) Find the sum of the series

$$\sum_{n=1}^{\infty} \frac{(-0.5)^n}{n}.$$

$$\sum_{n=1}^{\infty} \frac{(-0.5)^n}{n} = \sum_{n=1}^{\infty} \frac{(-1)(-1)^{n-1}(0.5)^n}{n}$$

$$= -\sum_{n=1}^{\infty} \frac{(-1)^{n-1}(0.5)^n}{n}.$$

The power series for $\ln(1+x)$

$$= \sum_{n=1}^{\infty} \frac{(-1)^{n-1}x^n}{n}.$$

Therefore,

$$\sum_{n=1}^{\infty} \frac{(-0.5)^n}{n} = -\ln(1+0.5) = -\ln 1.5$$

$$\approx -0.4054.$$

7) Find the infinite series expression for

$$\int_0^{1/2} \frac{1}{1-x^3}\,dx.$$

$$\frac{1}{1-x} = \sum_{n=0}^{\infty} x^n. \text{ Thus } \frac{1}{1-x^3} = \sum_{n=0}^{\infty} x^{3n}.$$

$$\int \frac{1}{1-x^3}\,dx = \int \sum_{n=0}^{\infty} x^{3n}\,dx$$

$$= \sum_{n=0}^{\infty} \frac{x^{3n+1}}{3n+1} + C.$$

$$\int_0^{1/2} \frac{1}{1-x^3}\,dx = \sum_{n=0}^{\infty} \frac{x^{3n+1}}{3n+1}\Bigg]_0^{1/2}$$

$$= \sum_{n=0}^{\infty} \frac{\left(\frac{1}{2}\right)^{3n+1}}{3n+1} - 0$$

$$= \sum_{n=0}^{\infty} \frac{1}{(6n+2)8^n}.$$

8) Find $\int_0^1 xe^{-x}\,dx$ using a series to within 0.01.

$$e^x = \sum_{n=0}^{\infty} \frac{x^n}{n!}, \; e^{-x} = \sum_{n=0}^{\infty} \frac{(-1)^n x^n}{n!},$$

so $xe^{-x} = \sum_{n=0}^{\infty} \frac{(-1)^n x^{n+1}}{n!}$. Thus,

$$\int_0^1 xe^{-x}\,dx = \int_0^1 \sum_{n=0}^{\infty} \frac{(-1)^n x^{n+1}}{n!}\,dx$$

$$= \sum_{n=0}^{\infty} \frac{(-1)^n x^{n+2}}{n!(n+2)}\Big]_0^1 = \sum_{n=0}^{\infty} \frac{(-1)^n}{n!(n+2)} - 0$$

$$= \sum_{n=0}^{\infty} \frac{(-1)^n}{n!(n+2)}$$

$$= \frac{1}{0!2} - \frac{1}{1!3} + \frac{1}{2!4} - \frac{1}{3!5} + \frac{1}{4!6} - \frac{1}{5!7} + \cdots$$

$$= \frac{1}{2} - \frac{1}{3} + \frac{1}{8} - \frac{1}{30} + \frac{1}{144} + \cdots.$$

This is an alternating series and the fifth term $\frac{1}{144}$ is less than 0.01. Thus, we may use

$$\int_0^1 xe^{-x}\,dx \approx \frac{1}{2} - \frac{1}{3} + \frac{1}{8} - \frac{1}{30} \approx 0.2583.$$

Note: We could use integration by parts:

$$\int_0^1 xe^{-x}\,dx = (-xe^{-x} - e^{-x})\Big] = 1 - \frac{2}{e}$$

$$\approx 0.2642$$

The approximation 0.2583 is within 0.01 of the actual value.

B. If $f^{(n)}(a)$ exists, the **Taylor polynomial of degree n for f about a** is the nth partial sum of the Taylor series:

Page 754

$$T_n(x) = f(a) + \frac{f'(a)}{1!}(x-a) + \frac{f''(a)}{2!}(x-a)^2 + \frac{f^{(3)}(a)}{3!}(x-a)^3 + \cdots \frac{f^{(n)}(a)}{n!}(x-a)^n.$$

The Taylor polynomial of degree one, $T_1(x) = f(a) + f'(a)(x-a)$, is the familiar equation for the tangent line to $y = f(x)$ at a point a. This means f and T_1 have the same functional value and same first derivative value at $x = a$: $f(a) = T_1(a)$ and $f'(a) = T_1'(a)$.

$T_2(x) = f(a) + f'(a)(x-a) + \frac{1}{2}f''(a)(x-a)^2$ is the "tangent parabola" to $y = f(x)$ at $x = a$. $T_2(x)$ has the same value, the same first derivative value, and the same second derivative value as does f at $x = a$.

$T_3(x)$ would be the "tangent cubic" and so on.

Since the first n derivatives of $T_n(x)$ at a are equal to the corresponding first n derivatives of $f(x)$ at a, $T_n(x)$ is a very good approximation for $f(x)$ if x is near a.

Example: Let $f(x) = -2 + 2x + \dfrac{1}{x}$. Then $f'(x) = 2 - \dfrac{1}{x^2}$, $f''(x) = \dfrac{2}{x^3}$, and

$f'''(x) = \dfrac{-6}{x^4}$. At $a = 1$, $f(1) = -2 + 2(1) + \dfrac{1}{1} = 1$, $f'(1) = 2 - \dfrac{1}{1^2} = 1$, $f''(1) = \dfrac{2}{1^3} = 2$,

and $f'''(1) = \dfrac{-6}{1^4} = -6$. Therefore, the first three Taylor polynomials for f are:

$$T_1(x) = 1 + \frac{1}{1!}(x-1) = 1 + (x-1)$$

$$T_2(x) = 1 + \frac{1}{1!}(x-1) + \frac{2}{2!}(x-1)^2 = 1 + (x-1) + (x-1)^2$$

$$T_3(x) = 1 + \frac{1}{1!}(x-1) + \frac{2}{2!}(x-1)^2 + \frac{-6}{3!}(x-1)^3 = 1 + (x-1) + (x-1)^2 - (x-1)^3.$$

The three figures below contain the graphs of f and T_1, of f and T_2, and f and T_3. Notice that near $a = 1$, the graph of T_n becomes an increasingly better approximation to the graph of $y = f(x)$ as n increases.

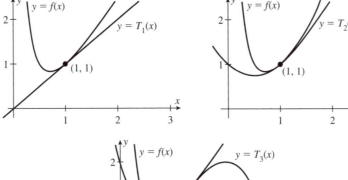

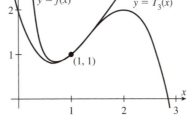

9) Let $f(x) = \sqrt{x}$. Construct the Taylor polynomial of degree 3 for $f(x)$ about $x = 1$.

$f(x) = x^{1/2}$ $\qquad$ $f(1) = 1$

$f'(x) = \frac{1}{2}x^{-1/2}$ $\qquad$ $f'(1) = \frac{1}{2}$

$f''(x) = -\frac{1}{4}x^{-3/2}$ $\qquad$ $f''(1) = -\frac{1}{4}$

$f^{(3)}(x) = \frac{3}{8}x^{-5/2}$ $\qquad$ $f^{(3)}(1) = \frac{3}{8}.$

Therefore, $T_3(x)$

$= 1 + \frac{1}{2}(x-1) - \frac{1/4}{2!}(x-1)^2 + \frac{3/8}{3!}(x-1)^3$

$= 1 + \frac{1}{2}(x-1) - \frac{1}{8}(x-1)^2 + \frac{1}{16}(x-1)^3.$

10) Find the Taylor polynomial of degree 6 for $f(x) = \cos x$ about $x = 0$.

$$f(x) = \cos x \qquad\qquad f(0) = 1$$
$$f'(x) = -\sin x \qquad\quad f'(0) = 0$$
$$f''(x) = -\cos x \qquad\quad f''(0) = -1$$
$$f^{(3)}(x) = \sin x \qquad\quad f^{(3)}(0) = 0$$
$$f^{(4)}(x) = \cos x \qquad\quad f^{(4)}(0) = 1$$
$$f^{(5)}(x) = -\sin x \qquad f^{(5)}(0) = 0$$
$$f^{(6)}(x) = -\cos x \qquad f^{(6)}(0) = -1.$$

$$T_6(x) = 1 + \tfrac{-1}{2!}x^2 + \tfrac{1}{4!}x^4 - \tfrac{1}{6!}x^6$$
$$= 1 - \tfrac{x^2}{2} + \tfrac{x^4}{24} - \tfrac{x^6}{720}.$$

11) a) Use your answer to question 9) to approximate $\sqrt{\tfrac{3}{2}}$ using the Taylor polynomial of degree 3.

$$\sqrt{\tfrac{3}{2}} \approx T_3\!\left(\tfrac{3}{2}\right) = 1 + \tfrac{1}{2}\!\left(\tfrac{1}{2}\right) - \tfrac{1}{8}\!\left(\tfrac{1}{2}\right)^2 + \tfrac{1}{16}\!\left(\tfrac{1}{2}\right)^3$$
$$= \tfrac{157}{128} \approx 1.2266.$$

(To 4 decimals, $\sqrt{\tfrac{3}{2}} = 1.2247$.)

b) Use your answer to question 10) to approximate $\cos 1$.

$$\cos 1 \approx T_6(1) = 1 - \tfrac{(1)^2}{2} + \tfrac{(1)^4}{24} - \tfrac{(1)^6}{720}$$
$$\approx 1 - 0.5 + 0.041667 - 0.001389$$
$$\approx 0.540278.$$

($\cos 1 = 0.540302$ to 6 decimal places.)

Page 754

C. If $f^{(n)}(x)$ exists for $|x - a| < d$, the **nth remainder of the Taylor series of degree n for f** is:

$$R_n(x) = f(x) - T_n(x).$$

(The R used in $R_n(x)$ should not be confused with the R used earlier in the chapter for the radius of convergence of an infinite series.)

$|R_n(x)|$ is the error resulting from using $T_n(x)$ to estimate $f(x)$.

Example: Let $f(x) = \cos 2x$. Then $f'(x) = -2 \sin 2x$, $f''(x) = -4 \cos 2x$, $f'''(x) = 8 \sin 2x$, and $f^{(4)}(x) = 16 \cos 2x$. At $a = 0, f(0) = 1$, $f'(0) = -2 \cdot 0 = 0$, $f''(0) = -4 \cdot 1 = -4$, $f'''(0) = 8 \cdot 0 = 0$, and $f^{(4)}(0) = 16 \cdot 1 = 16$.

The Taylor polynomial of degree 4 for f is

$$T_4(x) = 1 + \frac{0}{1!}(x - 0) + \frac{-4}{2!}(x - 0)^2 + \frac{0}{3!}(x - 0)^3 + \frac{16}{4!}(x - 0)^4$$

$$= 1 - 2x^2 + \tfrac{2}{3}x^4.$$

Choosing any x, say $x = 0.4$, $T_4(0.4) = 1 - 2 \cdot 0.4^2 + \tfrac{2}{3} \cdot 0.4^4 \approx 0.69707$ is an approximation of $f(0.4)$. To five decimals, $f(0.4) = \cos(0.8) = 0.69671$. Therefore, the error in this approximation (the remainder) is
$|R_3(0.4)| = |0.69671 - 0.69707| = 0.00036$.

In general, $R_n(x)$ will approach 0 as n gets larger. Taylor's Inequality, stated next, is a way to establish an upper bound for $|R_n(x)|$.

Taylor's Inequality: If $|f^{(n+1)}(x)| \le M$ for $|x - a| \le d$, then the nth remainder $R_n(x)$ of the Taylor series satisfies the inequality

$$|R_n(x)| \le \frac{M}{(n+1)!}|x - a|^{n+1} \text{ for } |x - a| \le d.$$

Example: From the pattern in the example on the previous page, the form of the nth derivative of $f(x) = \cos 2x$ is

$$f^{(n)}(x) = \pm 2^n \sin 2x \text{ or } \pm 2^n \cos 2x.$$

Since sine and cosine have ranges from -1 to 1, in each case we have $|f^{(n+1)}(x)| \le 2^{n+1}$ for all x. Therefore, using $M = 2^{n+1}$ in Taylor's Inequality, an upper bound for $|R_n(x)|$ for all x is

$$|R_n(x)| \le \frac{2^{n+1}}{(n+1)!}|x|^{n+1}.$$

In the specific case where $n = 4$ and $x = 0.4$,

$$|R_4(0.4)| \le \frac{2^5}{(4+1)!}|0.4|^5 \approx 0.00273.$$

This is certainly true since we know from the example that, to five decimals, $|R_4(0.4)|$ is 0.00036.

If f has derivatives of all orders and $\lim\limits_{n \to \infty} R_n(x) = 0$ for $|x - a| \le d$, then $f(x)$ is equal to its Taylor series for $|x - a| \le d$. This is one way to show that a function is equal to its Taylor series—show $\lim\limits_{n \to \infty} R_n(x) = 0$. You will often need to use this limit:

$$\lim_{n \to \infty} \frac{t^n}{n!} = 0 \text{ for all real numbers } t.$$

Example: In the example above, where $f(x) = \cos 2x$, $|R_n(x)| \le \dfrac{2^{n+1}}{(n+1)!}|x|^{n+1}$. We then have

$$0 \le \lim_{n \to \infty}|R_n(x)| \le \lim_{n \to \infty} \frac{2^{n+1}}{(n+1)!}|x|^{n+1} = |x|^{n+1}\left(\lim_{n \to \infty} \frac{2^{n+1}}{(n+1)!}\right) = |x|^{n+1}(0) = 0 \text{ for all } x.$$

By the Squeeze Theorem, $\lim\limits_{n \to \infty}|R_n(x)| = 0$, which proves for all x that

$f(x) = \cos 2x$ is equal to its Maclaurin series; that is,

$$\cos 2x = \sum_{n=0}^{\infty} \frac{(-1)^n 2^{2n}}{(2n)!} x^{2n} \text{ for all } x.$$

12) True or False:
In general, the larger the number n, the closer the Taylor polynomial approximation is to the actual functional value.

True.

13) Let $f(x) = 2^x$.

    a) Find an expression for the nth Taylor polynomial $T_n(x)$ for $f(x)$ at $x = 0$.

$$f(x) = 2^x \qquad\qquad f(0) = 1$$
$$f'(x) = (\ln 2)\, 2^x \qquad f'(0) = \ln 2$$
$$f''(x) = (\ln 2)^2\, 2^x \qquad f''(0) = (\ln 2)^2$$

and, in general,

$$f^n(x) = (\ln 2)^n\, 2^x \qquad f^n(0) = (\ln 2)^n.$$

$$T_n(x) = 1 + (\ln 2)x + \frac{(\ln 2)^2}{2!}x^2 + \dots$$
$$+ \frac{(\ln 2)^n}{n!}x^n.$$

    b) Find an upper bound for $\left|f^{(n+1)}(x)\right|$ for $|x| \le 1$.

For $|x| \le 1$,

$$\left|f^{(n+1)}(x)\right| = \left|(\ln 2)^{n+1} \cdot 2^x\right|$$
$$= 2^x(\ln 2)^{n+1} \le 2(\ln 2)^{n+1}$$

    c) Show that $f(x) = 2^x$ is equal to its Taylor series for $|x| < 1$.

We need to show that $\lim_{n \to \infty} R_n(x) = 0$ for $|x| \le 1$. From part **b)**

$$\left|f^{(n+1)}(x)\right| \le 2\,(\ln 2)^{n+1}.$$

By Taylor's Inequality

$$0 \le |R_n(x)| \le \frac{2(\ln 2)^{n+1}}{(n+1)!}|x|^{n+1} \le \frac{2(\ln 2)^{n+1}}{(n+1)!}$$

since $|x| \le 1$.

$$\lim_{n \to \infty} 2\frac{(\ln 2)^{n+1}}{(n+1)!} = 2\lim_{n \to \infty}\frac{(\ln 2)^{n+1}}{(n+1)!} = 2(0) = 0.$$

Therefore, by the Squeeze Theorem for limits at infinity, $\lim_{n \to \infty} R_n(x) = 0$.

Page 761

D. For any real number k and a non-negative integer n, the **binomial coefficient** is

$$\binom{k}{n} = \begin{cases} 1 & \text{if } n = 0 \\ \dfrac{k(k-1)(k-2)\ldots(k-n+1)}{n!} & \text{if } n \geq 0 \end{cases}.$$

Example: $\begin{pmatrix} \frac{3}{2} \\ 4 \end{pmatrix} = \dfrac{\frac{3}{2} \cdot \frac{1}{2} \cdot \frac{-1}{2} \cdot \frac{-3}{2}}{4!} = \dfrac{\frac{9}{16}}{24} = \dfrac{3}{128}.$

Example: Evaluate $\begin{pmatrix} 5 \\ n \end{pmatrix}$ for all non-negative integers n.

By definition, for $n = 0$, $\binom{5}{0} = 1$. For $n = 1, 2,$ and 3, $\binom{5}{1} = \dfrac{5}{1!} = 5$,

$\binom{5}{2} = \dfrac{5 \cdot 4}{2!} = \dfrac{5 \cdot 4}{2 \cdot 1} = 10$, and $\binom{5}{3} = \dfrac{5 \cdot 4 \cdot 3}{3!} = \dfrac{5 \cdot 4 \cdot 3}{3 \cdot 2 \cdot 1} = 10$. With similar

calculations, we see $\binom{5}{4} = 5$ and $\binom{5}{5} = 1$. For $n = 6$, $\binom{5}{6} = \dfrac{5 \cdot 4 \cdot 3 \cdot 2 \cdot 1 \cdot 0}{6!} = 0$.

In fact, $\binom{5}{n} = 0$ for all integers $n > 5$.

For any real number k the **binomial series for** $(1 + x)^k$ is the Maclaurin
series for $(1 + x)^k$:

$$(1 + x)^k = \sum_{n=0}^{\infty} \binom{k}{n} x^n = 1 + kx + \frac{k(k-1)}{2!} x^2 + \ldots + \binom{k}{n} x^n + \ldots .$$

Examples: a) We found above that $\begin{pmatrix} \frac{3}{2} \\ 4 \end{pmatrix} = \dfrac{3}{128}$ is the coefficient for x^4 in the

binomial series for $(1 + x)^{\frac{3}{2}}$. The binomial series is

$$(1 + x)^{\frac{3}{2}} = \sum_{n=0}^{\infty} \begin{pmatrix} \frac{3}{2} \\ n \end{pmatrix} x^n = 1 + \frac{3}{2}x + \frac{3}{16}x^2 - \frac{1}{16}x^3 + \frac{3}{128}x^4 + \ldots .$$

b) The binomial series for $(1 + x)^5$ is finite because $\binom{5}{n} = 0$ for all integers $n > 5$.

In fact, the series $(1 + x)^5 = \sum_{n=0}^{\infty} \binom{5}{n} x^n = 1 + 5x + 10x^2 + 10x^3 + 5x^4 + x^5$

is the binomial expansion of $(1 + x)^5$ you have seen in algebra.

For $k = 0, 1, 2, 3, \ldots$, the binomial series for $(1 + x)^k$ is a polynomial of degree k. When k is any number other than a non-negative integer, the binomial series for $(1 + x)^k$ is an infinite series.

The intervals of convergence of the binomial series for various values of k are given in the table at the right.

Condition on k	Interval of Convergence
$k \leq -1$	$(-1, 1)$
$-1 < k < 0$	$(-1, 1]$
$k = 0, 1, 2, 3, \ldots$	$(-\infty, \infty)$
$k > 0$, but not an integer	$[-1, 1]$

14) a) Find $\binom{6}{4}$.

$$\binom{6}{4} = \frac{6(5)(4)(3)}{4!} = 15.$$

b) Find $\begin{pmatrix} \frac{4}{3} \\ 5 \end{pmatrix}$.

$$\begin{pmatrix} \frac{4}{3} \\ 5 \end{pmatrix} = \frac{\left(\frac{4}{3}\right)\left(\frac{1}{3}\right)\left(-\frac{2}{3}\right)\left(-\frac{5}{3}\right)\left(-\frac{8}{3}\right)}{5 \cdot 4 \cdot 3 \cdot 2 \cdot 1} = \frac{-8}{3^6} = -\frac{8}{729}.$$

c) Expand $(1 + x)^6$.

$$\binom{6}{0} = 1, \binom{6}{1} = 6, \binom{6}{2} = 15, \binom{6}{3} = 20,$$
$$\binom{6}{4} = 15, \binom{6}{5} = 6, \binom{6}{6} = 1.$$

Thus $(1 + x)^6$
$$= 1 + 6x + 15x^2 + 20x^3 + 15x^4 + 6x^5 + x^6.$$

d) Find the coefficient of x^5 in the binomial expansion of $\sqrt{1 + x}$.

$$\sqrt{1 + x} = (1 + x)^{1/2}.$$
The coefficient of x^5 is

$$\begin{pmatrix} \frac{1}{2} \\ 5 \end{pmatrix} = \frac{\left(\frac{1}{2}\right)\left(-\frac{1}{2}\right)\left(-\frac{3}{2}\right)\left(-\frac{5}{2}\right)\left(-\frac{7}{2}\right)}{5 \cdot 4 \cdot 3 \cdot 2 \cdot 1}$$

$$= \frac{1(-1)(-3)(-5)(-7)}{2^5 \cdot 5 \cdot 4 \cdot 3 \cdot 2 \cdot 1}$$

$$= \frac{7}{2^6} = \frac{7}{64}.$$

Therefore,
$$\sqrt{1 + x} = 1 + \cdots + \frac{7}{64}x^5 + \cdots.$$

15) True or False:

If k is a non-negative integer, then the binomial series for $(1 + x)^k$ is a finite sum.

True, only the first $k + 1$ terms of the infinite series may have non-zero coefficients.

16) Find the binomial series for $\sqrt[3]{1 + x}$.

Since $\sqrt[3]{1 + x} = (1 + x)^{\frac{1}{3}}$, let $k = \frac{1}{3}$. Then

$$\sqrt[3]{1 + x} = \sum_{n=0}^{\infty} \binom{\frac{1}{3}}{n} x^n.$$

17) Find a power series for $f(x) = \sqrt{4 + x}$.

Let $k = \frac{1}{2}$. Then $\sqrt{4 + x}$

$$= 2\left(1 + \frac{x}{4}\right)^{1/2} = 2\sum_{n=0}^{\infty} \binom{\frac{1}{2}}{n}\left(\frac{x}{4}\right)^n$$

$$= 2\sum_{n=0}^{\infty} \binom{\frac{1}{2}}{n}\frac{x^n}{4^n}.$$

18) Find an infinite series for $\sin^{-1} x$.
(Hint: Start with $(1 + x)^{-1/2}$.)

$$\frac{1}{\sqrt{1 + x}} = (1 + x)^{-1/2} = \sum_{n=0}^{\infty} \binom{-\frac{1}{2}}{n} x^n.$$

Substitute $-x$ for x:

$$\frac{1}{\sqrt{1 - x}} = \sum_{n=0}^{\infty} \binom{-\frac{1}{2}}{n} (-x)^n$$

$$= \sum_{n=0}^{\infty} (-1)^n \binom{-\frac{1}{2}}{n} x^n.$$

Substitute x^2 for x:

$$\frac{1}{\sqrt{1 - x^2}} = \sum_{n=0}^{\infty} (-1)^n \binom{-\frac{1}{2}}{n} x^{2n}.$$

Since the derivative of $\sin^{-1} x$ is $\dfrac{1}{\sqrt{1 - x^2}}$,

integrate term by term:

$$\sin^{-1} x = \sum_{n=0}^{\infty} \frac{(-1)^n \binom{-\frac{1}{2}}{n} x^{2n+1}}{2n + 1} + C.$$

Page
763

E. Convergent series may be multiplied and divided just like polynomials. Often finding the first few terms of the result is sufficient.

Example: Find the first three terms of the Maclaurin series for $e^x \sin x$.

$$e^x = 1 + x + \frac{x^2}{2} + \frac{x^3}{6} + \cdots$$

$$\sin x = x - \frac{x^3}{3!} + \frac{x^5}{5!} - \cdots$$

$$e^x \sin x = x\left(1 + x + \frac{x^2}{2} + \frac{x^3}{6} + \cdots\right) - \frac{x^3}{6}\left(1 + x + \frac{x^2}{2} + \frac{x^3}{6} + \cdots\right) + \frac{x^5}{120}\left(1 + x + \frac{x^2}{2} + \frac{x^3}{6} + \cdots\right) + \cdots$$

$$= \left(x + x^2 + \frac{x^3}{2} + \frac{x^4}{6} + \cdots\right) - \left(\frac{x^3}{6} + \frac{x^4}{6} + \frac{x^5}{12} + \frac{x^6}{36} + \cdots\right) + \left(\frac{x^5}{120} + \frac{x^6}{120} + \frac{x^7}{240} + \cdots\right)$$

$$= x + x^2 + \frac{x^3}{3} + \cdots$$

19) Use the power series

$$\cos x = 1 - \frac{x^2}{2} + \frac{x^4}{24} - \cdots \text{ and}$$

$$e^x = 1 + x + \frac{x^2}{2} + \frac{x^3}{6} + \frac{x^4}{24} + \cdots$$

to find the first four terms of the power series for $e^{-x} \cos x$.

First, substitute $-x$ for x:

$$e^{-x} = 1 - x + \frac{x^2}{2} - \frac{x^3}{6} + \frac{x^4}{24} + \cdots.$$

$e^{-x} \cos x$

$$= 1\left(1 - \frac{x^2}{2} + \frac{x^4}{24} - \cdots\right) - x\left(1 - \frac{x^2}{2} + \frac{x^4}{24} - \cdots\right)$$

$$+ \frac{x^2}{2}\left(1 - \frac{x^2}{2} + \frac{x^4}{24} - \cdots\right) - \frac{x^3}{6}\left(1 - \frac{x^2}{2} + \frac{x^4}{24} - \cdots\right) + \cdots$$

$$= \left(1 - \frac{x^2}{2} + \frac{x^4}{24} - \cdots\right) - \left(x - \frac{x^3}{2} + \frac{x^5}{24} - \cdots\right)$$

$$+ \left(\frac{x^2}{2} - \frac{x^4}{4} + \frac{x^6}{48} - \cdots\right) - \left(\frac{x^3}{6} - \frac{x^5}{12} + \frac{x^7}{144} - \cdots\right) + \cdots$$

Therefore, $e^{-x} \cos x = 1 - x + \frac{1}{3}x^3 - \frac{5}{24}x^4 + \cdots$

20) Use the series in question **19)** to find the first three terms of the power series for $\dfrac{e^{-x}}{\cos x}$.

$$
\begin{array}{r}
1 - x \\
1 - \frac{x^2}{2} + \frac{x^4}{24} - \cdots \overline{\smash{\big)}\, 1 - x + \frac{x^2}{2} - \frac{x^3}{6} + \frac{x^4}{24} - \frac{x^5}{120} + \cdots} \\
1 \quad\quad - \frac{x^2}{2} \quad\quad + \frac{x^4}{24} \\
\hline
- x + x^2 - \frac{x^3}{6} + 0 - \frac{x^5}{120} + \cdots \\
- x \quad\quad + \frac{x^3}{2} \quad\quad - \frac{x^5}{24} + \cdots \\
\hline
x^2 - \frac{2x^3}{3} \quad\quad + \frac{5x^5}{30} + \cdots
\end{array}
$$

Therefore, $\dfrac{e^{-x}}{\cos x} = 1 - x + x^2 + \cdots.$

F. Technology Plus. Use a computer algebra system or a graphing calculator to solve.

T-1) On the same screen sketch a graph of $f(x)\ \dfrac{1}{\sqrt{x}}$ and its first 3 Taylor polynomials about $x = 1$.

$f(x) = x^{-1/2},\ f(1) = 1$

$f'(x) = -\frac{1}{2}x^{-3/2},\ f'(1) = -\frac{1}{2}$

$f''(x) = \frac{3}{4}x^{-5/2},\ f''(1) = \frac{3}{4}$

$f'''(x) = -\frac{15}{8}x^{-7/2},\ f'''(1) = -\frac{15}{8}.$

$T_1(x) = 1 + (-\frac{1}{2})(x-1) = 1 - \frac{1}{2}(x-1).$

$T_2(x) = 1 - \frac{1}{2}(x-1) + \dfrac{\frac{3}{4}}{2!}(x-1)^2$

$\quad\ = 1 - \frac{1}{2}(x-1) + \frac{3}{8}(x-1)^2.$

$T_3(x) = 1 - \frac{1}{2}(x-1) + \frac{3}{8}(x-1)^2$

$\qquad + \dfrac{-\frac{15}{8}}{3!}(x-1)^3$

$\quad\ = 1 - \frac{1}{2}(x-1) + \frac{3}{8}(x-1)^2 - \frac{5}{16}(x-1)^3.$

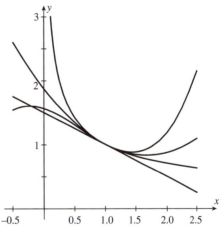

Section 11.11 Applications of Taylor Polynomials

This last section uses Taylor polynomials to estimate values of functions and uses Taylor's Inequality to measure the accuracy of such approximations.

Concepts to Master

A. Approximate functional values using a Taylor polynomial

B. Estimate the error of a Taylor polynomial approximation

Summary and Focus Questions

Page 768

A. The Taylor polynomial of degree n for f about a is

$$T_n(x) = f(a) + \frac{f'(a)}{1!}(x - a) + \frac{f''(a)}{2!}(x - a)^2 + \frac{f^{(3)}(a)}{3!}(x - a)^3 + \dots \frac{f^{(n)}(a)}{n!}(x - a)^n.$$

$T_n(x)$ may be used to approximate $f(x)$ for any given x in the interval of convergence for the Taylor series for f about a.

1) a) For $f(x) = \frac{1}{3 + x}$, find $T_3(x)$ where $a = 2$.

$$f(x) = (3 + x)^{-1} \qquad f(2) = \frac{1}{5}$$
$$f'(x) = -(3 + x)^{-2} \qquad f'(2) = \frac{-1}{25}$$
$$f''(x) = 2(3 + x)^{-3} \qquad f''(2) = \frac{2}{125}$$
$$f'''(x) = -6(3 + x)^{-4} \qquad f'''(2) = \frac{-6}{625}$$

$$T_3(x) = \frac{1}{5} - \frac{1}{25}(x - 2) + \frac{2/125}{2!}(x - 2)^2$$
$$\qquad - \frac{6/625}{3!}(x - 2)^3$$
$$= \frac{1}{5} - \frac{1}{25}(x - 2) + \frac{1}{125}(x - 2)^2$$
$$\qquad - \frac{1}{625}(x - 2)^3.$$

b) Estimate $f(3)$ with $T_3(3)$.

$$T_3(3) = \frac{1}{5} - \frac{(3 - 2)}{25} + \frac{(3 - 2)^2}{125} - \frac{(3 - 2)^3}{625}$$
$$= \frac{1}{5} - \frac{1}{25} + \frac{1}{125} - \frac{1}{625}$$
$$= \frac{104}{625} = 0.1664.$$

We note that this is a lot of work to approximate $f(3) = \frac{1}{6}$ but understanding the method is important.

2) **a)** For $f(x) = x^{5/2}$, find $T_3(x)$ where $a = 4$.

$f(x) = x^{5/2}$ $f(4) = 32$

$f'(x) = \frac{5}{2}x^{3/2}$ $f'(4) = 20$

$f''(x) = \frac{15}{4}x^{1/2}$ $f''(4) = \frac{15}{2}$

$f'''(x) = \frac{15}{8}x^{-1/2}$ $f'''(4) = \frac{15}{16}$.

$T_3(x) = 32 + 20(x - 4) + \frac{15/2}{2!}(x - 4)^2$
$\qquad + \frac{15/16}{3!}(x - 4)^3$

$\qquad = 32 + 20(x - 4) + \frac{15}{4}(x - 4)^2$
$\qquad + \frac{5}{32}(x - 4)^3$.

b) Estimate $(5)^{5/2}$ with $T_3(x)$.

$T_3(5) = 32 + 20(1)^1 + \frac{15}{4}(1)^2 + \frac{5}{32}(1)^3$

$\qquad = \frac{1789}{32} \approx 55.9063.$

To four decimals $(5)^{5/2}$ is approximately
55.9017.

B. The accuracy of the approximation $f(x) \approx T_n(x)$ will depend on the values of n, x, and a. In general:

Page
768

i) The larger the value of n, the better the estimate.

ii) The closer that x is to a, the better the estimate.

From the previous section, we may use Taylor's Inequality to estimate the error:

If $\left|f^{(n+1)}(x)\right| \le M$ for $|x - a| \le d$, then

$$\left|R_n(x)\right| \le \frac{M}{(n+1)!}|x - a|^{n+1} \text{ for } |x - a| \le d.$$

The approximation $T_n(x)$ will be accurate to within any given number ϵ by selecting n such that

$$\frac{M}{(n+1)!}|x - a|^{n+1} < \epsilon.$$

In the special case where the Taylor series is an alternating series, $\left|R_n(x)\right|$

may be estimated (by the Alternating Series Estimation Theorem) with

$\left|\frac{f^{(n+1)}(x)}{(n+1)!}(x - a)^{n+1}\right|$, the $(n + 1)^{\text{st}}$ term of the Taylor series for f.

Example: This example puts together the concepts from Section 11.10 and this section.

a) Find the Taylor polynomial of degree 3 about $x = 0$ for $f(x) = xe^{-x}$.

$f(x) = xe^{-x}$ $\qquad\qquad f(0) = 0 \cdot e^{-0} = 0$

$f'(x) = x(-e^{-x}) + e^{-x}(1) = -e^{-x}(x-1)$ $\qquad f'(0) = -e^{-0}(0-1) = 1$

$f''(x) = -(x-1)(-e^{-x}) + e^{-x}(-1) = e^{-x}(x-2)$ $\qquad f''(0) = e^{-0}(0-2) = -2$

$f'''(x) = (x-2)(-e^{-x}) + e^{-x}(1) = -e^{-x}(x-3)$ $\qquad f'''(0) = -e^{-0}(0-3) = 3.$

In general, $f^{(n)}(x) = (-1)^n e^{-x}(x-n)$.

Therefore, $T_3(x) = 0 + \dfrac{1}{1!}x + \dfrac{-2}{2!}x^2 + \dfrac{3}{3!}x^3 = x - x^2 + \dfrac{1}{2}x^3.$

Alternately, we could have modified the power series $e^x = \displaystyle\sum_{n=0}^{\infty} \dfrac{x^n}{n!}$ by

replacing x by $-x$: $e^{-x} = \displaystyle\sum_{n=0}^{\infty} \dfrac{(-x)^n}{n!} = \sum_{n=0}^{\infty} \dfrac{(-1)^n x^n}{n!},$

then multiply by x: $xe^{-x} = \displaystyle\sum_{n=0}^{\infty} \dfrac{(-1)^n x^n}{n!} x = \sum_{n=0}^{\infty} \dfrac{(-1)^n x^{n+1}}{n!}$

and selecting the first 3 terms of the series to obtain $T_3(x) = x - x^2 + \dfrac{1}{2}x^3.$

b) For $-1 < x < 1$, estimate the accuracy of $T_3(x)$ using Taylor's Inequality.

For $n = 3$, the error is $R_3(x) = xe^{-x} - \left(x - x^2 + \frac{1}{2}x^3\right).$

The graph of $y = |R_3(x)|$, given at the right, shows that for $-1 < x < 1$, $|R_3(x)|$ is small.

Let's see how small.

For any n and $-1 < x < 1$, we have

$0 < e^{-x} < e$ and $|x - n| < n + 1$. Thus

$|f^{(n)}(x)| = |(-1)^n e^{-x}(x-n)| \le e(n+1).$

Then $M = e(n+1)$ is an upper bound for the values of the nth derivative, and by Taylor's Inequality

$$|R_n(x)| \le \frac{e(n+1)}{(n+1)!} = \frac{e}{n!}.$$

Therefore, for $-1 < x < 1$, $|R_3(x)| \le \dfrac{e}{3!} \approx 0.4530.$

This means that for any x in $(-1, 1)$, the error from estimating $f(x)$ with $T_3(x)$ is at most 0.4530. Specifically, for $x = 0.8$,

$T_3(0.8) = 0.8 - 0.8^2 + 0.8^3/2 = 0.4160$ is an estimate for

$f(0.8) = 0.8e^{-0.8} \approx 0.3595.$ The actual error from this estimate is

$|R_3(0.8)| = |0.3595 - 0.4160| = 0.0565$, which is well within the error estimate of 0.4530.

c) Use the Alternating Series Estimation Theorem to estimate the error for $n = 3$.
For any n, the theorem says $|R_n(x)|$ is less than the $(n + 1)$st term. Thus

$$|R_n(x)| \le \left| \frac{f^{(n+1)}(0)}{(n + 1)!} x^{n+1} \right| = \left| \frac{(-1)^{n+1}(0 - (n + 1))e^{-0}}{(n + 1)!} x^{n+1} \right| = \frac{|n + 1|}{(n + 1)!} = \frac{1}{n!},$$

which is a smaller estimate for the error than the estimate found in part b).

In particular, for $n = 3$, $|R_3(x)| \le \dfrac{1}{3!} \approx 0.1667$ and for $x = 0.8$, $|R_3(0.8)| =$

0.0565 is within our error bound of 0.1667.

3) a) Use Taylor's Inequality to estimate the accuracy of $f(x) \approx T_3(x)$ for $1 < x < 4$ for question **1**).

From **1 a)**, $f^{(4)}(x) = 24(3 + x)^{-5}$.
If $1 \le x \le 4$, then $4 \le x + 3 \le 7$.
$$4^5 \le (x + 3)^5 \le 7^5$$

$$\frac{1}{7^5} \le (x + 3)^{-5} \le \frac{1}{4^5}$$

$$\frac{24}{7^5} \le 24(x + 3)^{-5} \le \frac{24}{4^5}.$$

Thus, $\left| f^{(4)}(x) \right| \le \dfrac{24}{4^5}$.

From $1 \le x \le 4$, $-1 \le x - 2 \le 2$.
$(x - 2)^4 \le 2^4 = 16$.
Finally,

$$|R_3(x)| \le \frac{\frac{24}{4^5}}{4!}(x - 2)^4$$

$$\le \frac{24}{4^5 \cdot 4!} \cdot 16 = 0.0156.$$

b) Check the accuracy of the estimate when $x = 3$.

$f(3) = \dfrac{1}{3 + 3} = \dfrac{1}{6} \approx 0.1667$.

$T_3(3) \approx 0.1664$ by **1 b)**.

$R_3(3) \approx 0.1667 - 0.1664 = 0.0003$,

which is within 0.0156.

4) a) For $f(x) = x^{5/2}$, estimate the error between $f(5)$ and $T_3(5)$. (See question **2**).

From **2 a)**, $f^{(4)}(x) = -\dfrac{15}{16}x^{-3/2}$.

For $|x - 4| \le 2$, $2 \le x \le 6$.

$6^{-3/2} \le x^{-3/2} \le 2^{-3/2}$.

$\left| f^{(4)}(x) \right| \le \dfrac{15}{16} \cdot 2^{-3/2} = \dfrac{15}{32\sqrt{2}}$.

$$|R_3(x)| \le \frac{15}{32\sqrt{2}}|x - 4|^4 \le \frac{\frac{15}{32\sqrt{2}}}{4!}(2)^4$$

$$\approx 0.221.$$

b) Check the accuracy of the estimate $f(5) \approx T_3(5)$.

$f(5) = 5^{5/2} \approx 55.9017.$
$T_3(5) \approx 55.9063$ (question **2b**)).
$R_3(5) \approx 55.9017 - 55.9063 = -0.0046$,
which is within 0.221.

5) a) What degree Maclaurin polynomial is needed to approximate cos 0.5 accurate to within 0.00001?

Rephrased, the question asks:
For what n is $|R_n(0.5)| < 0.00001$ when $f(x) = \cos x$ and $a = 0$?
Because $f^{(n)}(x) = \pm \sin x$ or $\pm \cos x$ for all n, $|f^{(n+1)}(x)| \leq 1$.

Thus, $|R_n(x)| \leq \frac{1}{(n+1)!}|x|^{n+1}$

$\quad = \frac{|x|^{n+1}}{(n+1)!} \leq \frac{1}{(n+1)!}$ if $|x| \leq 1$.

$\frac{1}{(n+1)!} \leq 0.00001$ means

$(n+1)! \geq 100{,}000.$
Therefore, $n = 8$, since $8! = 40{,}320$ and $9! = 362{,}800.$

b) Use your answer to part **a)** to estimate cos 0.5 to within 0.00001.

From the Taylor series for cosine

$$T_8(x) = 1 - \frac{x^2}{2} + \frac{x^4}{24} - \frac{x^6}{720} + \frac{x^8}{40320}.$$

$T_8(0.5)$
$= 1 - \frac{(0.5)^2}{2} + \frac{(0.5)^4}{24} - \frac{(0.5)^6}{720} + \frac{(0.5)^8}{40320}$
$\approx 0.877583.$

To 6 decimals, cos 0.5 = 0.877583 so the approximation is within 0.00001 as required by part **a)**.

Section 1.1

_____ **1.** True or False:
$x^2 + 6x + 2y = 1$ defines y as a function of x.

_____ **2.** True or False:
The graph at the right is the graph of a function.

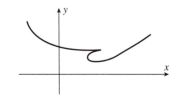

_____ **3.** The implied domain of $f(x) = \dfrac{1}{\sqrt{1-x}}$ is:
   a) $(1, \infty)$ c) $x \neq 1$
   b) $(-\infty, 1)$ d) $(-1, 1)$

_____ **4.** True or False:
The graph at the right is the graph of an even function.

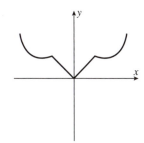

_____ **5.** Which graph represents $y = x + \dfrac{1}{x}$?

a)   b)   c)

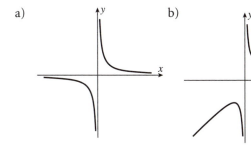 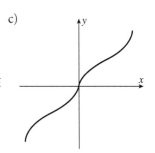

_____ **6.** True or False:
$f(x) = x^2$ is decreasing for $-10 \leq x \leq -1$.

____ **1.** True or False:

$f(x) = \dfrac{x^2 + \sqrt{x}}{2x + 1}$ is a rational function.

____ **2.** True or False:

The degree of $f(x) = 4x^3 + 7x^6 + 1$ is 7.

____ **3.** Which graph is best described by a linear model?

a) 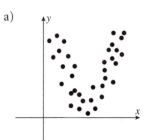 b) 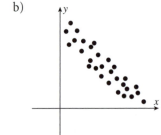 c)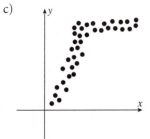

____ **4.** The best linear fit for the data at the right is $y = -3.4x + 39$.
An estimate for $x = 5$ is

a) 20 b) 22 c) 24 d) 25

x	y
1	36
2	32
3	28
4	26

Section 1.3

____ **1.** For $f(x) = 3x + 1$ and $g(x) = x$, $\dfrac{f}{g}(x) = $ ____?

    a) $\dfrac{3 + x}{x}$ b) $\dfrac{x}{3x + 1}$ c) $\dfrac{3x + 1}{x}$ d) $\dfrac{3x + 1}{x}$

____ **2.** For $f(x) = \sqrt{x^2 + x}$, we may write $f(x) = (h \circ g)(x)$, where:

    a) $h(x) = \sqrt{x}$ and $g(x) = x^2 + x$

    b) $h(x) = x^2 + x$ and $g(x) = \sqrt{x}$

    c) $h(x) = x^2$ and $g(x) = \sqrt{x}$

    d) $h(x) = x^2 + x$ and $g(x) = x^2$

____ **3.** Let $f(x) = 2 + \sqrt{x}$ and $g(x) = x + 3$. Then $(g \circ f)(x) = $ ____.

    a) $2 + \sqrt{x + 3}$ c) $3 + \sqrt{x + 2}$

    b) $2 + \sqrt{x} + 3$ d) $(x + 3)(2 + \sqrt{x})$

____ **4.** For $f(x) = x^2$ and $g(x) = 2x + 1$, $(f \circ g)(x) = $

    a) $2x^2 + 1$ b) $(2x)^2 + 1$ c) $(2x + 1)^2$ d) $x^2(2x + 1)$

____ **5.** For $f(x) = \cos(x^2 + 1)$ we may write $f(x) = (h \circ g)(x)$, where:

    a) $h(x) = \cos x^2$ and $g(x) = x + 1$

    b) $h(x) = \cos x$ and $g(x) = x^2 + 1$

    c) $h(x) = x^2 + 1$ and $g(x) = \cos x$

    d) $h(x) = x^2$ and $g(x) = \cos(x + 1)$

____ **6.** The graph of $y = f(3x)$ is obtained from the graph of $y = f(x)$ by:

    a) stretching vertically by a factor of 3.

    b) compressing vertically by a factor of 3.

    c) stretching horizontally by a factor of 3.

    d) compressing horizontally by a factor of 3.

____ **7.** Given the graph of $y = f(x)$, the graph to its right is the graph of:

    a) $y = -f(x)$

    b) $y = f(-x)$

    c) $y = -f(-x)$

    d) $y = f(|x|)$

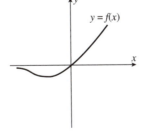
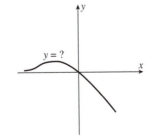

Section 1.4

_____ **1.** Which is the best viewing rectangle for the graph at the right?

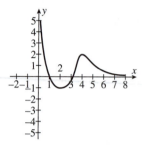

a) $[-6, 6]$ by $[-4, 4]$
b) $[-1, 6]$ by $[-2, 4]$
c) $[0, 8]$ by $[0, 4]$
d) $[-4, 4]$ by $[0, 8]$

_____ **2.** Which is the best viewing rectangle for $y = \dfrac{1}{x^2 - 9}$?

a) $[0, 10]$ by $[0, 10]$
b) $[-3, 3]$ by $[-10, 10]$
c) $[-5, 5]$ by $[-10, 10]$
d) $[-5, 5]$ by $[0, 10]$

_____ **3.** By graphing $y = \cos x$ and $y = 0.43x$ on a calculator, the number of solutions to $\cos x = 0.43x$ is

a) 0
b) 1
c) 2
d) 3

Section 1.5

____ **1.** Which is the best graph of $y = 1.5^x$? (The graph of $y = 2^x$ is shown as a dashed curve for comparison.)

a)

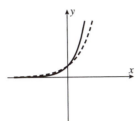

b)

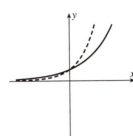

c)

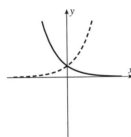

d)

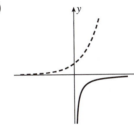

____ **2.** A mosquito population of 100 grows to 500 after two weeks. If the population follows an exponential growth model, how many mosquitoes are there after 5 weeks?

a) 1069   b) 1100   c) 5590   d) 31250

____ **3.** Which number is the best choice for the value of a in the following graph?

a) e b) 1   c) 2   d) 3

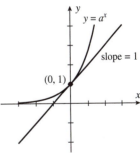

____ **4.** $e^x(e^x)^2 =$

a) e^{x^3} b) e^{3x} c) $3e^x$ d) e^{4x}

Section 1.6

_____ **1.** A function f is one-to-one means:

    a) if $x_1 = x_2$, then $f(x_1) = f(x_2)$
    b) if $x_1 \neq x_2$, then $f(x_1) = f(x_2)$
    c) if $x_1 \neq x_2$, then $f(x_1) \neq f(x_2)$
    d) if $f(x_1) \neq f(x_2)$, then $x_1 \neq x_2$

_____ **2.** True or False:

The function graphed at the right is one-to-one.

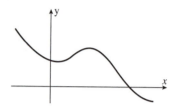

_____ **3.** If $f(x) = \sqrt[3]{x + 3}$, then $f^{-1}(x) =$

    a) $\dfrac{1}{\sqrt[3]{x + 3}}$ b) $x^3 + 3$ c) $(x + 3)^3$ d) $x^3 - 3$

_____ **4.** Sometime, Always, or Never:

If f is one-to-one and the point (a, b) is on the graph of $y = f(x)$ then (b, a)
is on the graph of $y = f^{-1}(x)$.

_____ **5.** $\log_{27} 9 =$

    a) $\dfrac{1}{3}$ b) $\dfrac{2}{3}$ c) $\dfrac{3}{2}$ d) $\dfrac{1}{2}$

_____ **6.** True or False:
$\ln(a + b) = \ln a + \ln b$.

_____ **7.** Simplified, $\log_3 9x^3$ is:
    a) $2 + 3 \log_3 x$ b) $6 \log_3 x$ c) $9 \log_3 x$ d) $9 + \log_3 x$

_____ **8.** Solve for x: $e^{2x-1} = 10$.

    a) $\frac{1}{2}(1 + 10^e)$ b) $3 + \ln 10$ c) $2 + \ln 10$ d) $\dfrac{1 + \ln 10}{2}$

_____ **9.** Solve for x: $\ln(e + x) = 1$.
    a) 0 b) 1 c) $-e$ d) $e^e - e$

Section 2.1

____ 1. For $f(x) = 5x^2 + 1$ the slope of the secant line between the points corresponding to $x = 3$ and $x = 4$ is:

    a) 7   b) 35   c) 81   d) 1

____ 2. Make a guess for the slope of the tangent line to $f(x) = 5x^2 + 1$ at $P(3, 46)$ based on the chart:

    a) 26

    b) 30

    c) 35

    d) 46

x	4	3.1	3.01	3.001	3
$f(x)$	81	49.05	46.3005	46.030005	46

____ 3. The distance a ship is from a lighthouse after t hours is given in the table. What is the average velocity between 2 and 4 hours?

    a) 20 km/hr

    b) 25 km/hr

    c) 27.5 km/hr

    d) 40 km/hr

t	d (km)
0	0
1	40
2	70
3	95
4	110
5	130
6	150

____ 4. True or False:
The slope of a tangent line may be interpreted as average velocity.

Section 2.2

_____ **1.** In the graph at the right;
$$\lim_{x \to 2} f(x) = \underline{\hspace{1cm}}.$$

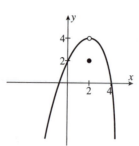

    a) 2
    b) 4
    c) 0
    d) does not exist

_____ **2.** $\lim_{x \to 6} (x^2 - 10x) = \underline{\hspace{1cm}}.$

    a) -24 b) 6   c) 26   d) does not exist

_____ **3.** In the graph at the right,
$$\lim_{x \to 1^-} f(x) = \underline{\hspace{1cm}}.$$

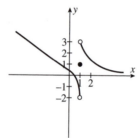

    a) -2
    b) 1
    c) 3
    d) does not exist

_____ **4.** Let $f(x) = \begin{cases} 2 - x^2 & \text{for } x > 1 \\ 3 & \text{for } x \le 1 \end{cases}$. Which limit does not exist?

    a) $\lim_{x \to 3} f(x)$ b) $\lim_{x \to 1^+} f(x)$ c) $\lim_{x \to 1^-} f(x)$ d) $\lim_{x \to 1} f(x)$

_____ **5.** $\lim_{x \to -1} \dfrac{-1}{(1 + x)^2} = \underline{\hspace{1cm}}.$

    a) ∞ b) $-\infty$ c) 1   d) -1

_____ **6.** $\lim_{x \to 0} \cot x =$

    a) 0   b) ∞ c) $-\infty$ d) does not exist

_____ **7.** True or False:

The graph in question 3 has a vertical asymptote at $x = 1$.

Section 2.3

_____ **1.** True or False:

If $h(x) = g(x)$ for all $x \neq a$ and $\lim_{x \to a} h(x) = L$, then $\lim_{x \to a} g(x) = L$.

_____ **2.** $\lim_{x \to 3} \frac{7x + 6}{5x - 12} =$ _____.

     a) $-\frac{1}{2}$ b) $\frac{7}{5}$ c) 9   d) 4

_____ **3.** Sometimes, Always, or Never:

$$\lim_{x \to a} \frac{f(x)}{g(x)} = \frac{\lim_{x \to a} f(x)}{\lim_{x \to a} g(x)}.$$

_____ **4.** $\lim_{x \to 2} \frac{x^2(x - 1)}{1 - x} =$ _____.

     a) -4 b) -1 c) 0   d) does not exist

_____ **5.** $\lim_{x \to 1} \frac{x^3 - x^2}{1 - x} =$ _____.

     a) 1   b) -1 c) 0   d) does not exist

_____ **6.** $\lim_{x \to 2} \frac{\frac{1}{2} - \frac{1}{x}}{2 - x} =$ _____.

     a) 0   b) $\frac{1}{4}$ c) $-\frac{1}{4}$ d) does not exist

_____ **7.** If $2x + 2 \leq f(x) \leq x^2 + 3$ for all x, then $\lim_{x \to 1} f(x) =$ _____.

     a) 1

     b) 4

     c) does not exist

     d) exists but cannot be determined from the information given

Section 2.4

_____ **1.** True or False:

In the definition of $\lim\limits_{x \to a} f(x) = L$, the expression $|f(x) - L| < \epsilon$
is interpreted as "the distance between $f(x)$ and L is less than ϵ."

_____ **2.** For a given $\epsilon > 0$, the corresponding δ in the definition of
$\lim\limits_{x \to 6} (3x + 5) = 23$ is:

a) $\frac{\epsilon}{6}$ b) $\frac{\epsilon}{3}$ c) $\frac{\epsilon}{5}$ d) $\frac{\epsilon}{23}$

_____ **3.** Let $\lim\limits_{x \to a} f(x) = L$. For $\epsilon > 0$ pictured
at the right, what is the best choice for δ in the
limit definition?

a) p

b) q

c) $a - p$

d) $q - a$

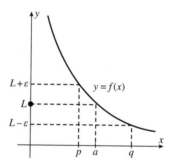

_____ **4.** For a given $M > 0$, the corresponding δ in the definition of $\lim\limits_{x \to 0} \dfrac{1}{2x^4} = \infty$ is:

a) $\sqrt[4]{2M}$ b) $2\sqrt[4]{M}$ c) $\dfrac{1}{2\sqrt[4]{M}}$ d) $\dfrac{1}{\sqrt[4]{2M}}$

Section 2.5

____ **1.** Sometimes, Always, or Never:

If $\lim\limits_{x \to a} f(x)$ and $f(a)$ both exist, then f is continuous at a.

____ **2.** $f(x) = \frac{x+2}{x(x-5)}$ is continuous for all x except:

a) $x = -2$
b) $x = -2, 0, 5$
c) $x = 0, 5$
d) f is continuous for all x.

____ **3.** True or False:

$f(x) = \frac{x^2(x+1)}{x}$ is continuous for all real numbers.

____ **4.** $f(x) = \begin{cases} 2x+1 & \text{if } x \le 1 \\ \frac{1}{x} & \text{if } 1 < x \le \pi \\ \sin x & \text{if } x > \pi \end{cases}$ is continuous:

a) for all x
b) for all x except $x = 1$
c) for all x except $x = \pi$
d) for all x except $x = 1$ and $x = \pi$.

____ **5.** For the continuous function at the right and the given value N, how many points c satisfy the conclusion of the Intermediate Value Theorem?

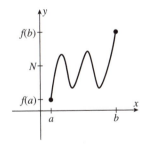

a) 1
b) 3
c) 5
d) none

____ **6.** True or False:

Let $f(x) = x^2$. The Intermediate Value Theorem guarantees there is $w \in (-1, 5)$ for which $f(w) = 0$.

On Your Own

Section 2.6

_____ **1.** Sometime, Always, or Never:

$$\lim_{x \to \infty} f(x) = \lim_{x \to -\infty} f(x).$$

_____ **2.** $\lim\limits_{x \to \infty} \dfrac{\sqrt{x^2 + 1}}{3x + 2} =$

    a) 0   b) 1   c) $\dfrac{1}{3}$ d) does not exist

_____ **3.** True or False:

If the three limits exist, then $\lim\limits_{x \to \infty} (f(x) + g(x)) = \lim\limits_{x \to \infty} f(x) + \lim\limits_{x \to \infty} g(x)$.

_____ **4.** The horizontal asymptote(s) for $f(x) = \dfrac{|x|}{x + 1}$ is (are):

    a) $y = 0$ b) $y = 1$
    c) $y = -1$ d) $y = 1$ and $y = -1$

_____ **5.** True or False:

$$\lim_{x \to \infty} \frac{6x^3 + 8x^2 - 9x + 7}{5x^3 - 4x + 30} = \lim_{x \to \infty} \frac{6x^3}{5x^3}.$$

_____ **6.** $\lim\limits_{x \to \infty} \dfrac{6 - x^3}{3 + x} =$

    a) ∞ b) $-\infty$ c) 0   d) 2

_____ **7.** Sometimes, Always, or Never:

$$\lim_{x \to -\infty} a^x = 0.$$

_____ **8.** $\lim\limits_{x \to -\infty} \left(\dfrac{1}{2}\right)^{x^2} =$

    a) 0   b) 1   c) ∞ d) does not exist

_____ **9.** $\lim\limits_{x \to \infty} 2^{x/(x + 1)} =$

    a) 0   b) 1   c) 2   d) ∞

Section 2.7

1. True or False:
$$\lim_{x \to a} \frac{f(x) - f(a)}{x - a} = \lim_{k \to 0} \frac{f(a + k) - f(a)}{k}.$$

2. The slope of the tangent line to $y = x^3$ at $x = 2$ is:

a) 18   b) 12   c) 6   d) 0

3. True or False:
$\dfrac{f(x) - f(a)}{x - a}$ may be interpreted as the instantaneous velocity of a particle at time a.

4. True or False:
Slope of a tangent line, instantaneous velocity, and instantaneous rate of change of a function are each all interpretations of the same limit concept.

5. The instantaneous rate of change of $y = x^3 + 3x$ at the point corresponding to $x = 2$ is:

a) 10   b) 15   c) 2   d) 60

6. Given the table of function values below, which of the following is the best estimate for the instantaneous rate of change of $y = f(x)$ at $x = 4$?

x	6	5	4.5	4.1	4.05	4.01	4
$f(x)$	17	13.3	11.7	10.32	10.16	10.031	10

a) 17   b) 10   c) 3   d) 0

(continued on next page)

7. True or False:

$$f'(a) = \lim_{a \to 0} \frac{f(a + h) - f(a)}{h}.$$

_____

8. For $f(x) = 10x^2, f'(3) =$ _____.

a) 10   b) 20   c) 30   d) 60

_____

9. $\lim\limits_{h \to 0} \dfrac{\sqrt{4 + h} - 2}{h}$ is the derivative of:

a) $f(x) = \sqrt{4 + x}$ at $x = 2$

b) $f(x) = \sqrt{x}$ at $x = 4$

c) $f(x) = \sqrt{x}$ at $x = 2$

d) $f(x) = \dfrac{\sqrt{4 + h} - 2}{h}$ at $x = 0$

_____

10. Which is the largest?

a) $f'(a)$

b) $f'(b)$

c) $f'(c)$

d) cannot tell from information given

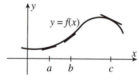

Section 2.8

_____ **1.** True or False:

$f(x) = \tan x$ is differentiable at $x = \frac{\pi}{2}$.

_____ **2.** True or False:

If $y = f(x)$ is not differentiable at a point $x = a$, then $y = f(x)$ is not continuous at $x = a$.

_____ **3.** Which is the best graph of $y = f'(x)$ for the given graph of $y = f(x)$?

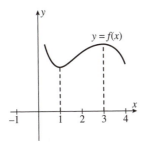

a)

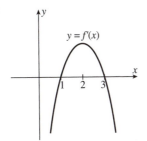

b)

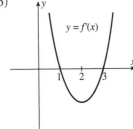

c)

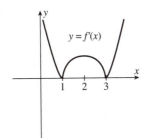

d)   None of these

_____ **4.** True or False:

$\frac{dm}{dv}$ stands for the derivative of the function m with respect to the variable v.

Section 3.1

____ **1.** For $f(x) = 6x^4$, $f'(x) =$ _____.

    a) $24x$ b) $6x^3$ c) $10x^3$ d) $24x^3$

____ **2.** For $y = \frac{1}{2}gt^2$ (where g is a constant), $\frac{dy}{dt} =$ _____.

    a) $\frac{1}{4}g^2$ b) $\frac{1}{2}g$ c) gt d) gt^2

____ **3.** For $f(x) = 5 + g(x)$, $f'(x) =$ _____.

    a) $5g'(x)$ b) $5 + g'(x)$ c) $0 \cdot g'(x)$ d) $g'(x)$

____ **4.** $\lim\limits_{h \to 0} \frac{a^h - 1}{h} = 1$ for $a =$

    a) 0 b) 1 c) e d) The limit is never 1.

____ **5.** For $f(x) = e^x - x$, $f'(x) =$

    a) $e^x - x$ b) $e^x - 1$ c) $e^1 - x$ d) $e^1 - 1$

____ **6.** A particle moves along an axis so that t seconds after it starts its position is $p(t) = 3t^2 - 4t + 1$ meters. The velocity at time $t = 3$ is

    a) 5 m/s   b) 14 m/s   c) 16 m/s   d) 22 m/s

Section 3.2

___ **1.** True or False:

$(f(x)g(x))' = f'(x)g'(x)$.

___ **2.** For $f(x) = (x^3 + 1)(x^4 + 1)$, $f'(x) =$

    a) $(3x^2)(4x^3)$

    b) $(x^3 + 1)(x^4) + (x^4 + 1)(x^3)$

    c) $(x^3 + 1)(4x^3) + (x^4 + 1)(3x^2)$

    d) $4x^6 + 3x^5 + 4x^3 + 3x^2$

___ **3.** For $f(x) = \frac{x}{x+1}$, $f'(x) =$ _____.

    a) 1   b) $\frac{x^2}{(x+1)^2}$ c) $\frac{-1}{(x+1)^2}$ d) $\frac{1}{(x+1)^2}$

___ **4.** For $f(x) = \frac{e^x}{x}$, $f'(x) =$

    a) e^x b) $\frac{xe^x - 1}{x^2}$ c) $\frac{e^x}{x} - \frac{e^x}{x^2}$ d) $\frac{xe^x - x}{x^2}$

Section 3.3

_____ **1.** $\lim\limits_{x\to 0} \dfrac{2 \sin x}{3x} =$ _____.

    a) 1   b) $\dfrac{2}{3}$ c) $\dfrac{3}{2}$ d) 0

_____ **2.** If $\lim\limits_{x\to 0} \dfrac{\sin x}{x} = 1$, the angle x must be measured in:

    a) radians   b) degrees   c) it does not matter

_____ **3.** For $y = x \sin x$, $y' =$ _____.

    a) $x \cos x$
    b) $x \cos x + 1$
    c) $\cos x$
    d) $x \cos x + \sin x$

_____ **4.** For $y = 3x + \tan x$, $y' =$ _____.

    a) $3 + \sec x$
    b) $3x \sec^2 x + 3 \tan x$
    c) $3 + \sec^2 x$
    d) $x \cos x + \sin x$

_____ **5.** For $y = \sin^2 x + \cos^2 x$, $y' =$ _____.

    a) $2 \sin x - 2 \cos x$
    b) $2 \sin x \cos x$
    c) $4 \sin x \cos x$
    d) 0

_____ **6.** What is the equation of the tangent line to $y = \cos x$ at $x = \dfrac{\pi}{3}$?

    a) $y = \dfrac{1}{2} - \dfrac{\sqrt{3}}{2}\left(x - \dfrac{\pi}{3}\right)$

    b) $y = -\dfrac{1}{2} + \dfrac{\sqrt{3}}{2}\left(x - \dfrac{\pi}{3}\right)$

    c) $y = \dfrac{1}{2} + \dfrac{\sqrt{3}}{2}\left(x - \dfrac{\pi}{3}\right)$

    d) $y = -\dfrac{1}{2} - \dfrac{\sqrt{3}}{2}\left(x - \dfrac{\pi}{3}\right)$

Section 3.4

_____ **1.** True or False:
If $y = f(g(x))$ then $y' = f'(g'(x))$.

_____ **2.** For $y = (x^2 + 1)^2, y' = $ _____.
a) $2(x^2 + 1)$ b) $4(x^2 + 1)$ c) $4x(x^2 + 1)$ d) $2x(x^2 + 1)$

_____ **3.** For $y = \sin 6x, \dfrac{dy}{dx} = $ _____.
a) $\cos 6x$ b) $6 \cos x$ c) $\cos 6$ d) $6 \cos 6x$

_____ **4.** For $y = \dfrac{4}{\sqrt{3 + 2x}}, y' = $ _____.

a) $-4(3 + 2x)^{-3/2}$

b) $4(3 + 2x)^{-3/2}$

c) $-2(3 + 2x)^{-1/2}$

d) $2(3 + 2x)^{-1/2}$

_____ **5.** True or False:
For $y = f(u)$ and $u = g(x), \dfrac{dy}{dx} = \dfrac{dy}{du} \cdot \dfrac{du}{dx}$.

_____ **6.** Sometimes, Always, or Never:
$(a^x)' = a^x$.

_____ **7.** If $f(x) = 2e^{3x}, f'(x) = $
a) $2e^{3x}$ b) $2xe^{3x}$ c) $6e^{3x}$ d) $6xe^{3x}$

Section 3.5

___ **1.** True or False:

$y = \sqrt{4 - \sqrt{x^2 + 1}}$ defines y implicitly as a function of x.

___ **2.** If y is defined as a function of x by $y^3 = x^2$, $y' = $ _____.

   a) $\dfrac{2x}{y^3}$　　　　b) $\dfrac{2x}{3y^2}$　　　　c) $\dfrac{x^2}{3y^2}$　　　　d) $\dfrac{x^2}{3y}$

___ **3.** The slope of the line tangent to the circle $x^2 + y^2 = 100$ at the point $(-6, 8)$ is:

   a) $\dfrac{3}{4}$　　　　b) $-\dfrac{3}{4}$　　　　c) $\dfrac{4}{3}$　　　　d) $-\dfrac{4}{3}$

___ **4.** The range of $f(x) = \cos^{-1} x$ is:

   a) $\left[-\dfrac{\pi}{2}, \dfrac{\pi}{2}\right]$

   b) $[0, \pi]$

   c) $\left[0, \dfrac{\pi}{2}\right] \cup \left[\pi, \dfrac{3\pi}{2}\right]$

   d) all reals

___ **5.** $\sec^{-1}\left(\dfrac{2}{\sqrt{3}}\right) = $

   a) $-\dfrac{\pi}{3}$　　　　b) $\dfrac{\pi}{3}$　　　　c) $-\dfrac{\pi}{6}$　　　　d) $\dfrac{\pi}{6}$

___ **6.** For $f(x) = \cos^{-1}(2x)$, $f'(x) = $

   a) $\dfrac{-1}{\sqrt{1 - 4x^2}}$

   b) $\dfrac{-2}{\sqrt{1 - 4x^2}}$

   c) $\dfrac{-4}{\sqrt{1 - 4x^2}}$

   d) $\dfrac{-8}{\sqrt{1 - 4x^2}}$

Section 3.6

_____ **1.** For $y = \ln(3x^2 + 1)$, $y' =$

    a) $\dfrac{1}{3x^2 + 1}$ b) $\dfrac{6}{x}$ c) $\dfrac{6x}{3x^2 + 1}$ d) $\dfrac{1}{6x}$

_____ **2.** For $y = x^3 \ln x$, $y' =$

    a) $x^2(1 + 3 \ln x)$ b) $x^3 + 3x^2(\ln x)$ c) $\dfrac{3}{x}$ d) $4x^2(\ln x)$

_____ **3.** For $y = 3^x$, $y' =$

    a) $3^x \log_3 e$ b) $3^x \ln 3$ c) $\dfrac{3x}{\log_3 e}$ d) $\dfrac{3x}{\ln 3}$

_____ **4.** By logarithmic differentiation, if $y = \dfrac{\sqrt{x}}{x + 1}$, then $y' =$

    a) $\dfrac{\sqrt{x}}{x + 1}\left(\dfrac{1}{\sqrt{x}} - \dfrac{1}{x + 1}\right)$ b) $\dfrac{\sqrt{x}}{x + 1}\left(\dfrac{1}{x} - \dfrac{1}{x + 1}\right)$

    c) $\dfrac{\sqrt{x}}{x + 1}\left(\dfrac{2}{x} - \dfrac{1}{x + 1}\right)$ d) $\dfrac{\sqrt{x}}{x + 1}\left(\dfrac{1}{2x} - \dfrac{1}{x + 1}\right)$

_____ **5.** $\lim\limits_{x \to 0^+} (1 + x)^{-1/x} =$

    a) e b) $-e$ c) $\dfrac{1}{e}$ d) $-\dfrac{1}{e}$

Section 3.7

_____ **1.** True or False:

$f'(t)$ is used to measure the average rate of change of f with respect to t.

_____ **2.** Air is being pumped into a chamber so that after t seconds the pressure in the chamber is $3 + 3x + x^2$ pounds/in². The rate of change of the pressure at $t = 2$ second is:

a) 7 pounds/in²/sec

b) $3\frac{1}{6}$ pounds/in²/sec

c) $3\frac{1}{2}$ pounds/in²/sec

d) $2\frac{5}{6}$ pounds/in²/sec

_____ **3.** A bacteria population grows in such a way that after t hours its population is $1000 + 10t + t^3$. The growth rate at $t = 3$ hours is:

a) 13 bacteria/hour

b) 37 bacteria/hour

c) 10 bacteria/hour

d) 1057 bacteria/hour

_____ **4.** A circle is increasing in area. The rate of change of area with respect to its radius r is:

a) $2\pi r$ b) πr^2 c) πr d) $2\pi r^2$

Section 3.8

_____ **1.** The general solution to $\frac{dy}{dt} = ky$ is:

   a) $y(t) = y(0)e^{kt}$

   b) $y(t) = y(k)e^t$

   c) $y(t) = y(t)e^k$

   d) $y(t) = e^{y(0)kt}$

_____ **2.** A bacteria culture starts with 50 organisms and after 2 hours there are 100. Assuming natural growth, how many will there be after 5 hours?

   a) $50 \ln 5$

   b) $50e^5$

   c) $200\sqrt{2}$

   d) 300

_____ **3.** If \$5,000 is deposited in an account that accumulates interest compounded continuously at 4% then after 6 years the amount accumulated is

   a) $5000 \, e^{(-0.04)6}$

   b) $5000 \, e^{(0.04)6}$

   c) $5000 \, e^{(-4)6}$

   d) $5000 \, e^{(4)6}$

Section 3.9

_____ **1.** A right triangle has one leg with constant length 8 cm. The length of the other leg is decreasing at 3 cm/sec. The rate of change of the hypotenuse when the variable leg is 6 cm is:

a) $-\frac{9}{5}$ cm/s

b) 3 cm/s

c) $-\frac{5}{3}$ cm/s

d) $\frac{5}{3}$ cm/s

_____ **2.** How fast is the angle between the hands of a clock increasing at 3:30 pm?

a) 2π radians/hr

b) $\frac{\pi}{6}$ radians/hr

c) $\frac{11\pi}{6}$ radians/hr

d) $\frac{5\pi}{6}$ radians/hr

Section 3.10

_____ **1.** For $f(x) = x^3 + 7x$, $df =$ _____.

    a) $(x^3 + 7x)\,dx$
    b) $(3x^2 + 7)\,dx$
    c) $3x^2 + 7$
    d) $x^3 + 7x + dx$

_____ **2.** The best approximation of Δy using differentials for $y = x^2 - 4x$ at $x = 6$ when $\Delta x = dx = 0.03$ is:

    a) 0.36   b) 0.24   c) 0.20   d) 0.30

_____ **3.** The linearization of $f(x) = \dfrac{1}{\sqrt{x}}$ at $x = 4$ is:

    a) $L(x) = \dfrac{1}{2} + \dfrac{1}{\sqrt{x}}(x - 2)$
    b) $L(x) = \dfrac{1}{4} + \dfrac{1}{2}(x - 2)$
    c) $L(x) = \dfrac{1}{2} + \dfrac{1}{2}(x - 4)$
    d) $L(x) = \dfrac{1}{2} - \dfrac{1}{16}(x - 4)$

_____ **4.** Using differentials, an approximation to $(3.04)^3$ is:

    a) 28.094464   b) 28.08   c) 28.04   d) 28

Section 3.11

_____ **1.** True or False:

$\sinh x = \frac{e^x + e^{-x}}{2}$.

_____ **2.** $\cosh 0 =$

    a)   0   b)   1   c) $\frac{e^2}{2}$ d)   is not defined

_____ **3.** For $f(x) = \sinh x$, $f''(x) =$

    a) $\cosh^2 x$ b) $\sinh^2 x$ c) $\tanh x$ d) $\sinh x$

_____ **4.** For $f(x) = \sqrt{x^2 + \sinh x}$, $f'(x) =$

    a) $\dfrac{2x + \cosh x}{2\sqrt{x^2 + \sinh x}}$

    b) $\dfrac{x + \cosh x}{\sqrt{x^2 + \sinh x}}$

    c) $\dfrac{1}{2\sqrt{2x + \cosh x}}$

    d) $\dfrac{x \cosh x}{\sqrt{x^2 + \sinh x}}$

_____ **5.** For $f(x) = \tanh^{-1} 2x$, $f'(x) =$

    a) $2(\operatorname{sech}^{-1} 2x)^2$

    b) $2(\operatorname{sech}^2 2x)^{-1}$

    c) $\dfrac{2}{1 - 4x^2}$

    d) $\dfrac{1}{1 - 4x^2}$

Section 4.1

_____ **1.** Sometimes, Always, or Never:
If c is a critical number, then $f'(c) = 0$.

_____ **2.** Sometimes, Always, or Never:
If $f'(c) = 0$, then c is a critical number.

_____ **3.** The critical numbers of $f(x) = 3x^4 + 20x^3 - 36x^2$ are:

    a) $0, 1, 6$ b) $0, -1, 6$ c) $0, 1, -6$ d) $0, -1, -6$

_____ **4.** True or False:
If $f'(c)$ exists and f has a local minimum at c, then $f'(c) = 0$.

_____ **5.** True or False:
The absolute extrema of a continuous function on a closed interval always exist.

_____ **6.** The graph at the right has a local maximum at $x =$ _____.

    a) $1, 3,$ and 5
    b) 2
    c) 4
    d) 2 and 4

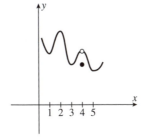

_____ **7.** The absolute maximum value of $f(x) = 6x - x^2$ on $[1, 7]$ is:

    a) 3 b) 9 c) 76 d) -7

Section 4.2

_____ **1.** Which of the following is not a hypothesis of Rolle's Theorem?

    a) f is continuous on $[a, b]$.
    b) $f(a) = f(b)$.
    c) $f(x) \geq 0$ for all $x \in (a, b)$.
    d) f is differentiable on (a, b).

_____ **2.** True or False:
All the hypotheses for the Mean Value Theorem hold for
$f(x) = 1 - |x|$ on $[-1, 2]$.

_____ **3.** A number c that satisfies the Mean Value Theorem for $f(x) = x^3$ on $[1, 4]$ is:

    a) $\sqrt{7}$ b) $\sqrt{21}$ c) $\sqrt{63}$ d) 63

_____ **4.** True or False:
If $f'(x) = g'(x)$ for all x then $f(x) = g(x)$.

_____ **5.** If $f'(x) = g'(x)$ for all $x \in (a, b)$, then _____.

    a) $f(a) = g(a)$ and $f(b) = g(b)$
    b) $f(x) = g(x)$
    c) $f(x) = g(x) + c$ for some constant c.
    d) $f(x) + g(x) = c$ for some constant c.

Section 4.3

_____ **1.** If 4 is a critical number for f and $f'(x) < 0$ for $x < 4$ and $f'(x) > 0$ for $x > 4$, then:
  a) f has a local maximum at 4.
  b) f has a local minimum at 4.
  c) 4 is either a local maximum or minimum, but not enough information is given to decide which.
  d) 4 is neither a local maximum nor local minimum.

_____ **2.** True or False:
  $f(x) = 10x - x^2$ is increasing on $(4, 8)$.

_____ **3.** True or False:
  $f(x) = 4x^2 - 8x + 1$ is increasing on $(0, 1)$.

_____ **4.** $f(x) = 12x - x^3$ has a local maximum at:
  a) $x = 0$
  b) $x = 2$
  c) $x = -2$
  d) f does not have a local maximum

_____ **5.** True or False:
  If a function is not increasing on an interval, then it is decreasing on the interval.

_____ **6.** True or False:
  $f(x) = 3x - x^3$ is concave down for $x > 1$.

_____ **7.** True or False:
  If $f''(c) = 0$, then c is a point of inflection for f.

_____ **8.** $f(x) = (x^2 - 3)^2$ has points of inflection at $x =$ _____.
  a) $0, \sqrt{3}, -\sqrt{3}$ b) $-1, 1$
  c) $0, -1, 1$ d) $\sqrt{3}, -\sqrt{3}$

_____ **9.** $f(x) = x^3 - 2x^2 + x - 1$ has a local maximum at:
  a) 0 b) 1 c) $\frac{1}{3}$ d) f has no local maxima

(continued on next page)

Section 4.3 continued

_____ **10.** What graph has $f'(2) > 0$ and $f''(2) < 0$?

a)
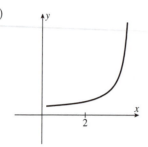

b)
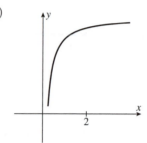

c)
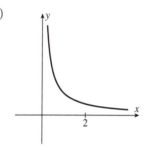

d)
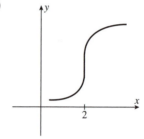

Section 4.4

_____ **1.** $\lim_{x\to 5} \dfrac{x^2 - 25}{x^2 - 9x + 20} =$

    a) 1   b) 10   c) ∞ d) does not exist

_____ **2.** $\lim_{x\to 0^+} \sqrt[3]{x}\,\ln x =$

    a) 0   b) 1   c) $-\infty$ d) does not exist

_____ **3.** $\lim_{x\to 1} \dfrac{1}{\ln x} - \dfrac{1}{x - 1} =$

    a) 0   b) $\frac{1}{2}$ c) 1   d) does not exist

_____ **4.** $\lim_{x\to\infty} (\ln x)^{1/x} =$

    a) 0   b) 1   c) e d) ∞

Section 4.5

_____ **1.** True or False:
The domain is all numbers x for which $f(x)$ is defined.

_____ **2.** True or False:
The x-intercepts are the values of x for which $f(x) = 0$.

_____ **3.** True or False:
$f(x)$ is symmetric about the origin if $f(x) = -f(x)$.

_____ **4.** True or False:
If $f'(x) > 0$ for all x in an interval I then f is increasing on I.

_____ **5.** True or False:
If $f'(c) = 0$ and $f''(c) < 0$, then c is a local minimum.

_____ **6.** True or False:
If $f''(x) > 0$ for $x < c$ and $f''(x) < 0$ for $x > c$, then c is a point of inflection.

_____ **7.** Which is the graph of $f(x) = \dfrac{x^2 - x}{x^2 - 1}$?

_____

a)

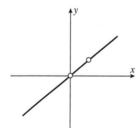

b)

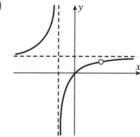

c)

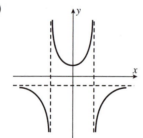

d)

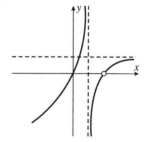

Section 4.6

_____ **1.** The graph of $f(x) = 3x^5 - 5x^3$ is:

a)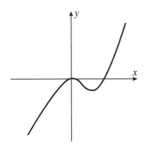

b)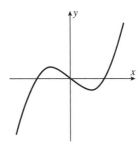

c)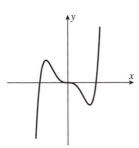

d)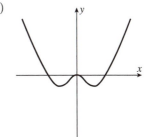

_____ **2.** For the family of functions $f(x) = x^2 - ax$, each function is:
  a) a parabola through $(0, 0)$ and $(a, 0)$.
  b) is concave up.
  c) both a) and b) are correct.

_____ **3.** $f(x) = \dfrac{2x^2 + 1}{x + 3}$ has slant asymptote(s):

  a) $y = 2x$
  b) $y = x$
  c) $y = 2x$ and $y = -2x$
  d) f does not have slant asymptotes.

____ 1. For two nonnegative numbers, twice the first plus the second is 12. What is the maximum product of two such numbers?

   a) 12
   b) 18
   c) 36
   d) There is no such maximum.

____ 2. A carpenter has a 10-foot-long board to mark off a triangular area on the floor in the corner of a room. See the figure. What is the function for the triangular area in terms of x that may be used to determine the maximum such area?

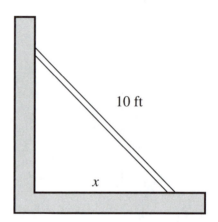

10 ft

x

   a) $A = \frac{1}{2}x^2$
   b) $A = \frac{1}{2}(x^2 + 10)$
   c) $A = \frac{1}{2}x\sqrt{100 - x^2}$
   d) $A = \frac{1}{2}x\sqrt{100 - 2x^2}$

____ 3. The maximum area in question 2) is

   a) $2\sqrt{50}$ ft^2 b) 20 ft^2 c) 25 ft^2 d) 50 ft^2

____ 4. Suppose the price (demand function) for an item is $p = 16 - 0.01x$, where x is the number of items. If the cost function is $C(x) = 1000 + 10x + 0.02x^2$, how many items should be made and sold to maximize profits?

   a) 50   b) 100   c) 200   d) 250

Section 4.8

_____ **1.** Sometimes, Always, or Never:

The value of x_2 in Newton's Method will be a closer approximation to the root than the initial value x_1.

_____ **2.** With an initial estimate of -1, use Newton's Method once to estimate a zero of $f(x) = 5x^2 + 15x + 9$.

a) $-\dfrac{6}{5}$ 　　　 b) $-\dfrac{4}{5}$ 　　　 c) $-\dfrac{7}{5}$ 　　　 d) $-\dfrac{3}{5}$

_____ **3.** Suppose in an estimation of a root of $f(x) = 0$ we make an initial guess x_1 and it turns out that $f'(x_1) = 0$. Which of the following will be true?

a) x_1 is a root of $f(x) = 0$.
b) Newton's Method will produce a sequence $x_1, x_2, x_3, \ldots$ that does approximate a zero.
c) Newton's Method will produce a sequence $x_1, x_2, x_3, \ldots$ such that $x_1 = x_2 = x_3 = \ldots$
d) Newton's Method cannot be used when $f'(x_1) = 0$.

Section 4.9

_____ **1.** True or False:
If $h'(x) = k(x)$, then $k(x)$ is an antiderivative of $h(x)$.

_____ **2.** Find $f(x)$ if $f'(x) = 10x^2 + \cos x$.

    a) $20x - \sin x + C$

    b) $20x - \cos x + C$

    c) $\frac{10}{3}x^3 - \cos x + C$

    d) $\frac{10}{3}x^3 + \sin x + C$

_____ **3.** Find $f(x)$ if $f'(x) = \frac{1}{x^2}$ and $f(2) = 0$.

    a) $-\frac{1}{x} + \frac{1}{2}$

    b) $-\frac{1}{x} - \frac{1}{2}$

    c) $-\frac{3}{x^3} + \frac{3}{8}$

    d) $-\frac{3}{x^3} - \frac{3}{8}$

_____ **4.** True or False:
If $f'(x) = g'(x)$, then $f(x) = g(x)$.

_____ **5.** If velocity is $v(t) = 4t + 4$ and $s(1) = 2$, then $s(t) =$

    a) $2t^2 + 4t - 4$

    b) $2t^2 + 4t + 6$

    c) $t^2 + 4$

    d) $t^2 - 4$

Section 5.1

_____ 1. Find the sum of the areas of approximating rectangles for the area under
$f(x) = 48 - x^2$, between $x = 1$ and $x = 5$. Use $\Delta x = 1$ and the right
endpoints of each subinterval for x_i*.

    a) 138   b) 99   c) 15   d) 192

_____ 2. True or False:
In general, better approximations to the area under the curve $y = f(x)$
between $x = a$ and $x = b$ are obtained by selection of partitions with smaller
values of Δx.

_____ 3. Selecting midpoints of subintervals for x_i* to approximate the area under
$f(x) = 10x + 5$ between $x = 1$ and $x = 4$ will generate an approximating
sum which is:

    a) greater than the actual area
    b) less than the actual area
    c) equal to the actual area

_____ 4. If the interval $[1, 3]$ is divided into n subintervals of length Δx, then the
shaded area at the right is

    a) $\lim\limits_{n \to \infty} \sum\limits_{i=1}^{n} \left((x_i^*)^2 + 1\right) \Delta x$

    b) $\lim\limits_{n \to \infty} \sum\limits_{i=1}^{n} \left((3 - 1)^2 + 1\right) \Delta x$

    c) $\lim\limits_{n \to \infty} \sum\limits_{i=1}^{n} \left(3(x_i^*)^2 + x_i^*\right) \Delta x$

    d) $\lim\limits_{n \to \infty} \sum\limits_{i=1}^{n} \left(\frac{(x_i^*)^3}{3} + x_i^*\right) \Delta x$

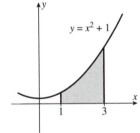

$y = x^2 + 1$

Section 5.2

_____ **1.** Sometimes, Always, or Never:

$\int_a^b f(x)\,dx$ equals the area between $y = f(x)$, the x-axis, $x = a$, and $x = b$.

_____ **2.** Sometimes, Always, or Never:

If $f(x)$ is continuous on $[a, b]$, then $\int_a^b f(x)\,dx$ exists.

_____ **3.** Using $n = 4$ and midpoints for x_i^*, then Riemann sum for $\int_{-1}^7 x^2\,dx$ is

    a) $\frac{344}{3}$ b) 72   c) 168   d) 112

_____ **4.** If $\int_1^3 f(x)\,dx = 10$ and $\int_1^3 g(x)\,dx = 6$, then $\int_1^3 (2f(x) - 3g(x))dx =$

    a) 2   b) 4   c) 18   d) 38

_____ **5.** If $f(x) \geq 5$ for all $x \in [2, 6]$ then $\int_2^6 f(x)\,dx \geq$ _____.

(Choose the best answer.)

    a) 4   b) 5   c) 20   d) 30

_____ **6.** $\int_3^5 6\,dx =$

    a) 2   b) 8   c) 6   d) 12

_____ **7.** $\sum_{i=1}^n (4i^2 + i) =$

    a) $4\sum_{i=1}^n (i^2 + i)$ c) $\sum_{i=1}^n 5i^2$

    b) $4\sum_{i=1}^n i^2 + \sum_{i=1}^n i$ d) $4\sum_{i=1}^n i^2 + 4\sum_{i=1}^n i$

_____ **8.** By determining the limit of Riemann sums, the exact value of $\int_{-1}^2 3x^2\,dx$ is

    a) 7   b) 9   c) $\frac{14}{3}$ d) 18

_____ **9.** True or False:

$\int_0^7 2\sqrt[3]{x}\,dx = 2\int_0^7 \sqrt[3]{x}\,dx.$

Section 5.3

_____ **1.** If $h(x) = \int_3^x (8t^3 + 2t)dt$, then $h'(x) =$

    a) $8t^3 + 2t$ b) $24x^2 + 2$ c) $8x^3 + 2x - 3$ d) $8x^3 + 2x$

_____ **2.** Evaluate $\int_0^3 4x \, dx$ using the Fundamental Theorem of Calculus.

    a) 6   b) 12   c) 18   d) 24

_____ **3.** Evaluate $\int_{-1}^2 3x^2 \, dx$ using the Fundamental Theorem of Calculus.

    a) 7   b) 9   c) $\frac{14}{3}$ d) 18

_____ **4.** True or False:

    If $f'(x) = F(x)$, then $\int_b^a f(x) \, dx = F(b) - F(a)$.

____ **1.** $\int f(x)\, dx = F(x)$ means

     a) $f'(x) = F(x)$
     b) $f(x) = F'(x)$
     c) $f(x) = F(b) - F(a)$
     d) $f(x) = F(x) + C$

____ **2.** $\int x^{-4} dx =$

     a) $-4x^{-5} + C$ b) $\frac{-4}{5} x^{-5} + C$ c) $-3x^{-3} + C$ d) $\frac{x^{-3}}{-3} + C$

____ **3.** $\int (x - \cos x) dx =$

     a) $1 - \sin x + C$

     b) $1 + \sin x + C$

     c) $\frac{x^2}{2} - \sin x + C$

     d) $\frac{x^2}{2} + \sin x + C$

____ **4.** Evaluate $\int_{-1}^{1} (x^2 - x^3) dx$

     a) $\frac{3}{2}$ b) $\frac{5}{6}$ c) $\frac{1}{2}$ d) $\frac{2}{3}$

____ **5.** If A_1 and A_2 are the areas at the right then $\int_{a}^{b} f(x)\, dx =$

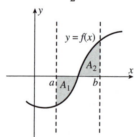

     a) $A_1 - A_2$
     b) $-A_1 + A_2$
     c) $A_1 + A_2$
     d) $A_1 \cdot A_2$

____ **6.** 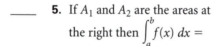 $\int \frac{1}{2x + 1}\, dx =$

     a) $\ln |2x + 1| + C$ b) $\frac{1}{2} \ln |2x + 1| + C$

     c) $2 \ln |2x + 1| + C$ d) $\ln \left| \frac{2x + 1}{2} \right| + C$

Section 5.5

_____ **1.** Suppose $\int f(x) = F(x) + C$. Then $\int f(g(x))g'(x)\,dx =$

a) $F(x)g(x) + C$

b) $f'(g(x)) + C$

c) $F(g'(x)) + C$

d) $F(g(x)) + C$

_____ **2.** What substitution should be made to evaluate $\int \sqrt{x^2 + 2x}\,(x + 1)\,dx$?

a) $u = \sqrt{x^2 + 2x}$

b) $u = x^2 + 2x$

c) $u = x + 1$

d) $u = \sqrt{x^2 + 2x}\,(x + 1)$

_____ **3.** What substitution should be made to evaluate $\int \sin^2 x \cos x\,dx$?

a) $u = \sin x$

b) $u = \cos x$

c) $u = \sin x \cos x$

d) $u = \sin^2 x \cos x$

_____ **4.** Evaluate $\int_0^1 (x^3 + 1)^2 x^2\,dx$.

a) $\frac{1}{9}$ b) $\frac{1}{3}$ c) $\frac{7}{9}$ d) $\frac{8}{9}$

_____ **5.** $\int \sin \frac{x}{2}\,dx =$

a) $2 \cos \frac{x}{2} + C$ b) $\frac{1}{2} \cos \frac{x}{2} + C$ c) $-2 \cos \frac{x}{2} + C$ d) $-\frac{1}{2} \cos \frac{x}{2} + C$

_____ **6.** True or False:

If $f(x) = f(-x)$ for all x then $\int_{-2}^{2} f(x) = 2\int_0^2 f(x)$.

_____ **7.** $\int_0^1 e^{2x}\,dx =$

a) $2e^2 - 1$ b) $e^2 - 1$ c) $\frac{1}{2}e^2$ d) $\frac{1}{2}(e^2 - 1)$

Section 6.1

_____ 1. A definite integral for the
area shaded at the right is:

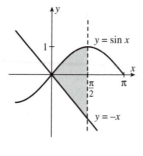

a) $\displaystyle\int_0^{\pi/2} (\sin x - x)\,dx$

b) $\displaystyle\int_0^{\pi/2} (\sin x + x)\,dx$

c) $\displaystyle\int_0^1 (\sin x - x)\,dx$

d) $\displaystyle\int_0^1 (\sin x + x)\,dx$

_____ 2. For the area of the shaded
region, which integral form,
$\int \dots dx$ or $\int \dots dy$, should be
used?

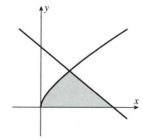

a) $\displaystyle\int \dots dx$

b) $\displaystyle\int \dots dy$

c) Either may be used with
equal ease.

_____ 3. A definite integral for the area of the region bounded by $y = 2 - x^2$
and $y = x^2$ is:

a) $\displaystyle\int_{-1}^1 (2 - 2x^2)\,dx$

b) $\displaystyle\int_{-1}^1 (2x^2 - 2)\,dx$

c) $\displaystyle\int_0^2 (2 - 2x^2)\,dx$

d) $\displaystyle\int_0^2 (2x^2 - 2)\,dx$

On Your Own

Section 6.2

____ **1.** Find a definite integral for the volume of a solid with a circular base of radius 4 inches such that the cross section of any slice perpendicular to a certain diameter in the base is a square.

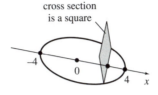
cross section is a square

a) $\displaystyle\int_{-4}^{4} (64 - 4x^2)\,dx$

b) $\displaystyle\int_{-4}^{4} 4x^2\,dx$

c) $\displaystyle\int_{-4}^{4} 4x^4\,dx$

d) $\displaystyle\int_{-4}^{4} (4 - x^2)^2\,dx$

____ **2.** An integral for the solid obtained by rotating the region at the right about the x-axis is:

a) $\displaystyle\int_{0}^{1} \pi(x^4 - x^2)\,dx$

b) $\displaystyle\int_{0}^{1} \pi(x^2 - x^4)\,dx$

c) $\displaystyle\int_{0}^{1} \pi(x - x^2)\,dx$

d) $\displaystyle\int_{0}^{1} \pi(x^2 - x)\,dx$

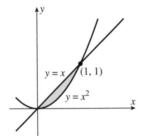
$y = x$ (1, 1) $y = x^2$

____ **3.** An integral for the solid obtained by rotating the region at the right about the y-axis is:

a) $\displaystyle\int_{0}^{4} \pi y\,dy$

b) $\displaystyle\int_{0}^{2} \pi y^2\,dy$

c) $\displaystyle\int_{0}^{4} \pi\sqrt{y}\,dy$

d) $\displaystyle\int_{0}^{2} \pi\sqrt{y}\,dy$

(2, 4) $y = x^2$

Section 6.3

____ **1.** Which definite integral is the volume of the solid obtained by revolving the region at the right about the *y*-axis using cylindrical shells?

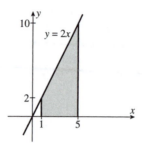

a) $\displaystyle\int_1^5 2\pi x^2\, dx$

b) $\displaystyle\int_1^5 4\pi x^2\, dx$

c) $\displaystyle\int_1^5 2\pi x\, dx$

d) $\displaystyle\int_2^{10} 4\pi x\, dx$

____ **2.** Which definite integral is the volume of the solid obtained by revolving the region at the right about the *x*-axis using cylindrical shells?

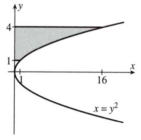

a) $\displaystyle\int_1^{16} 2\pi y^3\, dy$

b) $\displaystyle\int_1^{16} 2\pi(4 - \sqrt{x})\, dy$

c) $\displaystyle\int_1^4 2\pi y^2\, dy$

d) $\displaystyle\int_1^4 2\pi y^3\, dy$

Section 6.4

_____ **1.** How much work is done in lifting a 60-pound bag of bird seed 3 feet in the air?

  a)   20 ft-lbs
  b)   20g ft-lbs
  c)   180 ft-lbs
  d)   180g ft-lbs

_____ **2.** A particle moves along an x-axis from 2 m to 3 m pushed by a force of x^2 N (newtons). The definite integral that determines the amount of work done is:

  a) $\displaystyle\int_2^3 gx^3 \, dx$

  b) $\displaystyle\int_2^3 x \, dx$

  c) $\displaystyle\int_2^3 2\pi x^2 \, dx$

  d) $\displaystyle\int_2^3 x^2 \, dx$

Section 6.5

_____ **1.** The average value of $f(x) = 3x^2 + 1$ on the interval $[2, 4]$ is:

    a) 29   b) 66   c) 58   d) 36

_____ **2.** Which point x_1, x_2, x_3, or x_4, on the graph of $y = f(x)$ is the best choice to serve as the point guaranteed by the Mean Value Theorem for Integrals?

    a) x_1
    b) x_2
    c) x_3
    d) x_4

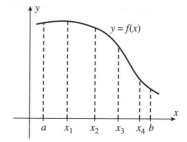

_____ **3.** Find a point c for the Mean Value Theorem for Integrals for $f(x) = x^2 - 2x$ on $[2, 5]$.

    a) 18   b) 6   c) $1 + \sqrt{7}$ d) no such c exists

_____ **4.** True or False:
The Mean Value Theorem for Integrals is a restatement of the Mean Value Theorem (of Chapter 4) but stated in terms of a definite integral instead of in terms of a derivative.

Section 7.1

____ **1.** Find the correct u and dv for integration by parts of $\int x^2 \cos 2x\, dx$.

   a) $u = x^2, dv = \cos 2x\, dx$
   b) $u = \cos 2x, dv = x^2\, dx$
   c) $u = x^2, dv = 2x\, dx$
   d) $u = x \cos 2x, dv = x\, dx$

____ **2.** Evaluate $\int 16t^3 \ln t\, dt$.

   a) $4t^4 + C$
   b) $4t^3 \ln t - 16t^3 + C$
   c) $4t^4 - t^3 \ln t + C$
   d) $4t^4 \ln t - t^4 + C$

____ **3.** Evaluate $\int_1^{e^2} \ln \sqrt{x}\, dx$.

   a) $e^2 + \dfrac{1}{2}$

   b) $\dfrac{e^2 + 1}{2}$

   c) $e + 1$

   d) $e + \dfrac{1}{2}$

____ **4.** The integration by parts rule corresponds to which differentiation formula?

   a) $(u + v)' = u' + v'$
   b) $(uv)' = u'v'$
   c) $(uv)' = uv' + u'v$
   d) $u\, dv = v\, du$

____ 1. By the methods of trigonometric integrals, $\int \sin^3 x \cos^2 x \, dx$ should be rewritten as:

    a) $\int \sin^3 x \, (1 - \sin^2 x) dx$

    b) $\int (1 - \cos^2 x)^{3/2} \cos^2 x \, dx$

    c) $\int (\sin^3 x)(\cos x) \cos x \, dx$

    d) $\int (1 - \cos^2 x) \cos^2 x \sin x \, dx$

____ 2. Using $\tan^2 x = \sec^2 x - 1$, $\int \tan^3 x \, dx =$

    a) $\frac{3}{2} \ln |\sec^2 x - 1| + C$

    b) $\frac{1}{2} \tan^2 x - \ln |\sec x| + C$

    c) $\frac{1}{2} \tan^2 x - \frac{1}{4} \sec^2 x + C$

    d) $\frac{1}{4} \tan^2 x + \ln |\sec x| + C$

____ 3. The first step to evaluate $\int \cos^4 x \, dx$ is to rewrite the integral as

    a) $\int \left(\frac{1 + \cos 2x}{2} \right)^2 dx$

    b) $\int \cos^3 x \cdot \cos x \, dx$

    c) $\int (1 - \sin^4 x) \, dx$

    d) $\int \cos^2 x \, (1 - \sin^2 x) \, dx$

____ 4. $\int \sin 3x \cos 6x \, dx =$

    a) $\int \frac{1}{2}(\sin(-3x) + \cos 9x) \, dx$

    b) $\int \frac{1}{2}(\sin(-3x) + \sin 9x) \, dx$

    c) $\int \frac{1}{2}(\cos(-3x) + \sin 9x) \, dx$

    d) $\int \frac{1}{2}(\cos(-3x) + \cos 9x) \, dx$

Section 7.3

_____ **1.** What trigonometric substitution should be made for $\int \frac{x^2}{\sqrt{x^2-25}}\, dx$?

a) $x = 5 \sin \theta$
b) $x = 5 \tan \theta$
c) $x = 5 \sec \theta$
d) No such substitution will evaluate the integral.

_____ **2.** Rewrite $\int \frac{\sqrt{10-x^2}}{x}\, dx$ using a trigonometric substitution.

a) $\int \frac{\cos \theta}{\sqrt{10} \sin \theta}\, d\theta$

b) $\int \frac{1}{\sqrt{10}} \sin \theta \cos^2 \theta\, d\theta$

c) $\int \frac{\sqrt{10}\ \cos^2 \theta}{\sin \theta}\, d\theta$

d) $\int \sqrt{10} \sin \theta \cos \theta\, d\theta$

_____ **3.** True or False:

A trigonometric substitution is necessary to evaluate $\int \frac{x}{\sqrt{x^2+10}}\, dx$.

_____ **4.** Evaluate: $\int_{1}^{\sqrt{2}} \frac{\sqrt{x^2-1}}{x}\, dx$.

a) $\frac{\sqrt{3}}{2} - \frac{\pi}{2}$

b) $\frac{\pi}{2} + \sqrt{2} - 1$

c) $\frac{\pi}{4} + \sqrt{2}$

d) $1 - \frac{\pi}{4}$

Section 7.4

_____ **1.** What is the partial fraction form of $\dfrac{2x + 3}{(x - 2)(x + 2)}$?

     a) $\dfrac{Ax}{x - 2} + \dfrac{B}{x + 2}$

     b) $\dfrac{A}{x - 2} + \dfrac{B}{x + 2}$

     c) $\dfrac{Ax}{x - 2} + \dfrac{Bx}{x + 2}$

     d) $\dfrac{Ax + B}{x - 2} + \dfrac{Cx + D}{x + 2}$

_____ **2.** What is the partial fraction form of $\dfrac{2x + 3}{(x^2 + 1)^2}$?

     a) $\dfrac{Ax + B}{(x^2 + 1)^2}$

     b) $\dfrac{A}{x^2 + 1} + \dfrac{B}{(x^2 + 1)^2}$

     c) $\dfrac{A}{x^2 + 1} + \dfrac{Bx + C}{(x^2 + 1)^2}$

     d) $\dfrac{Ax + B}{x^2 + 1} + \dfrac{Cx + D}{(x^2 + 1)^2}$

_____ **3.** Evaluate $\displaystyle\int \dfrac{2}{(x + 1)(x + 2)}\, dx$ using partial fractions.

     a) $2 \ln |x + 1| - 2 \ln |x + 2| + C$
     b) $\ln |x + 1| + 3 \ln |x + 2| + C$
     c) $3 \ln |x + 1| - \ln |x + 2| + C$
     d) $\ln |x + 1| + \ln |x + 2| + C$

_____ **4.** What rationalizing substitution should be made for $\displaystyle\int \dfrac{\sqrt[3]{x} + 2}{\sqrt[3]{x} + 1}\, dx$?

     a) $u = x$

     b) $u = \sqrt[3]{x}$

     c) $u = \sqrt[3]{x} + 2$

     d) $u = \sqrt[3]{x} + 1$

_____ **5.** What rationalizing substitution should be made for $\displaystyle\int \dfrac{1}{\sqrt{x} + \sqrt[5]{x}}\, dx$?

     a) $u = x$

     b) $u = \sqrt{x}$

     c) $u = \sqrt[5]{x}$

     d) $u = \sqrt[10]{x}$

Section 7.5

_____ **1.** True or False:

$\int x^2 \sqrt{x^2 - 4}\, dx$ should be evaluated using integration by parts.

_____ **2.** True or False:

A straight (simple) substitution may be used to evaluate $\int \frac{2 + \ln x}{x}\, dx$.

_____ **3.** True or False:

Partial fractions may be used for $\int \frac{1}{\sqrt{x} + 2\sqrt{x + 3}}\, dx$.

_____ **4.** True or False:

$\int \frac{\sqrt{x^2 + 1}}{x^3}\, dx$ may be solved using a trigonometric substitution.

_____ **5.** Sometimes, Always, or Never:

The sum of two elementary functions is an elementary function.

_____ **6.** Sometimes, Always, or Never:

The antiderivative of an elementary function is elementary.

Section 7.6

_____ **1.** What is the number of the integral formula in your text's Table of Integrals that may be used to evaluate $\int \dfrac{dx}{x\sqrt{2 - x^2}}$?

    a) #35   b) #36   c) #18   d) #24

_____ **2.** What is the number of the integral formula in your text's Table of Integrals that may be used to evaluate $\int \dfrac{x^2 + 2x + 1}{\sqrt{x^2 + 2x + 5}}\,dx$?

    a) #37   b) #38   c) #26   d) #46

_____ **3.** Use a Table of Integrals to evaluate $\int \dfrac{\sqrt{x^2 - 4}}{x^2}\,dx$.

    a) $-\dfrac{\sqrt{x^2 - 4}}{x} + \ln\left|x + \sqrt{x^2 - 4}\right| + C$

    b) $\ln\left|x + \sqrt{x^2 - 4}\right| + C$

    c) $\dfrac{\sqrt{x^2 - 4}}{4x} + C$

    d) $\dfrac{x}{2}\sqrt{x^2 - 4} - 2\ln\left|x + \sqrt{x^2 - 4}\right| + C$

_____ **4.** Use a Table of Integrals to evaluate $\displaystyle\int_0^1 x\tan^{-1} x\,dx$.

    a) $2 - \dfrac{\pi}{4}$ b) $2 + \dfrac{\pi}{4}$ c) $\dfrac{\pi}{4} - \dfrac{1}{2}$ d) $\dfrac{\pi}{4} + 1$

Section 7.7

_____ **1.** Estimate $\int_{-1}^{4} (x^3 - x^2)\, dx$ using the Left Endpoint Rule and $n = 5$.

a) $\frac{20}{3}$ b) 20   c) 24   d) 30

_____ **2.** If $f(x)$ is integrable and concave upward on $[a, b]$, then an estimate of $\int_{a}^{b} f(x)\, dx$ using the Trapezoidal Rule will always be:

a) too large
b) too small
c) exact
d) within $\frac{b - a}{12n}$ of the exact value

_____ **3.** An estimate of $\int_{-1}^{3} x^4\, dx$ using Simpson's Rule with $n = 4$ gives:

a) $\frac{242}{5}$ b) $\frac{148}{3}$ c) $\frac{152}{3}$ d) $\frac{244}{5}$

_____ **4.** Using $\frac{K(b - a)^3}{24n^2}$ as an estimator, find the maximum error in estimating $\int_{-1}^{3} x^4\, dx$ with the Midpoint Rule and 8 subintervals.

a) $\frac{1}{2}$ b) 9   c) $\frac{9}{2}$ d) $\frac{9}{8}$

_____ **1.** True or False:

$\int_{-2}^{3} |x|\, dx$ is an improper integral.

_____ **2.** By definition the improper integral $\int_{1}^{e} \frac{1}{x \ln x}\, dx =$

    a) $\lim_{t \to 1^+} \int_{t}^{e} \frac{1}{x \ln x}\, dx$

    b) $\lim_{t \to e^-} \int_{1}^{t} \frac{1}{x \ln x}\, dx$

    c) $\int_{1}^{2} \frac{1}{x \ln x}\, dx + \int_{2}^{e} \frac{1}{x \ln x}\, dx$

    d) It is not an improper integral.

_____ **3.** $\int_{0}^{2} \frac{x-2}{\sqrt{4x-x^2}}\, dx =$

    a) 0   b) 2   c) -2 d) does not exist

_____ **4.** Suppose we know $\int_{1}^{\infty} f(x)\, dx$ diverges and that $f(x) \geq g(x) \geq 0$ for all $x \geq 1$.

What conclusion can be made about $\int_{1}^{\infty} g(x)dx$?

    a) it converges to the value of $\int_{1}^{\infty} f(x)dx$

    b) it converges, but we don't know to what value

    c) it diverges

    d) no conclusion (it could diverge or converge)

Section 8.1

___ **1.** A definite integral for the length of $y = x^3$, $1 \le x \le 2$ is:

a) $\displaystyle\int_1^2 \sqrt{1 + x^3}\, dx$

b) $\displaystyle\int_1^2 \sqrt{1 + x^6}\, dx$

c) $\displaystyle\int_1^2 \sqrt{1 + 9x^4}\, dx$

d) $\displaystyle\int_1^2 \sqrt{1 + 3x^2}\, dx$

___ **2.** The arc length for the function $f(x) = 3x + 2$ from $x = 1$ to $x = 5$ is

a) $2\sqrt{10}$
b) $5\sqrt{2}$
c) $10\sqrt{2}$
d) $4\sqrt{10}$

___ **3.** The arc length function for $y = x^2$, $1 \le x \le 3$ is $s(x) =$

a) $\displaystyle\int_1^x \sqrt{1 + 4x^2}\, dx$

b) $\displaystyle\int_1^x \sqrt{1 + 4t^2}\, dt$

c) $\displaystyle\int_1^x \sqrt{1 + t^4}\, dt$

d) $\displaystyle\int_1^x \sqrt{1 + 4t^4}\, dt$

Section 8.2

___ **1.** An integral for the surface area obtained by rotating $y = \sin x + \cos x$, $0 \le x \le \frac{\pi}{2}$, about the x-axis is:

a) $\displaystyle\int_0^{\pi/2} 2\pi(\sin x + \cos x)\sqrt{2 - 2\sin x \cos x}\ dx$

b) $\displaystyle\int_0^{\pi/2} 2\pi(\sin x + \cos x)\sqrt{1 + \sin^2 x - \cos^2 x}\ dx$

c) $\displaystyle\int_0^{\pi/2} 2\pi(\cos x - \sin x)\sqrt{1 + \sin^2 x - \cos^2 x}\ dx$

d) $\displaystyle\int_0^{\pi/2} 2\pi(\cos x - \sin x)\sqrt{2 - 2\sin x \cos x}\ dx$

___ **2.** A hollow cylinder with no ends of radius 3 in. and height 4 in. is a surface of revolution. Its surface area may be expressed as:

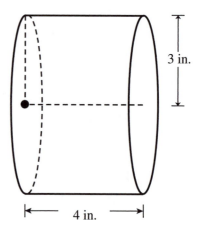

3 in.

4 in.

a) $\displaystyle\int_0^3 2\pi(4)\sqrt{1 + 0}\ dx$

b) $\displaystyle\int_0^3 2\pi(3)\sqrt{1 + 0}\ dx$

c) $\displaystyle\int_0^4 2\pi(3)\sqrt{1 + 0}\ dx$

d) $\displaystyle\int_0^4 2\pi(4)\sqrt{1 + 0}\ dx$

Section 8.3

_____ **1.** Find a definite integral for the hydrostatic force exerted by a liquid with density 800 kg/m³ on the semicircular plate in the center of the figure.

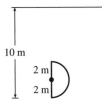

a) $\displaystyle\int_{-2}^{2} 800(9.8)[8 - y]\left[2\sqrt{4 - y^2}\right] dy$

b) $\displaystyle\int_{-2}^{2} 800(9.8)[8 - y]\sqrt{4 - y^2}\, dy$

c) $\displaystyle\int_{-2}^{2} 800(9.8)[10 - y]\left[2\sqrt{4 - y^2}\right] dy$

d) $\displaystyle\int_{-2}^{2} 800(9.8)[10 - y]\sqrt{4 - y^2}\, dy$

_____ **2.** The x-coordinate of the center of mass of the region bounded by $y = \frac{1}{x}$, $x = 1, x = 2, y = 0$ is:

a) $\ln 2$ b) $\dfrac{1}{\ln 2}$ c) 1 d) $2\ln 2$

_____ **3.** The y-coordinate of the center of mass of the region bounded by $y = \sqrt{x - 1}, x = 1, x = 5, y = 0$ is:

a) $\dfrac{3}{14}$ b) $\dfrac{3}{4}$ c) $\dfrac{14}{3}$ d) 4

_____ **4.** A lamina with center of mass $\left(\frac{3}{8}, \frac{6}{5}\right)$ and uniform density ρ is at the right. The moments about the x- and y-axes are, respectively,

_____ and _____.

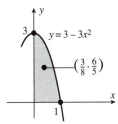

a) $\dfrac{5}{12}\rho$ and $\dfrac{4}{3}\rho$ b) $\dfrac{4}{3}\rho$ and $\dfrac{5}{12}\rho$

c) $\dfrac{12}{5}\rho$ and $\dfrac{3}{4}\rho$ d) $\dfrac{3}{4}\rho$ and $\dfrac{12}{5}\rho$

Section 8.4

_____ **1.** True or False:
In many applications of definite integrals, the integral is used to compute the total amount of a varying quantity.

_____ **2.** Find a definite integral for the consumer surplus if 180 units are available and the demand function is $p(x) = 3000 - 10x$.

a) $\int_0^{180} (3000 - 10x)dx$ b) $\int_0^{180} (1800 - 10x)dx$

c) $\int_0^{180} (1200 - 10x)dx$ d) $\int_0^{180} (180 - 10x)dx$

_____ **3.** The dye dilution method is used to measure the cardiac output of a pig's heart with 10 mg of dye. After t seconds the dye concentration is modeled by $c(t) = -\frac{1}{2}t^2 + 4t$ for $0 \le t \le 6$ seconds. Find the cardiac output.

a) $\frac{5}{18}$ L/s b) $\frac{3}{4}$ L/s

c) $\frac{10}{39}$ L/s d) 36 L/s

Section 8.5

_____ 1. True or False:
The function whose graph is given
at the right is a probability density
function.

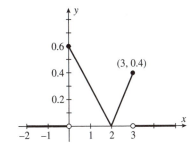

_____ 2. For the probability density function

$$f(x) = \begin{cases} 0 & \text{for } x < 0 \\ 3\,e^{-3x} & \text{for } x \geq 0 \end{cases}$$

the mean is

a) $\displaystyle\int_0^\infty 3e^{-3x}dx$ b) $\displaystyle\int_0^\infty 3e^{-3x+1}dx$ c) $\displaystyle\int_0^\infty 3e^{-3x-1}dx$ d) $\displaystyle\int_0^\infty 3xe^{-3x}dx$

_____ 3. For a normal probability density function $f(x) = \dfrac{1}{\sigma\sqrt{2\pi}}\,e^{-(x-\mu)^2/(2\sigma^2)}$,
the probability that $x \geq \mu$ is:

a) 1   b) $\dfrac{1}{2}$ c) $\dfrac{1}{\sigma}$ d) $\dfrac{1}{2\sigma^2}$

_____ 4. For the probability density function in question 2, the probability of
observing a value of greater than 2 is:

a) $\displaystyle\int_0^2 3e^{-3x}\,dx$ b) $\displaystyle\int_0^\infty 3e^{-3x}\,dx$ c) $\displaystyle\int_2^\infty 3e^{-3x}\,dx$ d) $\displaystyle\int_{-2}^2 3e^{-3x}\,dx$

_____ **1.** $(y''')^2 + 6xy - 2y' + 7x = 0$ is a differential equation of order _____.

    a) 1   b) 2   c) 3   d) 4

_____ **2.** True or False:

    $y = xe^x$ is a solution to $y' = y + \dfrac{y}{x}$.

_____ **3.** The general solution to $y' = \dfrac{2x}{y}$ is $y = \sqrt{2x^2 + C}$. Find a solution to the initial value problem $y' = \dfrac{2x}{y}$, $y(0) = 4$.

    a) $y = 4\sqrt{2x^2 + 1}$

    b) $y = \sqrt{2x^2 + 4}$

    c) $y = 2\sqrt{2x^2 + 4}$

    d) $y = \sqrt{2x^2 + 16}$

_____ **4.** True or False:

    The function $y = f(x)$ graphed at the right could be a solution to $y' = x + y$.

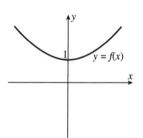

Section 9.2

_____ **1.** True or False:
The figure at the right is the
graph of the direction field
for $y' = (y - 1)x^2$.

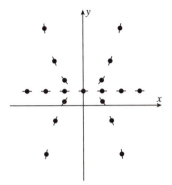

_____ **2.** The graph at the right
is the direction field for:

a) $y' = x - y$
b) $y' = xy$
c) $y' = x + y$
d) $y' = \frac{x}{y}$

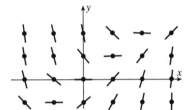

_____ **3.** The first point determined by Euler's method for $\frac{dy}{dx} = x - 4y$, $y(11) = 2$
with step size 0.5 is:

a) $(11.0, 3.5)$ b) $(11.5, 3.5)$ c) $(11.0, 3.75)$ d) $(11.5, 3.75)$

Section 9.3

_____ **1.** A differential equation is separable if it can be written in the form:

    a) $\dfrac{dy}{dx} = f(x) + g(y)$ b) $\dfrac{dy}{dx} = f(g(x, y))$

    c) $\dfrac{dy}{dx} = \dfrac{f(x)}{g(y)}$ d) $\dfrac{dy}{dx} = f(x)^{g(y)}$

_____ **2.** True or False:
    $y' + xe^y = e^{x+y}$ is separable.

_____ **3.** The general solution to $\dfrac{dy}{dx} = \dfrac{1}{xy}$ (for $x,\ y > 0$) is:

    a) $y = \ln x + C$ b) $y = \sqrt{\ln x + C}$

    c) $y = \ln(\ln x + C)$ d) $y = \sqrt{\ln x^2 + C}$

_____ **4.** The solution to $y' = -y^2$ with $y(1) = \frac{1}{3}$ is:

    a) $y = \dfrac{1}{x + 2}$ b) $y = \ln x + \dfrac{1}{3}$

    c) $y = \dfrac{1}{2x^2} - \dfrac{1}{6}$ d) $y = \dfrac{1}{2x} - \dfrac{1}{6}$

Section 9.4

_____ **1.** The solution to $\frac{dy}{dt} = ky$, where k is a constant and P_0 is the value of y when $t = 0$, is
   a) $y = e^{kt}$
   b) $y = xe^{kt}$
   c) $y = ke^{P_0 t}$
   d) $y = P_0 e^{kt}$

_____ **2.** The 2000 Census showed that the town of Grandville has a population of 1020 residents. The 2010 Census was 1170. Assuming natural growth of the population of Grandville, what is the population in the year 2025?
   a) 1395
   b) 1410
   c) 1437
   d) 1503

_____ **3.** True or False:
   A population that follows a logistic model will never reach its carrying capacity.

_____ **4.** A population of ants on an island is modeled by $P(t) = \dfrac{80000}{1 + 2000e^{-0.5t}}$, where t is the number of years since 1970. How long did it take the population to reach half that of the island's carrying capacity?

   a) 7.6 years   b) 15.2 years   c) 22.8 years   d) 30.4 years

Section 9.5

____ **1.** True or False:

$xy' + 6x^2y = 10 - x^3$ is linear.

____ **2.** What is the integrating factor for $xy' + 6x^2y = 10 - x^3$?

a) e^{3x^2} b) e^{6x} c) e^{3x^3} d) e^{6x^2}

____ **3.** The solution to $x\dfrac{dy}{dx} - 2y = x^3$ is:

a) $y = e^{x^3} + Cx^2$ b) $y = e^{x^3} + x^2 + C$

c) $y = x^3 + x^2 + C$ d) $y = x^3 + Cx^2$

Section 9.6

____ **1.** True or False:

Predator-prey population models are the solutions to two differential equations.

____ **2.** The populations of sheep and wolves on an island is governed by a predator-prey model with constants $k = 0.03$, $r = 0.2$, $a = 0.0005$, and $b = 0.0001$. What is the equilibrium solution for these populations?

   a) 60 wolves and 2000 sheep
   b) 30 wolves and 3000 sheep
   c) 30 wolves and 2000 sheep
   d) 60 wolves and 3000 sheep

____ **3.** If the predator population increases in a predator-prey model, the prey population will

   a) decrease
   b) increase
   c) remain constant

Section 10.1

_____ **1.** The graph of $x = 2 + 3t,\ y = 4 - t$ is a:

    a) circle   b) ellipse

    c) line   d) parabola

_____ **2.** The graph of $x = \cos t,\ y = \sin^2 t$ is:

    a)

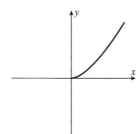

    b)

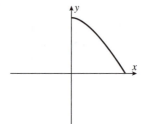

    c)

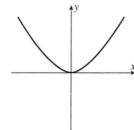

    d)

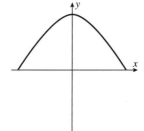

_____ **3.** Elimination of the parameter in $x = 2t^{3/2}, y = t^{2/3}$ gives:

    a) $x^4 = 16y^9$ b) $16x^4 = y^4$

    c) $x^3 = 8y^4$ d) $8x^3 = y^4$

Section 10.2

_____ **1.** Find $\dfrac{dy}{dx}$ if $x = \sqrt{t},\ y = \sin 2t$.

     a) $\dfrac{4 \cos t}{\sqrt{t}}$ b) $\dfrac{\cos 2t}{\sqrt{t}}$ c) $\dfrac{\cos 2t}{2\sqrt{t}}$ d) $4\sqrt{t} \cos 2t$

_____ **2.** Find $\dfrac{d^2y}{dx^2}$ for $x = 3t^2 + 1,\ y = t^6 + 6t^5$.

     a) $t^4 + 5t^3$ b) $4t^3 + 15t^2$

     c) $\dfrac{2}{3}t^2 + \dfrac{5}{2}t$ d) $t^3 + \dfrac{1}{2}t^2$

_____ **3.** The slope of the tangent line at the point where $t = \dfrac{\pi}{6}$ to the curve
     $y = \sin 2t,\ x = \cos 3t$ is:

     a) $\dfrac{1}{3}$ b) $-\dfrac{1}{3}$ c) 3 d) -3

_____ **4.** A definite integral for the area under the curve described by $x = t^2 + 1$,
     $y = 2t,\ 0 \le t \le 1$ is:

     a) $\displaystyle\int_0^1 (2t^3 + 2t)\,dt$ b) $\displaystyle\int_0^1 4t^2\,dt$

     c) $\displaystyle\int_0^1 (2t^2 + 2)\,dt$ d) $\displaystyle\int_0^1 4t\,dt$

_____ **5.** The length of the curve given by $x = 3t^2 + 2,\ y = 2t^3,\ t \in [0, 1]$ is:

     a) $4\sqrt{2} - 2$ b) $8\sqrt{2} - 1$

     c) $\dfrac{2}{3}\left(2\sqrt{2} - 1\right)$ d) $\sqrt{2} - 1$

_____ **6.** Find a definite integral for the area of the surface of revolution about the
     x-axis obtained by rotating the curve $y = t^2, x = 1 + 3t, 0 \le t \le 2$.

     a) $\displaystyle\int_0^2 2\pi t^2 \sqrt{t^4 + 9t^2 + 6t + 1}\,dt$

     b) $\displaystyle\int_0^2 2\pi t^2 \sqrt{4t^2 + 9}\,dt$

     c) $\displaystyle\int_0^2 2\pi(2t)\sqrt{t^4 + 9t^2 + 6t + 1}\,dt$

     d) $\displaystyle\int_0^2 2\pi(2t)\sqrt{4t^4 + 9}\,dt$

Section 10.3

_____ **1.** The polar coordinates of the
point P in the figure at the right are:

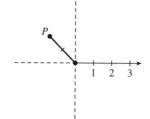

a) $\left(-2, \frac{\pi}{4}\right)$　　　　b) $\left(-2, \frac{3\pi}{4}\right)$

c) $\left(2, -\frac{\pi}{4}\right)$　　　　d) $\left(2, \frac{3\pi}{4}\right)$

_____ **2.** Polar coordinates of the point with rectangular coordinates $(5, 5)$ are:

a) $(25, 0)$　　　　b) $(5, \frac{\pi}{4})$　　　　c) $(5\sqrt{2}, \frac{\pi}{4})$　　　　d) $(50, -\frac{\pi}{4})$

_____ **3.** Rectangular coordinates of the point with polar coordinates $\left(-1, \frac{3\pi}{2}\right)$ are:

a) $(-1, 0)$　　　　b) $(0, 1)$　　　　c) $(0, -1)$　　　　d) $(1, 0)$

_____ **4.** The graph of $\theta = 2$ in polar coordinates is a:

a) circle　　　　b) line　　　　c) spiral　　　　d) 3-leaved rose

_____ **5.** Which is the best graph of $r = 1 - \sin \theta$ for $0 \leq \theta \leq \pi$?

a)

b)

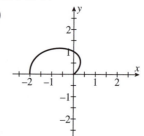

c)

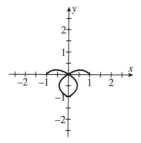

d)

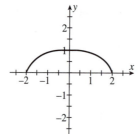

_____ **6.** The slope of the tangent line to $r = \cos \theta$ at $\theta = \frac{\pi}{3}$ is:

a) $\sqrt{3}$　　　　b) $\frac{1}{\sqrt{3}}$　　　　c) $-\sqrt{3}$　　　　d) $-\frac{1}{\sqrt{3}}$

Section 10.4

____ **1.** The area of the region bounded by $\theta = \frac{\pi}{3}$, $\theta = \frac{\pi}{4}$, and $r = \sec\theta$ is:

a) $\frac{1}{2}\left(\sqrt{3} - 1\right)$ b) $\sqrt{3}$

c) $2(\sqrt{3} - 1)$ d) $2\sqrt{3}$

____ **2.** The area of the shaded region is given by:

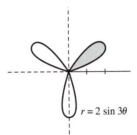
$r = 2\sin 3\theta$

a) $\displaystyle\int_0^{\pi/2} \sin 3\theta \, d\theta$

b) $\displaystyle\int_0^{\pi/2} 2\sin^2 3\theta \, d\theta$

c) $\displaystyle\int_0^{\pi/3} \sin 3\theta \, d\theta$

d) $\displaystyle\int_0^{\pi/3} 2\sin^2 3\theta \, d\theta$

____ **3.** The length of the arc $r = e^\theta$ for $0 \le \theta \le \pi$ is given by:

a) $\displaystyle\int_0^{\pi} \sqrt{e^\theta} \, d\theta$

b) $\displaystyle\int_0^{\pi} 2e^{2\theta} \, d\theta$

c) $\displaystyle\int_0^{\pi} \sqrt{2e^\theta} \, d\theta$

d) $\displaystyle\int_0^{\pi} \sqrt{2}e^\theta \, d\theta$

1. The graph of $\frac{y^2}{16} = 1 + \frac{x^2}{25}$ is:

a)

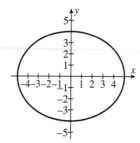

b)

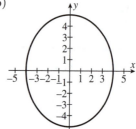

c)

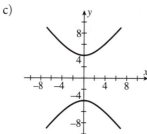

d)

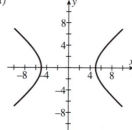

2. The center of $4x^2 + 8x - 3y^2 + 12y + 4 = 0$ is

    a) $(1, -2)$ b) $(-1, 2)$ c) $(2, -1)$ d) $(-2, 1)$

3. The conic section whose equation is $y(3 - y) + 4x^2 = 2x(1 + 2x) - y$ is a

    a) parabola   b) ellipse   c) hyperbola   d) None of these

_____ **1.** The figure at the right shows one point P on a conic and the distances of P to the focus and the directrix of a conic. What type of conic is it?

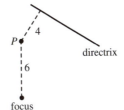

   a) parabola
   b) ellipse
   c) hyperbola
   d) not enough information is provided to answer the question.

_____ **2.** The polar equation of the conic with eccentricity 3 and directrix $x = -7$ is:

   a) $r = \dfrac{21}{1 + 3\cos\theta}$

   b) $r = \dfrac{21}{1 - 3\cos\theta}$

   c) $r = \dfrac{21}{1 + 3\sin\theta}$

   d) $r = \dfrac{21}{1 - 3\sin\theta}$

_____ **3.** The directrix of the conic given by $r = \dfrac{6}{2 + 10\sin\theta}$ is:

   a) $x = \dfrac{5}{3}$ b) $x = \dfrac{3}{5}$ c) $y = \dfrac{5}{3}$ d) $y = \dfrac{3}{5}$

_____ **4.** The polar form for the graph at the right is:

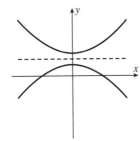

   a) $r = \dfrac{ed}{1 + e\cos\theta}$

   b) $r = \dfrac{ed}{1 - e\cos\theta}$

   c) $r = \dfrac{ed}{1 + e\sin\theta}$

   d) $r = \dfrac{ed}{1 - e\sin\theta}$

Section 11.1

_____ **1.** $\lim\limits_{n\to\infty} \dfrac{n^2 + 3n}{2n^2 + n + 1} =$

    a)   0   b) $\dfrac{1}{2}$ c)   1   d) ∞

_____ **2.** Sometimes, Always, or Never:
If $a_n \geq b_n \geq 0$ for all n and $\{b_n\}$ diverges, then $\{a_n\}$ diverges.

_____ **3.** Sometimes, Always, or Never:
If $a_n \geq b_n \geq c_n$ for all n and both $\{a_n\}$ and $\{c_n\}$ converge, then $\{b_n\}$ converges.

_____ **4.** Sometimes, Always, or Never:
If $\{a_n\}$ is increasing and bounded above, then $\{a_n\}$ converges.

_____ **5.** True or False:
$a_n = \dfrac{(-1)^n}{n^2}$ is monotonic.

_____ **6.** $\lim\limits_{n\to\infty} \dfrac{\arctan n}{2} =$

    a) $\dfrac{\pi}{4}$ b) $\dfrac{\pi}{2}$ c) π d)   does not exist

_____ **7.** $\lim\limits_{n\to\infty} \dfrac{2^n}{n!} =$

    a)   2   b)   1   c)   0   d)   does not exist

_____ **8.** The fourth term of $\{a_n\}$ defined by $a_1 = 3$, and $a_{n+1} = \dfrac{2}{3}a_n$ for $n = 1, 2, 3, \dots$ is:

    a) $\dfrac{16}{81}$ b) $\dfrac{16}{27}$ c) $\dfrac{8}{27}$ d) $\dfrac{8}{9}$

Section 11.2

_____ **1.** The third partial sum of $\sum_{n=1}^{\infty} \frac{n}{n^2+1}$ is:

     a) $\frac{3}{10}$ b) $\frac{1}{2}$ c) $\frac{3}{5}$ d) $\frac{6}{5}$

_____ **2.** True or False:

     $\sum_{n=1}^{\infty} a_n$ converges mean $\lim_{k \to \infty} \sum_{n=1}^{k} a_n$ exists.

_____ **3.** True or False:

     $\sum_{n=1}^{\infty} a_n$ converges if and only if $\lim_{n \to \infty} a_n = 0$.

_____ **4.** True or False:

     If $\sum_{n=1}^{\infty} a_n$ converges and $\sum_{n=1}^{\infty} b_n$ converges, then $\sum_{n=1}^{\infty} (a_n - b_n)$ converges.

_____ **5.** The harmonic series is:

     a) $1 + 2 + 3 + 4 + \dots$ b) $1 + \frac{1}{2} + \frac{1}{3} + \frac{1}{4} + \dots$

     c) $1 + \frac{1}{2} + \frac{1}{4} + \frac{1}{8} + \dots$ d) $1 - \frac{1}{2} + \frac{1}{4} - \frac{1}{8} + \dots$

_____ **6.** True or False:

     $\sum_{n=1}^{\infty} \frac{n}{n+1}$ converges.

_____ **7.** $\sum_{n=1}^{\infty} 2\left(\frac{1}{4}\right)^n$ converges to:

     a) $\frac{9}{4}$ b) 2 c) $\frac{2}{3}$ d) the series diverges

_____ **8.** True or False:

     $-3 + 1 - \frac{1}{3} + \frac{1}{9} - \frac{1}{27} + \dots$ is a geometric series.

Section 11.3

_____ **1.** For what values of p does the series $\sum_{n=1}^{\infty} \frac{1}{(n^2)^p}$ converge?

     a) $p > -\frac{1}{2}$ b) $p < -\frac{1}{2}$ c) $p > \frac{1}{2}$ d) $p < \frac{1}{2}$

_____ **2.** True or False:

If $f(x)$ is continuous and decreasing, $f(n) = a_n$ for all $n = 1, 2, 3, \ldots,$
and $\int_{1}^{\infty} f(x)\,dx = M$, then $\sum_{n=1}^{\infty} a_n = M$.

_____ **3.** True or False:

By the Integral Test, $\sum_{n=2}^{\infty} \frac{1}{n \ln n}$ converges.

_____ **4.** True or False:

$\sum_{n=1}^{\infty} \frac{1}{n^{2/3}}$ is a convergent series.

_____ **5.** True or False:

$\sum_{n=1}^{\infty} \frac{n + 2}{(n^2 + 4n + 1)^2}$ is a convergent series.

_____ **6.** For $s = \sum_{n=1}^{\infty} \frac{1}{n^3}$, an upper bound estimate for $s - s_6$ (where s_6 is the sixth
partial sum) is:

     a) $\int_{1}^{\infty} x^{-3}\,dx$ b) $\int_{6}^{\infty} x^{-3}\,dx$

     c) $\int_{7}^{\infty} x^{-3}\,dx$ d)   the series does not converge

Section 11.4

_____ **1.** Sometimes, Always, or Never:

If $0 \leq a_n \leq b_n$ for all n and $\sum_{n=1}^{\infty} a_n$ diverges, then $\sum_{n=1}^{\infty} b_n$ converges.

_____ **2.** True or False:

$\sum_{n=1}^{\infty} \frac{n+1}{n^3}$ is a convergent series.

_____ **3.** True or False:

$\sum_{n=1}^{\infty} \frac{\sqrt{n} + \sqrt[3]{n}}{n^{2/3} + n^{3/2} + 1}$ is a convergent series.

_____ **4.** True or False:

$\sum_{n=1}^{\infty} \frac{\cos^2(2^n)}{2^n}$ is a convergent series.

_____ **5.** $s = \sum_{n=1}^{\infty} \frac{1}{n \cdot 2^n}$ converges by the Comparison Test, comparing it to $\sum_{n=1}^{\infty} \frac{1}{2^n}$.

Using this information, make an estimate of the difference between s and its third partial sum.

a) $\frac{1}{16}$ b) $\frac{1}{8}$ c) $\frac{1}{32}$ d) 1

Section 11.5

_____ **1.** True or False:

$$\sum_{n=1}^{\infty} \frac{(-1)^{n+1} \ln n}{n^2} \text{ is a convergent series.}$$

_____ **2.** True or False:

$$\sum_{n=1}^{\infty} \frac{(-1)^n}{\sqrt[4]{n+1}} \text{ is a convergent series.}$$

_____ **3.** For what value of n is the nth partial sum within 0.01 of the value of $\sum_{n=1}^{\infty} \frac{(-1)^n}{2^n}$? (Choose the smallest such n.)

    a) $n = 4$ b) $n = 6$ c) $n = 8$ d) $n = 10$

_____ **4.** True or False:

The Alternating Series Test may be applied to determine the convergence of $\sum_{n=1}^{\infty} \frac{2 + (-1)^n}{2n^2}$.

_____ **1.** True or False:

If $\sum_{n=1}^{\infty} a_n$ converges absolutely, then it converges conditionally.

_____ **2.** True or False:

If $\lim_{n \to \infty} \left| \frac{a_n}{a_{n+1}} \right| = 3$, then $\sum_{n=1}^{\infty} a_n$ converge absolutely.

_____ **3.** True or False:

Every series must do one of these: converge absolutely, converge conditionally, or diverge.

_____ **4.** Which is true about the series $\sum_{n=1}^{\infty} 2^{-n} n!$?

   a) diverges
   b) converges absolutely
   c) converges conditionally
   d) converges, but not absolutely and not conditionally

_____ **5.** Which is true about the series $\sum_{n=1}^{\infty} \frac{(-5)^{n+1}}{n^n}$?

   a) diverges
   b) converges absolutely
   c) converges conditionally
   d) converges, but not absolutely and not conditionally

_____ **6.** True or False:

The series $\sum_{n=1}^{\infty} \frac{(-1)^n}{n^3}$ converges to a number that we will call s.

All rearrangements of $\sum_{n=1}^{\infty} \frac{(-1)^n}{n^3}$ also converge to s.

Section 11.7

____ **1.** True or False:
$$\sum_{n=2}^{\infty} \frac{1}{(\ln n)^n} \text{ converges.}$$

____ **2.** True or False:
$$\sum_{n=1}^{\infty} \frac{6}{7n + 8} \text{ converges.}$$

____ **3.** True or False:
$$\sum_{n=1}^{\infty} \frac{(-1)^n \sqrt{n}}{n + 3} \text{ converges.}$$

____ **4.** True or False:
$$\sum_{n=1}^{\infty} \frac{e^n}{n!} \text{ converges.}$$

____ **5.** True or False:
$$\sum_{n=1}^{\infty} \frac{(-1)^n}{\sqrt[4]{n}} \text{ converges conditionally.}$$

____ **6.** True or False:
$$\sum_{n=1}^{\infty} \frac{(-1)^n}{(1.1)^n} \text{ converges absolutely.}$$

Section 11.8

_____ **1.** Sometimes, Always, or Never:

The interval of convergence of a power series $\sum_{n=0}^{\infty} a_n(x - c)^n$ is an open interval $(c - R, c + R)$. (When $R = 0$ we mean $\{c\}$ and when $R = \infty$ we mean $(-\infty, \infty)$.)

_____ **2.** True or False:

If a number p is in the interval of convergence of $\sum_{n=0}^{\infty} a_n x^n$ then so is the number $\frac{p}{2}$.

_____ **3.** For $f(x) = \sum_{n=0}^{\infty} \frac{(x - 1)^n}{3^n}, f(3) =$

    a) 0   b) 2

    c) 3   d) $f(3)$ does not exist

_____ **4.** The interval of convergence of $\sum_{n=1}^{\infty} \frac{x^n}{\sqrt{n}}$ is:

    a) $[-1, 1]$ b) $[-1, 1)$ c) $(-1, 1]$ d) $(-1, 1)$

_____ **5.** The radius of convergence of $\sum_{n=0}^{\infty} \frac{n(x - 5)^n}{3^n}$ is:

    a) $\frac{1}{3}$ b) 1   c) 3   d) ∞

___ **1.** Given that $e^x = \sum_{n=0}^{\infty} \frac{x^n}{n!}$ for all x, a power series for xe^{x^2} is:

a) $\sum_{n=0}^{\infty} \frac{x^2}{n!}$

b) $\sum_{n=0}^{\infty} \frac{x^{2n}}{n!}$

c) $\sum_{n=0}^{\infty} \frac{x^{2n+1}}{n!}$

d) $\sum_{n=0}^{\infty} \frac{x^{2n}}{(n+1)!}$

___ **2.** For $f(x) = \sum_{n=0}^{\infty} \frac{x^{2n}}{n!}$, $f'(x) =$

a) $\sum_{n=1}^{\infty} \frac{x^{2n-1}}{n!}$

b) $\sum_{n=1}^{\infty} \frac{2x^{2n-1}}{(n-1)!}$

c) $\sum_{n=1}^{\infty} \frac{2^n x^{2n-1}}{n!}$

d) $\sum_{n=1}^{\infty} \frac{(2n-1)x^{2n-1}}{(n-1)!}$

___ **3.** Using $\frac{1}{1-x} = \sum_{n=0}^{\infty} x^n$ for $|x| < 1$, $\int \frac{x}{1-x^2} dx =$

a) $\sum_{n=0}^{\infty} \frac{x^{2n}}{2n}$

b) $\sum_{n=0}^{\infty} \frac{x^{2n+1}}{2n+1}$

c) $\sum_{n=0}^{\infty} \frac{x^{2n+2}}{2n+2}$

d) $\sum_{n=0}^{\infty} x^{2n+1}$

___ **4.** Using $\frac{1}{1-x} = \sum_{n=0}^{\infty} x^n$ for $|x| < 1$ and differentiation, find a power series for $\frac{1}{(1+x)^2}$.

a) $\sum_{n=1}^{\infty} \frac{x^{n-1}}{n}$

b) $\sum_{n=1}^{\infty} -n\, x^{n-1}$

c) $\sum_{n=1}^{\infty} (-n)^n x^{n-1}$

d) $\sum_{n=1}^{\infty} n(-1)^n x^{n-1}$

___ **5.** From $\frac{1}{1-x} = \sum_{n=0}^{\infty} x^n$ for $|x| < 1$ and substituting $4x^2$ for x, the resulting power

series $\sum_{n=0}^{\infty} (4x^2)^n$ has interval of convergence

a) $\left(-\frac{1}{4}, \frac{1}{4}\right)$

b) $\left(-\frac{1}{2}, \frac{1}{2}\right)$

c) $(-4, 4)$

d) $(-2, 2)$

____ **1.** Given the Taylor Series $e^x = \sum\limits_{n=0}^{\infty} \frac{x^n}{n!}$, a Taylor series for $e^{x/2}$ is:

a) $\sum\limits_{n=0}^{\infty} \frac{2^n x^n}{n!}$

b) $\sum\limits_{n=0}^{\infty} \frac{2x^n}{n!}$

c) $\sum\limits_{n=0}^{\infty} \frac{x^n}{2^n n!}$

d) $\sum\limits_{n=0}^{\infty} \frac{x^n}{2n!}$

____ **2.** Using the power series for $\cos x$, the sum of the series $\sum\limits_{n=0}^{\infty} \frac{(-1)^n}{(2n)!}(0.25)^n$ is:

a) $\cos \sqrt{0.5}$
b) $\cos(0.0625)$
c) $\cos(0.25)$
d) $\cos(0.5)$

____ **3.** The nth term in the Taylor series centered at 1 for $f(x) = x^{-2}$ is:
a) $(-1)^{n+1}(n+1)!(x-1)^n$
b) $(-1)^n n!(x-1)^n$
c) $(-1)^n (n+1)(x-1)^n$
d) $(-1)^n n(x-1)^n$

____ **4.** The Taylor polynomial of degree 3 for $f(x) = x(\ln x - 1)$ about 1 is $T_3(x) =$

a) $-1 + \frac{(x-1)^2}{2} - \frac{(x-1)^3}{6}$

b) $-2 + x + \frac{(x-1)^2}{2} - \frac{(x-1)^3}{6}$

c) $-1 + x - x^2 + x^3$

d) $-1 - x + x^2 - x^3$

____ **5.** Use the Taylor polynomial of degree 2 for $f(x) = \sqrt{x}$ centered at 1 to estimate $\sqrt{1.6}$.

a) 1.250
b) 1.255
c) 1.265
d) 1.270

____ **6.** True or False:

If $T_n(x)$ is the nth Taylor polynomial for $f(x)$ centered at c, then $T_n^{(k)}(c) = f^{(k)}(c)$ for $k = 0, 1, \ldots, n$.

____ **7.** If $f(x) = \sin 2x$, an upper bound for $|f^{(n+1)}(x)|$ is

a) 2
b) 2^n
c) 2^{n+1}
d) 2^{2n}

____ **8.** If we know that $\left| f^{(n+1)}(x) \right| \le M$ for all $|x - a| \le d$, then the absolute value of the nth remainder of the Taylor series for $y = f(x)$ about a for $|x - a| \le d$ is less than or equal to

a) $\frac{M}{n!}|x - a|^n$

b) $\frac{M}{(n+1)!}|x - a|^{n+1}$

c) $\frac{M}{n!}|x - a|^{n+1}$

d) $\frac{M}{(n+1)!}|x - a|^n$

(continued on next page)

Section 11.10 continued

—— **9.** $\dbinom{\frac{1}{2}}{3} =$

     a) $\dfrac{5}{16}$ b) $\dfrac{1}{16}$ c) $\dfrac{5}{8}$ d) $-\dfrac{5}{8}$

—— **10.** $\displaystyle\sum_{n=0}^{\infty}\dbinom{\frac{2}{3}}{n}x^n$ is the binomial series for:

     a) $(1+x)^{2/3}$ b) $(1+x)^{-2/3}$

     c) $(1+x)^{3/2}$ d) $(1+x)^{-3/2}$

—— **11.** Using a binomial series, the Maclaurin series for $\dfrac{1}{1+x^2}$ is:

     a) $\displaystyle\sum_{n=0}^{\infty}\dbinom{1}{n}x^n$ b) $\displaystyle\sum_{n=0}^{\infty}\dbinom{-1}{n}x^n$

     c) $\displaystyle\sum_{n=0}^{\infty}\dbinom{1}{n}x^{2n}$ d) $\displaystyle\sum_{n=0}^{\infty}\dbinom{-1}{n}x^{2n}$

—— **12.** From the Maclaurin series for $\ln(1+x)$ and $\sin x$, the first three terms of the Maclaurin series for $(\ln(1+x))(\sin x)$ is

     a) $x^2 - \dfrac{x^3}{2} - \dfrac{x^4}{6}$ b) $x^2 - \dfrac{x^3}{2} + \dfrac{x^4}{6}$

     c) $x - \dfrac{x^2}{2} - \dfrac{x^3}{6}$ d) $x - \dfrac{x^2}{2} + \dfrac{x^3}{6}$

Section 11.11

____ **1.** The quadratic approximation for $f(x) = x^6$ at $a = 1$ is:

   a) $P(x) = 6(x - 1) + 30(x - 1)^2$

   b) $P(x) = 1 + 6(x - 1) + 30(x - 1)^2$

   c) $P(x) = 6(x - 1) + 15(x - 1)^2$

   d) $P(x) = 1 + 6(x - 1) + 15(x - 1)^2$

____ **2.** If the Maclaurin polynomial of degree 2 for $f(x) = e^x$ is used to approximate $e^{0.2}$ then the best estimate for the error with $0 < x < 1$ is:

   a) $\frac{1}{24}e$ b) $\frac{1}{6}e$ c) $\frac{1}{2}e$ d) e

____ **3.** What degree Taylor polynomial about $a = 1$ is needed to approximate $e^{1.05}$ accurate to within 0.0001?

   a) $n = 2$ b) $n = 3$ c) $n = 4$ d) $n = 5$

Answers

■ CHAPTER 1

Section 1.1
1. True
2. False
3. b
4. True
5. b
6. True

Section 1.2
1. False
2. False
3. b
4. b

Section 1.3
1. c
2. a

Section 1.4
1. b
2. c
3. b

3. a
4. c
5. b
6. d
7. a

Section 1.5
1. b
2. c
3. a
4. b

Section 1.6
1. c
2. False

3. d
4. Always
5. b
6. False
7. a
8. d
9. a

■ CHAPTER 2

Section 2.1
1. b
2. b
3. a
4. False

Section 2.2
1. b
2. a
3. a
4. d
5. b
6. d
7. False

Section 2.3
1. True
2. c
3. Sometimes
4. a
5. b
6. c
7. b

Section 2.4
1. True
2. b
3. c
4. d

Section 2.5
1. Sometimes
2. c
3. False
4. d
5. c
6. False

Section 2.6
1. Sometimes
2. c
3. True
4. d
5. True

6. b
7. Sometimes
8. a
9. c

Section 2.7
1. True
2. b
3. False
4. True
5. b
6. c
7. False
8. d

9. b
10. b

Section 2.8
1. False
2. False
3. a
4. True

■ CHAPTER 3

Section 3.1
1. d
2. c
3. d
4. c
5. b
6. b

Section 3.2
1. False
2. c
3. d
4. c

Section 3.3
1. b
2. a
3. d
4. c
5. d
6. a

Section 3.4
1. False
2. c
3. d
4. a
5. True
6. Sometimes
7. c

Section 3.5
1. False
2. b
3. a
4. b
5. d
6. b

Section 3.6
1. c
2. a
3. b
4. d
5. c

Section 3.7
1. False
2. a
3. b
4. a

Section 3.8
1. a
2. c
3. b

Section 3.9
1. a
2. c

Section 3.10
1. b
2. b
3. d
4. b

Section 3.11
1. False
2. b
3. d
4. a
5. c

Answers

■ CHAPTER 4

Section 4.1
1. Sometimes
2. Always
3. c
4. True
5. True
6. b
7. b

Section 4.2
1. c
2. False
3. a

4. False
5. c

Section 4.3
1. b
2. False
3. False
4. b
5. False
6. True
7. False
8. b

9. c
10. b

Section 4.4
1. b
2. a
3. b
4. b

Section 4.5
1. True
2. True
3. False
4. True

5. False
6. True
7. b

Section 4.6
1. c
2. c
3. a

Section 4.7
1. b
2. c
3. c
4. b

Section 4.8
1. Sometimes
2. b
3. d

Section 4.9
1. False
2. d
3. a
4. False
5. a

■ CHAPTER 5

Section 5.1
1. a
2. True
3. c
4. a

Section 5.2
1. Sometimes
2. Always

3. d
4. a
5. c
6. d
7. b
8. b
9. True

Section 5.3
1. d
2. c
3. b
4. False

Section 5.4
1. b
2. d
3. c
4. d
5. b
6. b

Section 5.5
1. d
2. b
3. a
4. c
5. c
6. True
7. d

■ CHAPTER 6

Section 6.1
1. b
2. b
3. a

Section 6.2
1. a
2. b
3. a

Section 6.3
1. b
2. d

Section 6.4
1. c
2. d

Section 6.5
1. a
2. c
3. c
4. True

■ CHAPTER 7

Section 7.1
1. a
2. d
3. b
4. c

Section 7.2
1. d
2. b
3. a
4. b

Section 7.3
1. c
2. c
3. False
4. d

Section 7.4
1. b
2. d
3. a

4. b
5. d

Section 7.5
1. False
2. True
3. False
4. True
5. Always
6. Sometimes

Section 7.6
1. a
2. c
3. a
4. c

Section 7.7
1. b
2. a
3. b
4. c

Section 7.8
1. False
2. a
3. c
4. d

Answers

■ CHAPTER 8

Section 8.1
1. c
2. d
3. b

Section 8.2
1. a
2. c

Section 8.3
1. b
2. b
3. b
4. c

Section 8.4
1. True
2. b
3. a

Section 8.5
1. False
2. d
3. b
4. c

■ CHAPTER 9

Section 9.1
1. c
2. True
3. d
4. False

Section 9.2
1. False
2. a
3. b

Section 9.3
1. c
2. True

3. d
4. a

Section 9.4
1. d
2. c
3. True
4. b

Section 9.5
1. True
2. a
3. d

Section 9.6
1. True
2. a
3. a

■ CHAPTER 10

Section 10.1
1. c
2. d
3. a

Section 10.2
1. d
2. c
3. b
4. b
5. a
6. b

Section 10.3
1. d
2. c
3. b
4. b
5. d
6. b

Section 10.4
1. a
2. d
3. d

Section 10.5
1. c
2. b
3. a

Section 10.6
1. c
2. b
3. d
4. c

■ CHAPTER 11

Section 11.1
1. b
2. Always
3. Sometimes
4. Always
5. False
6. a
7. c
8. d

Section 11.2
1. d
2. True
3. False
4. True
5. b
6. False

7. c
8. True

Section 11.3
1. c
2. False
3. False
4. False
5. True
6. b

Section 11.4
1. Never
2. True
3. False
4. True
5. b

Section 11.5
1. True
2. True
3. b
4. False

Section 11.6
1. False
2. True
3. True
4. a
5. b
6. True

Section 11.7
1. True
2. False
3. True

4. True
5. True
6. True

Section 11.8
1. Sometimes
2. True
3. c
4. b
5. c

Section 11.9
1. c
2. c
3. c
4. d
5. a

Section 11.10
1. c
2. d
3. c
4. a
5. b
6. True
7. c
8. b
9. b
10. a
11. d
12. b

Section 11.11
1. d
2. b
3. b